SCIENCE FACT

SCIENCE FACT

Astounding and Exciting Developments That Will Transform Your Life

Edited by Prof. Frank George
Foreword by Desmond Morris
Produced under the direction of
Colin Rose

 **STERLING** PUBLISHING CO., INC.

NEW YORK

CONTENTS

Frank George is Professor of Cybernetics at Brunel University, and Chairman of the Bureau of Information Science and the Institution of Computer Science. He has also been consultant to NATO, the Treasury, the Department of Industry, the Ministry of Defence and various other organizations.

He was an Exhibitioner at Sidney Sussex College, Cambridge, where he read the Mathematical and Moral Science Triposes. After getting a 'double first' at Cambridge, he was appointed to the staff of the University of Bristol and subsequently took the Chair at Brunel. He has been a visiting Professor at McGill University, Princeton University, Stanford University, and Georgia 'Tech', as well as being a visitor at many other institutions, including UCLA.

Among his many previous books, the best known are perhaps *The Brain as a Computer* (1961), *Models of Thinking* (1969), *Science and the Crisis in Society* (1970), *Cybernetics – Science of the Environment* (1977) and *Man the Machine* (1977).

Foreword
By Desmond Morris

The origin of this foreword is a little unusual and rather appropriate, for it is based not only on my impressions of *Science Fact* itself but on questions I was subsequently able to put to the editor, Professor Frank George, over a 6,000-mile transatlantic phone call.

Science Fact sums up for me all the exciting new developments that are taking place in almost all the major areas of science and technology. I am grateful to the book for this because, like most scientists, I do not know a great deal about progress in all the other fields of science – it is a highly specialized world we live in.

Additionally, the book is nicely balanced between the fascination of the almost unbelievable scientific developments themselves and the often controversial implications of this scientific genius. Perhaps nowhere is this more evident than in the section on computers. The fact that you cannot only talk to computers, but that they will be capable of independent thought and could be self reproducing, is startling.

While I see the obvious potential threat, I also see an exciting opportunity. It was no accident that the great advances in thought, philosophy and ideas in Ancient Greece were based on the slave state. Certain, albeit privileged, people were free to devote themselves to creative thinking precisely because they were freed from the day to day tedium. Perhaps the computer offers us a slave machine – a return to a new Greece.

In this aspect and many others I find *Science Fact* also an optimistic book. Certainly this is very true of the developments in medicine. Some, such as the post-operative freezing of nerves to eliminate pain, will affect millions. Others will affect less people but are no less astonishing. I was especially intrigued by the discovery that a blind man can be made to see by having the electronic signals from a miniature TV camera transmitted from his *stomach* muscles to the visual center of his brain.

Inevitably, not all the developments have uniformly desirable implications. In reading about the possibility of using my telephone linked to my TV linked to my local supermarket in order to "go shopping" without leaving my armchair, I was struck that the world could become an increasingly private place – and perhaps too inward-looking. This thought was reinforced when I later read that almost any experience, including sexual, can be artificially reproduced by stimulating the brain electronically. An extraordinary toy but one open to exploitation.

There are other developments surveyed by *Science Fact* that I believe should be pursued because, while their ultimate chances of success are smaller, the implications of that success if achieved are so profound. I would put the NASA program to search for extra terrestrial life in this category. While I think that on the laws of chance alone there must be other intelligent life, the chances of finding it are remote – but if we did . . . . !

Similarly, I am glad the section on Parapsychology was included. Some of the experiments are almost literally incredible (the Russian research on moving objects with mind power alone, for example) but when we can find a universally accepted standard of proof for the paranormal, the implications of certain findings will necessitate some profound rethinking.

Finally, I felt that Professor George made an important point at the end of our transatlantic conversation. "Science," he commented, "is neutral. It can be used for good or evil. It is people who must decide."

To make a good decision you need a clear understanding of the issues. *Science Fact* makes a contribution to that understanding because the sheer range of its survey allows us to see the implications of apparently unconnected developments.

INTRODUCTION

This book was planned with a straightforward aim in mind. It was to describe the very latest, most dramatic and most significant developments in each of the major fields of science. The undertaking was a formidable one, but was motivated both by a sense of curiosity and sheer enjoyment at discovery. We wanted to know what was happening and what is about to happen in our world. This plan also involved some apprehension since the rate of change already observable is so obviously dramatic in itself.

The changes happening are sometimes bizarre to a degree, almost always astonishing and usually extremely exciting. But one has to admit there is a sort of social sense of preservation involved. We want to know – indeed have a right to know – the way that events are trending.

There is also a justification involved. Scientists spend most of their lives studying minutiae, and almost always studying their specialist subject. We feel the need, from time to time, to get away from such narrow confines and see the world in perspective. This allows us to see the interrelatedness of events, only possible in a broad survey of our subject – and our subject is virtually all of factual knowledge. In fact, we explicitly exclude the man-made socially determined knowledge, such as the professionalism of the accountant and the solicitor. We had to try to narrow down the field of activity somehow, even though when we limit ourselves to science our canvas is still enormous. It is after all science that is changing our whole life.

The method adopted was to invite experts in each field to describe the *latest* developments in that field. The resulting book is therefore a compendium of viewpoints in which the main emphasis has been on 'facts' and speculation about 'likely facts'. It is impossible not to be elated, enormously impressed and slightly frightened by examples both of human ingenuity and the complexity of our universe – still, and perhaps for ever, a vastly 'mysterious universe'.

With all these feelings in mind, feelings which are ambivalent to say the least, we start our pilgrimage. One thing is certain, by the end of the book it will be clear to everyone that nothing can ever be quite the same again.

A major consideration in our survey refers to the educational process. This applies to all civilization and is especially related to specialization, and the consequent privacy, of so much of science. It is difficult, with the exponential growth of knowledge, for any one person or even group of persons to know exactly what is happening in the world and to the world. This is why this book could not have been written by any one person. We have tried to bring together experts from all the most vital fields of human endeavor and get them to tell their own story.

One of the worst dangers of specialization is the fact that it is difficult for a person concerned with great precision of detail to be able to see the consequences of his discoveries. This is one reason why he is reticent to communicate and often extremely conservative in his forecasts. He is not aware of the dangers that may exist in his work because he does not see its implications and he does not see the implications because he is concentrating on just one small piece of the total jigsaw puzzle.

Confronted now with the specific problem of predicting the future, doing it in a responsible way is a forbidding prospect. Prediction has always been difficult – as Arthur C. Clarke has said, 'advanced technology is indistinguishable from magic' and magic is notoriously unpredictable.

Only a few years before the first nuclear bomb was exploded, Lord Rutherford, the *founder* of modern nuclear physics ('The man who split the atom'), was of the opinion that his own work would have no profound practical or political applications.

Very few people would have predicted the full implications of the computer's development, and so it has been with most scientific discoveries – even those most closely involved often cannot see the wood for the trees.

There are a vast number of other cases where eminent people were unable to anticipate the implications of their own research. Alexander Graham Bell's work on the telephone was not originally seen by him or by others as the enormously significant step it has turned out to be. Early work on X-rays by Roentgen was thought to be of some interest but no one was able to envisage the complete transformation that his work has brought about in the field of medicine.

Less surprising but more sinister than the failure of scientists to see the significance of their own work, is the failure of the profession as a whole – the general public can reasonably be excused – to anticipate even the bulk of the most important discoveries and inventions.

To my knowledge, one eminent professor of chemistry regarded the development of artificial organs for body transplants and the transplant of 'real' bodily organs as 'absurd and fanciful' even within a few years of it becoming a reality.

So conservative indeed are scientists as a group that they are loath to commit themselves in a manner that most people would regard as prim and almost childlike, but much of the explanation lies in the fact of specialization. Specialization and concern for detail seems to dull the imagination.

It is not of course only that scientists become highly specialized and sometimes get bound up in the narrowest of fields. We are almost at the stage where a specialist in 'the middle ear' is not too well informed on 'the outer ear' and more important still he does not communicate what he is doing.

The time of course is sometimes ripe for certain developments, as is shown by the fact that Sir Isaac Newton and Leibniz independently developed the Infinitesimal Calculus – this is a real sign of the needs of the times and such things do occasionally occur – in retrospect it is clear that the Special Theory of Relativity was also very much a child of its time, as all the signposts of physics pointed in that direction, but it took an Einstein to realize it.

So our first message to the reader is clearly a plea to keep an 'open mind'. We cannot, for example, be sure whether paranormal activities are of minor importance – even irrelevant – or a central feature of our future development. A prejudgement could be damaging. We cannot be certain that we will mine and farm most of the minerals of the sea bed although we probably could. While not all developments will fulfil their promise, we submit it is easier to be wrong by being too conservative than by being too imaginative.

If you bear in mind that we, the human species, have learned in the last fifty years more than we learned previously in the whole history of man, then the rate of growth of our knowledge becomes dramatically clear, and makes it even more difficult to decide what *will* happen : let alone what *can* happen.

The time scale involved in events is not always easy to grasp. If you take *one step* only towards the study wall and call it the scale of time of the 2000 years since Christ, then the beginning of human life is as far away from Boston as Nantucket Island, and the beginning of the world would take us millions of steps to the middle of the Atlantic. Civilized life on earth is an incredibly recent development!

The development of our computers helps again to bring out the point. Mathematical work on the tides of the St. Lawrence seaway in Canada would, it was reckoned, use every available mathematician in Canada and keep them all busy for a hundred years. The needed hydrodynamic computations were performed by the University of Toronto computer in just over a week – this was twenty years ago. Today the same work would take a modern computer some few seconds!

First cousin to this problem of dimensions where times and distances (consider going to Neptune so many 'light years' away) are almost impossible to envisage, is *exponential growth*. If the population of some community is 100,000 people and it doubled every five years then in a single generation say 25 years) it would become 3,200,000 – from Oxford to Chicago in a single generation! The population of the world is growing almost as fast as that, so the consequences could be alarming in the very near future – exponentially near!

Exponential growth is like compound interest, it is not only something growing but something growing at an increasing rate. Two simple stories illustrate the point.

The first example I have drawn from Professor Jay Forrester's *World Dynamics*. The following table shows the population growth for 700 years with a doubling time of 50 years:

Year	Millions of people
0	1
50	2
100	4
150	8
200	16
250	32
300	64
350	128
400	256
450	512
500	1024
550	2048
600	4096
650	8192
700	16,384

Other crucial variables like wealth and production tend to follow similar exponential paths.

A more abstract example is given by taking a sheet of paper, necessarily large (!), but only 1/16th of an inch thick and folding it over and over. If we fold it thirty times it is so thick that it has one side in Oxford and the other side in Kansas City. Fold it six more times and it will nearly fill the space from the earth to the moon. One can see why the sheet of paper had to be large, and one also sees why exponential growth is so sensational!

It is the exponential nature of growth which makes prediction more difficult precisely because the speed of development, although exponential, has a gradient that is difficult to estimate. This can all be seen clearly in the way in which 'breakthroughs' tend to occur. The build-up in say the field of artificial intelligence is complicated and yet can suddenly fall into shape. An analogy with problem solving is to the point. It takes a great deal of work to program a digital computer *to learn to play* even a simple game like 'tic-tac-toe', but when done it is then that much easier to get it to play three-dimensional tic-tac-toe. Other games like parchesi and checkers are progressively easier still. All this is a reminder that what a human being learns is a set of concepts which once grasped can apply to a whole range of activities. So what is initially difficult becomes cumulative and therefore progressively easier.

Conversely the making of one development, discovery or invention leads to another. This represents the cumulative nature of knowledge. It is natural, as in the case of a jigsaw puzzle, that the addition of one new piece leads to the possibility of the next piece (or pieces) and gradually the whole picture is built up.

An industrialized society 'takes off' slowly, as Britain did with the Industrial Revolution, and gathers momentum and then it takes a long time for a competitor to do the same thing and to catch up. Once they do so, of course, they have the advantage of being newer, having been planned as a whole and produced from scratch. They also have the advantage of learning from earlier mistakes, as indeed do the next group who come along and go one better still. We see just this sort of thing occurring in the present day. After Britain (where the Industrial Revolution started) came the other Western countries – particularly USA – who have now surpassed Britain. After them come the under-developed countries and it can only be a matter of a short time now before they too surpass Britain, unless something more fundamental stands in their way.

The 'more fundamental' obstacle which may or may not stand in their way is knowledge and education. The other western countries had only a small gap to bridge here; the undeveloped countries have a much bigger gap. It is for this reason that so much money is now being spent on education and new projects in such countries : their aim is to gain 'first division status'.

The exponential growth process can be just as fundamental in reverse – in the exponential decay process. We are finding that our resources are disappearing at enormous rates, much, much faster than is generally realized. We shall see that the majority of resources on which we depend may only last a few years, and this sort of threat could totally change our planning and our priorities.

In the next few years, as this books bears witness, we are going to see some fairly staggering developments in all aspects of science and society and we ought to have some glimmering of what to expect, otherwise we will be unprepared for the sort of change that will hit us like a tidal wave. We need to know and we need to know as much as possible in order to adapt.

I do not wish to anticipate the experts' reports in the text, but in my own subject covered by Dr Christopher Evans, we may expect to see the emergence of a 'machine species' and the only question is as to how quickly such a species will emerge.

Medicine, with spare parts surgery, new techniques and its relations with genetic developments and genetic control will allow us in turn to control far more of our biological world than was once thought possible. This could lead to our machine species growing from seeds and being protoplasmic rather than metallic in their construction. It is beginning to look as if for the first time a new species will evolve not from the body of its predecessor but from his mind !

Then there is the question of transportation. How long before we can travel to Mars and Venus? I suppose these days one will be even more concerned with the cost per passenger than the implications for the colonization of space.

So it goes on. Brain control by neurological and psychological means – and this includes brain washing and more effective advertising – will change the nature of our society. Paranormal research is now being carried out at a greater rate than ever before. In horse racing terminology this is perhaps an outsider – certainly all the others are hot favorites – but if it 'comes in', like any outsider, it will have the most dramatic effect of all.

We probably will mine the sea, we will seek new energy sources, we shall find new pesticides, new cures for disease, and this last, coupled with spare part surgery, will eventually hold over us the threat of immortality and that could cause the biggest social upheaval of all time.

We must be clear that all scientific and technological advance is not always for the better. The slowness with which social change is achieved makes it difficult for society to accept massive scientific advances and technical developments. To this extent we shall see all humanity, but particularly Western society, put to the biggest test of all times in the near future. The biggest threats must be of loss of freedom, either by anarchy or totalitarianism, and eventually subjugation by a new species – but there are other threats as this book will make clear.

The first step towards dealing with the threat of the future is to understand the nature of the future. This book attempts to do just that and our plea to the reader is 'approach the whole thing with an open mind' and give yourself time to absorb the implications of what could be the most dramatic sort of change – nothing it seems is impossible!

Of course, much that will happen comes more easily into perspective with the present. We know that there will be architectural changes and that we will be able – sooner or later – to control the weather. These examples alone have vitally important consequences. They make clear the *interdependent nature* of scientific advance, since if, for example, we can control our weather then we can let that influence our architecture. We could – at one extreme – have completely flat roofs if we decreed that no rain should fall in that area. Or if that suggests too discriminating a degree of weather control, then we could still arrange for special roofing – super-plastic coverings – to put over only when we allow rain to fall.

The design of houses though will also be influenced by the emergence of new heating systems and new building fabrics, and we might well see a return to 'throw away' designs reminiscent of 'pre-fabs', but at a totally different level of efficiency and comfort.

Automation has already sparked people's imagination, so the pilotless aircraft and the factory empty of human workers will not occasion quite as much surprise today as it would have only a few years ago. The interdependent nature of things though allows us to see automation change our ideas on house design. We can envisage the preparation of food and the consumption of food all occurring in one room (kitchens and dining rooms integrated) and conveyor belts replacing tables and food cooked only as a result of the cook's deft controlling of the dials of the computer console, or not even that if the market research is controlled too by automation.

The architectural and transportational changes will mean the development of new sorts of cities – probably large linear ones that stretch from Aberdeen to Naples, or New York to Los Angeles. The communities away from the linear city could be communal sites made up of chains of houses with central facilities, and as a result the new society may well be simple and on a small tribal pattern, de-centralized and yet with some sort of continental and eventually international representation and control.

But now we are 'jumping the gun'. Let us sit back and bear the contextual and social problems in mind, as we weigh and assess the forecast developments in each and every branch of science and technology. And once more we must say it – keep an open mind and remember that there seems to be no limit on the possibilities inherent in human genius.

What we somehow have to decide is which possibilities to encourage and which to discourage, which to put our money into and which not. We can only make this sort of decision if we can clearly see the implications of the plans and decisions we ourselves make.

COMPUTERS AND ARTIFICIAL INTELLIGENCE

Dr Chris Evans

Dr Chris Evans is a psychologist and computer scientist with over 100 scientific and technical publications to his name. He is also an author and broadcaster and liaises science programs for radio and TV in Britain and the U.S.A. He is well known for his theory which likens dreaming to computer program clearance.

Introduction

The boundary lines between the apparently imaginary worlds of science fiction and the down-to-earth worlds of science fact are becoming increasingly blurred. The future – even the kind of future predicted by the most extravagant prophets of paperback SF – is rolling up on us with astonishing rapidity. And in no other field perhaps is it rolling up so fast as in the field of computers, where even the computer scientists themselves are frequently taken by surprise at the technical developments occurring at their elbows. Furthermore, there is every indication that the pace will accelerate over the next decade or so with consequences so tremendous that our view of the world, even of our nature, will be transformed. For Man is about to create a new companion for himself on his native planet, a companion who will rival and vie with him not for natural resources, but for intellectual supremacy. Man, undisputed master of the earth, may before too long have to step gracefully aside and yield the reins of power to beings of his own creation. Contrary to most current scenarios, the 20th century will not be remembered as the era when space was conquered, or the power of the atom harnessed, but that in which there appeared the first machines with minds.

The concept of a machine with a mind – a thinking machine – is not an easy one to come to terms with, though it has been a familiar one to science-fiction fans ever since the first computer-based fantasies began to emerge about 25 years ago. But the vast majority of people remained largely untroubled by such eerie possibilities until Stanley Kubrick's sensational movie *2001* rocked audiences across the world in the mid 1960s. Part of the success of *2001* was its great technical authenticity, but the central image of the film was undoubtedly HAL, the talking, thinking and reasoning computer controlling all systems on board a manned spaceship

en route to the planet Jupiter. The human crew who had
little to do except plot the occasional entry in the ship's log,
spent most of their time chatting to the computer, who not
only understood anything that the crew said to him, but also
chatted back to them in conversational English. This led to
a peculiar intimacy between man and machine that was
chilling enough even before the computer began to suffer
from programming problems and made up its mind to kill
its human crew-members. Needless to say the humans got
the better of it – but only just! – and HAL 'died' a melo-
dramatic death, regressing to childish babbling as it did so.

'Human' computers

Whenever people discover that I am involved in the development of computers and in particular in improving communications between man and computer, one of the first
questions they ask is, 'Could it really be like that?' To this
I reply that if 'like that' means could computers come to
understand human speech, and could they have voices of
their own, and could they exhibit powers of reason and even
the rudiments of personality, then my answer is a definite
Yes. And if they then ask, 'and could these things come about
by the year 2001?', again my answer is that they could, *and
perhaps sooner.* This opinion is shared by a growing number of hard-headed scientists involved in the systematic
creation of artificial intelligence – a topic being actively
studied in some of the world's leading scientific institutions
and which is already occasioning scientific controversy.
Much of this controversy incidentally does not concern itself with *whether* artificial intelligence is worth studying, but
whether it *ought to be* studied because of its possible long-
term consequences. But to understand the significance of this
topic requires some background knowledge of the origins of
computers. Before hoping to comprehend the future one must
come to know something of the past.

The origin of computers

Many people put down automatic shutters when they hear the word 'computers', presumably feeling that as they are far too complicated to understand, it's not even worth trying to do so. Attitudes like this have contributed to a weird array of myths and superstitions which have permeated the media so intensively that no story or feature on computers is complete until reference has been made to people who receive million-dollar electricity bills, or who have been denied credit or threatened with legal action by some computer. The truth of the matter is that while technical details of computers – the precise structure of their interior architecture – is indeed complicated, and reflects the magnificent achievement of electronic engineers who designed and built them, the principles of computer operation are simple. Computers are in fact nothing more than elaborate calculating machines, capable in principle of solving any kind of mathematical problem.

If the idea of a machine solving problems seems incomprehensible, then all one has to do is accept the fact that numbers can be represented by physical things or by *states* of things. To give the simplest and most venerable example, the abacus is a device which represents numbers as beads, and which allows calculations to be made by manipulating the positions of the beads on a frame. The abacus is in fact an ultra-primitive computer, which must of course be operated by hand, and which can only perform one or two mathematical operations – e.g. addition, subtraction and rudimentary multiplication. Just as numbers or values can be represented and manipulated by beads, so they can be represented or manipulated by other physical objects – cogwheels, gears or even simple switches. A kind of cogwheel calculator was in fact constructed in 1642 by the French mathematician Gabriel Pascal. It could add, subtract, and rather laboriously multiply and he used it to help his tax-collector father and was much resented by clerks of the time who feared it would put them out of work. A far more elaborate and imaginative design for a calculator, but still using mechanical components of the cogs and wheels variety, came almost two centuries later when the mathematical genius, Charles Babbage, set out to build a 'general purpose calculator'. This would be a machine capable of solving *any* kind of mathematical problem. One moment one could use it for routine multiplication and division, another moment for solving polynomial equations, another time for something even more complex, with the task of the moment being determined by a set of instructions – today we could call them programs – fed into the machine. Unfortunately, when he came to construct the device – he called it the 'analytical engine' – his grand assembly of cogs and wheels, all of which had to be hand-turned out of brass, simply could not be made to fit together with sufficient precision. The main difficulty was that the representations of numbers in decimal form – i.e. all cogs had to have ten teeth – made the calculating machinery hopelessly complicated and prone to endless mechanical error.

It's fairly clear that Babbage realized that his analytical engine would never really work in practice, even though its design was flawless. Had he known the principles of binary arithmetic however and had he been able to construct his computer using binary units then it might have been a different story. This system of mathematics relies on the principle that all numbers, no matter how large, can be expressed in terms of different sequences of the numbers 0 and 1, and thus, if you think about it they can also be expressed in terms of anything which can be in one of two states. This allows something like a switch, which may simply be on or off, to act as the basic unit in a calculating device, which leads to a logically simpler design than any, like Babbage's, which relied on the decimal system. Nor of course did Babbage have electricity as a power source, and it wasn't until these two factors – electric power and binary counting – were combined together that really fast, automatic calculation became possible. Indeed, the world had to wait until the early 1940s, when Howard Aiken of IBM built a giant programmable computer using telephone relays as the basic binary switching units, that Babbage's dream was fulfilled.

Modern day computers arrive

Aiken's device worked very well, and was hundreds of times faster at calculating than any human being, but the telephone relays were large, and because they were metal switches, took a relatively long time in changing their state from on to off or back again, which after all was the essential mechanism of computation. A major advance came with the employment of valves or vacuum tubes to replace the relays, which because they were electrical and not mechanical could switch hundreds and even thousands of times a second. Later, in the 1950s, came transistors, smaller, faster and much more reliable than valves, which, with the development of what is known as the 'stored program', paved the way for really powerful modern computing.

Computer power

Now programs are simply instructions to the computer telling it what to do next – how to use its brain if you like – and in the early days they were fed to the machine on punched cards or tape. As *stored* programs, they became part of the computer itself, an array of separate sets of instructions built into its memory, and able to interact, so that one program can call up another, that in turn can call up yet another and thus set the computer on a long and complex course of actions. This made computers progressively more capable of acting independently, and performing more and more complex tasks as more and more sophisticated programs were developed. Gradually computers became more and more like the brains of the humans who had designed them. For example, human beings are characterized by having special devices capable of recognizing information (the senses), a system capable of storing this information (memory), a system capable of processing this information or acting on it (the cerebral cortex), and devices capable of feeding back information to the outside world and controlling it in some way (the psychomotor system).

Today's computers too can be looked at as a set of subsystems rather similar to the human brain and body. They have input terminals which gather data from the environment, various types of memory including short-term or buffer stores and longer term, central or archive stores, they have a main control unit, the central processor, and they have peripheral terminals which either print out the result of their calculations or which control the environment in some way. In one important respect, however, computers and human brains were pulling apart even by the 1950s, and that is that computers were actually quicker at calculating than humans – in other words their processing was already faster than thought. This huge gain in computing speed was helped by the development of exceedingly fast memories, and a major breakthrough came with the discovery that information could be stored in the form of minute electrical signals on any magnetic substance. Access time (the time taken to get information out of the memory) shrank from thousandths to ten, even a hundred, thousandths of a second, and the storage space taken up by these minute stores shrank by an equivalent amount. The result was that less than 15 years after the first production computers were made, their offspring were many times more powerful and at the same time took up less than a hundredth of the space.

Yet another important development was the advent of time-sharing, a technique which allows dozens – even hundreds of thousands – of people to use a big computer system at the same time. Companies sprang up which installed powerful computers, serviced them and maintained them, and then rented time out on them on an hourly basis to any user; all these users needed was a suitable terminal, a telephone to ring up the system and a special decoding device to allow the computer information to be transferred down the telephone lines. Suddenly the vast power of computers became available to anyone with enough space for a typewriter keyboard, and the money to pay for computer time.

Conversation with a computer?

In the past almost all computing had been concerned with what in computer jargon is known as number crunching, but with the cost of computing down to a few pounds an hour, one could start thinking about totally different applications. Before any real novel applications could be developed, however, one had to have some way of making computers comprehensible and usable by normal people – that is to say people not specifically trained in high-powered programming. Now programming is, in a nutshell, the art of making computers do what one wants them to, but to do this it had always been necessary to learn the computer's rather abstruse language. Well, if computer languages were too abstruse then the solution must be not to teach people how to talk to computers, but to teach computers *how to talk to people*. The result was the development of a number of languages, conversational programming languages as they are called, which consist largely of English words and phrases, all with an obvious meaning and role, but which the computer can also understand. These languages are easy to learn, and one communicates with the computer by typing in whatever commands one wants on a typewriter or similar keyboard terminal. Furthermore, using them one can arrange for the computer to communicate back to you using English (or any other language for that matter) words or phrases. Having such conversations is an eerie experience – one gets a sense of rapport with the machine which is not easy to brush off – a sense of there being a 'presence' at the other end of the cable or telephone line. The effect is particularly powerful when the computer has been programmed to interact with one on a meaningful matter – for example when the computer is trying to teach you, or when it is simulating a doctor and asking about the state of your health. Of course the feeling is illusory – one infers the intelligence and the computer is simply generating a sequence of questions. But the creation of such programs is one of the first steps in the simulation and ultimately the creation of artificial intelligence.

Computers shrink in size

In the past year or so have come even further reductions in the size of the basic components of which computers are made, and hence, of the computers themselves. Transistors ceased to be individual units wired together, but were reduced first to small slivers of silicon mounted on ceramic bases, and then in the process known as large-scale integration, became microscopic particles engraved into minute plates. The scale of reduction in size has been almost incredible. The central processors which in the big systems of the 1950s might fill a small room, shrank first to desk size, then to suitcase size, then to pocket size and then down to the size of a thumbnail. Memories too got more and more compact; the magnetic units are now so small you need a microscope to see them. By the mid 1970s mini-computers gave way to *micro*-computers, and at the time of writing the process of shrinkage has progressed so far that the current equivalent of the first giant valve computers, costing hundreds of thousands of pounds to buy and needing hundreds of kilowatts of power to feed them, could be reduced into something about the size of a matchbox and requiring only a tiny battery to power it. But if computers have evolved to this stage in so short a time, what are we likely to see taking place in the second quarter-century of their existence?

The short-term future (1977–1982)

In plotting the future of computers it's vital to realize the enormous pace of development in this aspect of technology. The suddenness with which pocket calculators came upon the public is a good example of the kind of change that can take place, as is the way in which the digital watch made centuries of mechanical watchmaking expertise redundant almost overnight. The most striking feature of the immediate future will be that computer technology will continue to become more common, more powerful, more miniaturized and progressively cheaper. This will first be felt by the increasing use of miniature calculators, most of them programmable, which before 1980 will be quite as common as watches today. In fact, to continue an existing trend, many of them will be, and are now being, incorporated into watches and pens!

A further interesting development will concern the displays on the calculators which at present involve numbers only, but which will soon offer letters as well. Initially this will have a gimmick value only, but it will ultimately be helpful in programming as pocket computers themselves become more powerful. Some pocket computers will also have what is known as hard copy – paper printouts of strings of numbers, program listings. Once alphanumeric display and hard copy facilities become available the pocket computer will have a new dimension of use as a teaching aid, and programs will be available to teach all kinds of topics from mathematics to history and foreign languages. Programs will even be available to *teach the user how to program.*

Another development likely to occur within the next five years concerns visual displays. Ever since the early 1960s it has been possible for computers to present their output to the user on cathode ray tubes – essentially TV sets – generally in the form of luminous letters coming up on the screen. Using a modified version of such a screen with a 'memory' built into it, a much more dramatic form of interaction is possible with the computer drawing a pattern on the screen. The pattern can vary in complexity from a geometrical shape to highly complex patterns and these patterns can be manipulated by the computer according to the whim of the user. Apart from the fact that this produces very pretty pictures, computer graphics as it is called, it has real practical value. Take for example the design of a motor car. When limited to pencil and paper the designer must sketch out the car on paper, and if he wants to look at it from different angles he must produce a drawing of it from each of these angles. By designing the car using interactive graphics and programming its dimensions and spatial characteristics into the system he can produce instant pictures of it from any angle, even moving back and to across the screen, turning round, driving away and so on. Furthermore if he fancies a minute style change all he has to do is to specify the dimensional change – this can be typed into the computer – and lo! the modified car is there for his inspection.

A similar area for computer aided design (CAD) is in architecture where architects can rough out a building and get instant feedback as to the heat loss from window space, the effects of altering the dimensions of rooms, etc., or in electrical engineering the design of electronic circuits. While interactions of this kind have been possible since the early 1960s they were only possible at vast cost. But as memories become smaller and cheaper and the sheer guts of the computers increase, dynamic interactive graphics is likely to become very common and may well become available, within the next five years, on specifically modified TV sets. Actually crude interactive graphics is already commonplace – in the TV games of table tennis, basketball, etc., that are controlled by a small but non-programmable computer or 'chip', and will be commonplace 'extras' when one buys a TV set by the end of this decade. Anyone who has played the present generation of games will realize how compelling they are and how doubly, triply compelling succeeding generations of the games are likely to be. Within the next five years for example the first TV games will appear in which the micro-computer in the TV set actually competes with the player. For instance, it might set itself up as the partner in the table tennis game, give one interesting mazes to solve, engage one in general knowledge quizzes, or set up a game of draughts or checkers. In case you feel that nothing could be easier than to beat a TV set at checkers, then you're in for a big surprise – a computer program is already the world checker champion. I suspect that before the turn of the decade anyone will have the option of taking a hiding from the world champion checkers player in the comfort of his own home and for the price of a few units of electricity. Chess playing programs too will be available, though in the early stages they will play only novice-like games. Again before 1980 expect to see small calculators equipped with miniature TV displays offering all the game-playing facilities plus miniaturized computer graphics so that one can design one's car, hi-fi circuit or new house while on the bus, train, car – or even in bed if one is so moved.

Of course the availability of all this gadgetry will not necessarily mean that it will be applied usefully and in the early days people will buy these devices almost entirely because of their gimmick value. Actually one gimmick in connection with watches will certainly appear shortly, and that will be a talking watch which when its button is pushed, or perhaps even when it is spoken to, will speak the time. This, seemingly ridiculous, development will arise because of recent developments in the synthesis of speech by computers – and in particular in the miniaturization of the speech-producing devices down to microscopic size. Crude synthesis, featuring flat nasal speech, has been practically possible for some time, but it has always involved large and expensive computers to generate it. Already an excellent talking calculator now exists for blind people – as you punch the keys the machine repeats what you have punched, and finally when you push the equals sign it instantly speaks the total.

It is important to distinguish incidentally between synthetic speech and recorded speech – the latter consists simply of a miniaturized tape recording, and is therefore limited in range and flexibility, while the former is the machine electronically generating the speechlike sounds itself. This of course is theoretically limitless and given a clever enough program the machine can generate long, variable strings of words and even sentences. For the wider, deeper significance of speech synthesis we will need to look ahead to the medium-term future, but over the next few years at any rate more and more machines will talk to one – although what they have to say will at first be pretty routine and uninteresting, or trivial in the case, for example, of their applications in toys. Increasingly we will see toys which respond – in early days just by stopping and going – to voice command and dolls will be equipped with a wide range of synthetic speech vocabularies. Really mind-blowing toys – dolls which move about like miniaturized people under microcomputer control – are not likely to appear until the late 1980s however.

Microprocessors

Somewhat less gimmicky, though still a rather trivial use of computer power, will be the introduction of microprocessors into motor cars. Initially they will be used for monitoring fuel supply – supremely efficient carburation – which will have marked effects on fuel consumption. They will also control instrumentation and draw the driver's attention – often by using synthetic voice – to the state of his battery, gas tank, brake-linings, etc., and ultimate insult of all, to the fact that he is exceeding the speed limit. Similar microprocessors are likely to find a quick and useful role in aviation – advising air traffic controllers, for example, assisting pilots and controlling complex transportation networks. Here, however, their arrival will not be so welcome, as in these roles the computers are in fact beginning to take over some of the functions of very highly trained and skilled human beings. The horrendous fact that many prized human skills are rather surprisingly easy to simulate by quite simple computers will not penetrate human consciousness overnight, but once it does there are likely to be fireworks.

In the home microprocessors will appear in most household labor-saving devices – cookers, washers and driers – and in the elaborate security systems which many people will employ to protect their homes. Perhaps one of the most dramatic features will be the disappearance of the key – an object which has served a common function for thousands of years and which will be replaced by pushbutton locks whose combination is known only to the owner, and later on a door which responds to the owner's voice, and to his or hers only.

No doubt criminals will find a way round these new systems, but they will need increasing sophistication and knowhow to do so, and they will of course find themselves confronted by a police force whose efficiency and power has been greatly amplified by complicated computer backup.

Computer teachers

Fascinating though these developments might be, nobody could say that they were of world-shaking importance, and if this were all that computer science had to show for itself after 25 years one could only marvel at the speed with which it had brought super-sophisticated but totally trivial gadgetry into an already gadget-ridden world. But this gadgetry represents only the first public impact of an enormous industry, and the more significant effects of computerization will be slower to materialize and less easy to identify. Nevertheless, the next five years will see an increasing impact of computers in two vital areas of society where experimental work has been slowly grinding ahead for the past decade – medicine and teaching.

Both are areas where a vast public need is served by a drastically limited supply of experts, all of whom need to be specially selected and expensively and lengthily trained. Until recently experiments employing computers as substitutes for doctors or teachers have been handicapped by the high initial capital costs of experimenting, but within the last year or two, however, a tremendous computer-assisted learning project, Project Plato, has sprung into action in the State of Illinois. Here, operating from a central computer in Urbana, in the southern part of the state, hundreds of schools participate in a scheme in which a significant proportion of their curriculum is actually taught by the computer. Pupils sit at terminals of the TV-set variety which can display in addition to text and diagrams, colored slides and even dynamic graphic displays. Some of the terminals even have the facility for outputting sound, and on a recent visit to Urbana I saw a class full of ten-year-olds learning the principles of musical notation with a number of these terminals, the computer generating a tone, the children identifying it and, depending upon their response, correcting them, or leading them on to the next note. Most significantly the computer program to teach the lesson had been written by the school's music teacher herself – a woman who had no special training in computer science. PLATO is the world's first major CAL project, but others are not far behind. For example, the ALOHA network links a chain of remote Pacific islands to a computer in Hawaii, the information being beamed to the schools by satellite, and the Indian government has a similar plan to link some of their rural areas to a central teaching system. Despite the success of these projects, which would have seemed like way-out science fiction if they had been spoken of 20 years ago, in a curious way they are almost already out of date. The reason is that their pioneers were unable to foresee just how rapidly miniaturization and cost cutting would progress. In many cases microprocessors attached to terminals, or even to small boxes barely bigger than today's calculators, will in due course provide all the power needed for the majority of interactive teaching, with of course all the added convenience of immediate availability.

Your doctor could be a computer?

In medicine the most promising developments to date have been in the use of computers for automatic interviewing of patients in hospitals. Computers, programmed with friendly, tolerant personalities, now chat to patients in numerous hospitals in Britain and the USA, the patient responding by pushing buttons indicating his or her answers – or even, in some very recent experiments, speaking their answers to the machine. Biggest surprise of all perhaps is that the patients, far from disliking the experience in the way that the computer's 'Big Brother' image would seem to suggest, often say that they prefer the computer to the doctor. There are a number of reasons for this preference, one of which is that the computer always exhibits friendliness, tolerance and above all, patience, thus scoring over the average overworked hospital doctor. The second is that the computer has a definite personality (albeit one programmed into it) which heightens the sense of communication and rapport struck up with the user. Carrying this point a bit further more recent experiments have given the computer a voice in the form of a computer-controlled tape recorder, and a 'visual personality' in the form of a computer-controlled videotape recorder in which a moving picture of the doctor comes up on the screen and chats to the patient. Again patient acceptability is very high, leading one to wonder whether the next generation of computers with personalities might not be even more rivetingly effective. Certainly, the patience of the computer whether in a teaching or medical role is a major bonus. Perhaps that is why computer-aided instruction has scored some of its major early successes in the teaching of handicapped people, especially children, where the computer's limitless reserves of tolerance are invaluable.

But the most intriguing developments in teaching by computer are going to come not from an expansion of existing teaching methods, but from a fundamental overhaul of the whole principle and theory of education. One of the pioneers of CAL Professor Seymour Papert of MIT, insists that the real power of the computer is that it teaches people not *facts*, but *how to think*. He has shown that young children, some aged only six or seven, can quickly learn to write computer programs, some of considerable sophistication, and that by learning to program they are teaching themselves the principles of logical thought. The long-term effects of giving children access to computers at an early age aren't obvious, but they are unlikely to be anything but beneficial. Incidentally large question marks are raised at the spread of sophisticated calculators in schools. Will mental arithmetic go out altogether, and if so does it really matter? Will constant playing with calculators allow children's grasp of mathematics to flower or to atrophy? The most likely bet is that mathematical sophistication will become in principle available to all – programmable calculators will cost less than a carton of cigarettes within a year or so. Of course some people will avail themselves of this opportunity while others will not, but for those youngsters who steep themselves in mathematics and explore its endless permutations with the fabulous calculating tools at their fingertips a new dimension of understanding is likely to emerge.

To summarize, the five years ahead will be mainly a period in which the fruits of computerization are trivial and gimmicky, designed to make life more entertaining and titillating, rather than to change its quality in any significant way. But the growth of projects to tap and amplify the power of computers for education and social purposes, already under way, will speed up around 1980 and with these will come the first signs of a massive transformation of man's society.

The medium-term future (1982–2000)

Making predictions about technological developments more than five years ahead is a chancy business, particularly in a realm where change is accelerating. In the short-term, one is largely predicting improvements in existing gadgetry, but looking further into the future requires one to imagine gadgetry which may not yet have been invented. But there are some moderately confident predictions one can venture about the period 1982–2000. To begin with, it will see the first widespread use of robots – not metal men cast in the traditional mode of science fiction, for such devices would have more curiosity than practical value, but public service vehicles performing their tasks without human guidance. Initially these will perform menial or unpleasant jobs like garbage collection, ditch digging, road laying, etc., and will also look after public transportation. By the end of the period most main highways in developed countries will have built-in vehicle guidance devices, smaller than the cats' eyes of today, which will control the speed and course of robot trucks and cars. Aircraft, both commercial and private, will do most of their flying under computer control, with all navigation, takeoffs and landings absolved from human fallibility. Most maritime transport will equally be computer controlled. Robots too will work in factories, though the old idea of an army of metal slaves clanking away tightening up nuts and bolts is a bit passé. Tomorrow's factories will be more properly looked on as each a kind of giant robot, the whole building being controlled by a central computer and with minimal involvement from humans. In the service industries all booking systems from hotels to transportation will be handled by computers; telephone orders for any goods will be universally accepted – or more likely orders via one's home TV-computer console – for the central computer will check back with the home computer and no doubt the bank computer and cause whatever financial transfer is required to automatically take place. Money in fact will almost totally disappear with credit cards, probably linked for security purposes to one's voice or fingerprints, being the prime channel of financial exchange. The disappearance of money will inevitably lead to a shift in the patterns of crime – when credit cards supply all one's needs and are the only way of obtaining goods or services, and when these transactions are instantly inspected by a central computer capable of determining whether the card has been stolen or not, old style robbery will become pointless. In the short-term thefts of material goods and even food from homes, with weapons employed to gain violent entry will become more common. But police forces will make increasing use of computer aids, including instant checks on data banks, and, very probably, constant surveillance of public roads and buildings by miniature TV scanners linked to central computers. Records featuring personal files on all individuals will probably be kept in these central computers, and no doubt there will be public dismay at this development. Some of this uneasiness will be quelled by allowing any individual immediate access to his own record at any time to check for errors and inconsistencies, and many people will feel that the presence of these banks will be a small price to pay for the dramatic reduction in crime and corruption which will result. Such computerization of police records and law enforcement will, however, be unavoidable, otherwise society will be overwhelmed by a wave of crime as criminals employ the power of computer technology against a hamstrung police force.

If this seems to us to be a depersonalized, automatized world, then to some extent that feeling is justified, and there is little doubt that our social environment and life space will change markedly along a number of dimensions. An examination of these dimensions will give some idea of the wide scope of change.

The three-day working week arrives

The first and most obvious dimension of change is that of patterns of work – indeed the whole fabric of industrial society. The Luddites of the 18th century took spades and crowbars to smash the spinning jennies which they feared were going to ruin them. In fact the mechanization of all forms of production which heralded the industrial revolution erased some jobs, but created work for millions and sent the standard of living of the whole industrial world leaping. Computers, whether in the form of industrial or service robots or of information-processing devices exhibiting ever greater degrees of intelligence, are likely to have an even more radical effect on a society than did the mechanical shovels and looms of the 19th century. But they are unlikely to create more jobs. Inevitably, as world markets fail to expand exponentially, even the increased demand for more and more desirable mechanical gimmicks will not be able to match the limitless ease with which the factories of the future will be able to manufacture them, these factories steadily requiring fewer and fewer human beings to man and control them. Like it or not, our medium-term future will see fewer and fewer jobs on offer, and shorter and shorter working hours. Nor will there be any point in individual countries attempting to preserve the work-patterns of the past by resisting automation and computerization, for any country which did would overnight become hopelessly uncompetitive. At first thought this might seem to confirm the worst possible predictions of the doom merchants – millions on a shrinking dole while factories roll day and night controlled by a single computer. At second, or rather at deeper thought, the problem is nowhere nearly so hopeless. Jobs, after all, are (in quite the majority of cases) merely strategies adopted by people to earn money in order to keep themselves alive, and purchase the desirable products manufactured by other people. The increasing wealth of the industrialized world has been steadily leading to a decreasing workload, or put another way, increasing leisure time. This pattern of increasing wealth must continue and the medium-term future therefore is likely to witness a sensational reduction in the length of the working week and even of the working life, which may last as little as 10 years in all. If the objection is raised that people hardly know what to do with the leisure time they have at their disposal already, then the rejoinder should be that people must learn to enjoy leisure, sustained leisure, in the way that some people enjoy sustained work. Man is an active, creative being, born to explore the universe around him and equipped with a fabulous brain to help him. There is no hint of a suggestion that his genetic structure, his psychological needs and drives, or even his social heritage demands that he spend a vast chunk of his life poring over figures in a neon-lit office, driving a diesel engine through smoky streets, or watching a million bottles a day pass him by on a conveyor belt. As we move towards the end of this century therefore expect to find man throwing off the shackles of compulsory mindless labor, and embracing the new intellectual riches which the computer will provide. And the key to these riches will come from the second dimension of transformation caused by computerization – the universality of learning and knowledge.

Since the time when social reform was considered to be a recognizable and worthy goal it has been realized that of all the chains that bind man, ignorance is the most cruel and the most unyielding. Once, education and the benefits it brings, was considered to be the prerogative only of wealthy privileged classes. Mass education methods which got under way in Western society only a century ago have been hampered by three factors : (1) the enormous cost of providing even sketchy group tuition, (2) the explosive population growth which constantly outdated the school system and, (3) the lack of any real understanding of how teaching should be done. The 1980s and 1990s are going to see, with an accelerating suddenness, an end to these three barriers, roughly in the order they are listed above. First, the cost of microprocessors with an almost infinite range of powerful interactive teaching programs will plummet to the point where these devices will be readily available through the state educational system, and effectively give personal tuition to everyone. What's more, the personal tuition will be graded precisely to suit the pace of the learner's development, and will be capable of testing him, highlighting his weaknesses, guiding and motivating him. Second, the rising tide of people to be educated will be more than matched by the spread of computer-aided teaching, and as so much of education will be achievable at home – or anywhere that the pupil requires – the problem of housing the student population will cease to be significant. The middle of the 1990s will probably see the disappearance of formal, group education along traditional lines, and the transformation of the present school system will be apparent a decade earlier. The third great barrier to universal education – the lack of any true understanding as to how teaching should be most efficiently done – will also fall. In the first instance the development of the computer tutors will force their designers to consider how best these devices can be used, and thus in turn force them to concentrate effort on discovering the optimum strategies of teaching. Later, as computers advance in intelligence, they will themselves be put to the task of researching and discovering new teaching methods with the inevitable outcome that for the first time man will come to understand the nature of teaching, learning and memory. With this revolution in the theory and practice of education the cultural, artistic and scientific heritage of mankind which has hitherto been in the hands of a minority, will be available freely for all. Thus the barriers which divide man so bitterly at present and which are substantially cultural and educational will also collapse.

The third, perhaps rather broader dimension concerns the

structure of social systems, political and economic. I have already indicated that patterns of work and leisure will be profoundly altered and, one can hardly expect traditional political themes, whether Capitalist or Communist, to survive. The convulsions of the old political systems as they realize their approaching demise may well shake society, and repressive and authoritarian action on the parts of some factions may occur. In the early stages opposition, possibly militant, may come from trade and labor unions who, incorrectly, or out of an inbred conservatism, see the computer only in terms of its threat to work and weekly income. Later, after these fears have been overcome, a further burst of opposition – possibly more intense – may come from the extreme right wing who see – correctly – the end of Capitalism in its exploitive form, and from those who, for one reason or another, seek to preserve the rights of a privileged minority or 'ruling class'.

Another radical development will be the increasing role that young people will play in society – not in their traditional guise as agitators and protestors but in holding political office and executive or management positions in industry. In our present society age carries with it one major bonus only, knowledge and experience, and it is this which tends to hand all top decision-making posts to older people. As we move towards the end of this century, the computer will provide more and more 'instant knowledge' at people's fingertips, and the sole advantage of having an old brain will slowly be eroded. This tendency for 'youthful power' will be encouraged by the rapid shift towards early retirement which is already under way in the 1970s, and which will become very marked in the medium-term future. By 1990, for example, most people will be retiring at the age of 40 or thereabouts.

Computer war generals

Most political, social and probably personal planning will be done on the basis of computer modelling – feeding relevant data to the computer (more likely the computer will gather its own data routinely) and getting it to map out the likely consequences of various decisions. Tremendous weight will be attached to such predictive modelling, because of the accuracy with which the super-computers of the 1980s and the even more super-computers of the 1990s will be able to predict the future – at least in comparison with the hasty hunches generated by even the most solid and well-informed of human 'experts'. Anyone who doubts the potential of computer modelling to influence public opinion and action should remember that the whole of the modern surge to 'ecology' and the rejection of throw-away capitalism arose because of the first large-scale computer simulations – the warning from the Club of Rome that continued profligacy with natural resources would lead to the destruction of modern society. The increasing confidence that all people in decision-making positions will attach to computer modelling will have far-reaching effects, but none perhaps so important as those involving militaristic planning. By the 1980s almost all military and tactical decisions will be computer-based, a dehumanization of warfare which will be entirely beneficial as the increasing computerization of weaponry will lead to a gradual reduction of human involvement – as weapons will fight weapons, with weapons the major casualties. More important, however, will be the fact that computer modelling will almost certainly eliminate the risk of any giant international conflagration. Any major power contemplating war will find the computer always spelling out the same grim message of nuclear holocaust, with destruction to both attacker and attacked. For the first time in history man will be obliged to take the advice of a truly independent arbiter when it comes to deciding whether to fight or not, and the computer's cold calculations will not go unheeded. What will, at long last, be rejected will be the obsessive hunches of dominating megalo-maniacs, the Hitlers and Strangeloves, and on a lesser scale of their blindly optimistic generals. Indeed, computer sim-ulation is already in full swing in the war rooms of all major powers, and it is worth speculating that the world already owes its continued existence, and the fact that no nuclear weapons have been used in anger since 1945, to the fatal prognostications of the Pentagon's computers.

The fourth dimension of change in the medium-term future in which the computer will have an overpowering effect will be that of science, for as science is essentially a philosophy and set of techniques for acquiring knowledge about the universe, the power of the computer will be particularly well applied. At the immediate and obvious level, advances will occur in medical science which may lead to the elimination of diseases and viral infections, the eradication of cancer and just about any non-accidental disruption of the body's processes. The consequence of this will be a rapid rise in the mean life expectancy, which will probably be 100 before the end of the medium-term future. On a more profound note, and one far less easy to assess, the computer will enhance our understanding of the universe and its basic laws to a remarkable degree. This may not necessarily be the blessing that it seems – already the universe is revealing itself to be a disturbingly complex and mysterious place, and it seems quite possible that the extra information about the universe which tomorrow's scientists will amass may make the universe seem even more complicated, more mysterious and more inhospitable. The psychological consequences of this may not be all that easy to handle, but by then we shall have computers around to help us in our moments of uneasy insight.

While the first four dimensions – patterns of work, education, social and political structure, and science – will all be changing sharply from about 1980 onwards, and will peak in the 1990s – the fifth dimension will begin slowly and really have its most radical effects in the long-term future. Nevertheless it will begin to impinge upon mankind and his society in the next decade or so. This is the dimension featuring the evolution of intelligence in computers – the creation of artificial minds.

The meaning of intelligence

Before developing this point, some discussion of what is meant by the term 'intelligence' is required. Most people assume that intelligence is a property of human beings only, and are often surprised at the suggestion that animals could be said to exhibit intelligence, and incredulous at the idea that machines might. This is based largely on a misunderstanding of what the word means. Intelligence in fact refers to the ability of any system, biological or mechanical, to adjust itself to changes in the environment. The more versatile its adjustment the more intelligent it is. To give an extreme example, a rock or piece of dead wood is purely at the mercy of environmental forces and cannot 'do anything' to escape from anything that 'threatens it'. If one had to make such a judgement, one would therefore rate a rock's intelligence as precisely zero. On the other hand a simple living thing such as an amoeba can avoid irritating substances, move towards 'pleasant' ones, and reproduce itself. Thus it has some degree of versatility in response to changes in the environment, and therefore a rudimentary degree of intelligence. Further up the scale in the case of, say, a beetle, the versatility of adjustment is greater, for the creature can do a largish range of things, all related in some practical way to changes taking place in the world about it. It can even learn simple things – store useful information in its brain which allows it to increase its versatility. Right up at the end of the existing scale of course comes man, with his stupendous capacity for learning, reasoning, problem solving, linguistic communication. Of course comparing man and beetle, not to say amoeba, on the same intelligence scale is rather absurd, but a crudely graphic approach might be to say that if a rock has an absolute intelligence of zero, an amoeba might be said to score say 10, a beetle, say 100 and man, say, 1,000,000. Anyway it is clear that intelligence is not just a function of man's brain and of no other type of brain, but of *any* system which is capable of responding selectively to environmental changes. From this the way is open to accepting that machines can exhibit intelligence, and are in practice already exhibiting evidence of this. For example, most computers can already make numerical calculations thousands, perhaps millions of times faster than any single human being. Their memories are also already vaster in terms of their basic capacity than any individual human's. Of course these are just aspects of intelligence, and at the present time computers are not particularly good at learning (i.e. modifying their own programs on the basis of their previous experience) though some degree of learning is present in some specially designed machines; nor are they particularly good at problem-solving, though they can be programmed to do so; nor are they particularly good at understanding or using human language, though some machines can (just) understand a small vocabulary of spoken words, and some are now very good at reading printed text, and even handwriting. But they are getting steadily better at all these tasks, and the pace at which their intellectual capacity improves is quickening. Furthermore, effort is already being directed at a growing number of universities and research establishments at enhancing this intelligence. Big projects are afoot at the key research centers of MIT in Boston and Stanford University in California, at Sussex and Edinburgh, in Russia and in Japan.

The main thrust in artificial intelligence at the moment is towards making computers which can replicate routine, but still at heart highly complex tasks – assembling gearboxes, reading columns of figures, and even recognizing particular types of human blood cells under the microscope. But the most intriguing research is devoted to controlling computers by 'natural language', and here some fairly remarkable advances have already been made. A celebrated computer program developed by the young American Terry Winograd, for example, has been taught everything there is to know about its own private universe. This universe consists of a small room with a number of blocks and shapes scattered about in it. It can tell where the blocks and shapes are at any time and can answer questions about them when these questions are typed in – in plain English – on its terminal. A good example of the spooky conversations one can have with this computer program is given in note 1. Of course the universe it understands so exhaustively is a microcosm of the universe that humans inhabit and are aware of, and is even minute in comparison with the kind of universe 'known' by, say, a beetle. Nevertheless it exhibits behavior which indicates that it does have at least rudimentary intelligent knowledge and it can perceive the relationships between things and grasp simple concepts – things are triangular, square, large, small, etc.

Is the next chess master a computer?

Another potentially fruitful area of research is in developing computer programs to play chess. Once again progress has been slow, but much quicker than anyone expected 25 years ago when it was argued that no computer could possibly play anything but the most childish game of chess since all that it could do would be to scan through all the many millions of possibilities after any particular move and try to work out which would be the most successful. This process would take so long that hundreds of years would pass before the computer would make its second move. This kind of argument had to be abandoned when programs began to appear which instead of searching through every possibility in an almost supremely unintelligent way, adopted strategies and tactics. Less than 20 years later good programs now play more than a passable club game and computer programs are in fact already the chess champions of a number of university clubs in the USA. On occasions they have beaten people up to master standard. But the important point here is that what is clearly an intellectual task – chess playing – embodying many of the rules of reasoning, conceptualization and even thinking, is already being tackled in a genuinely impressive way by machines. At this point the objection is usually raised that this is all very well but of course the machine had to be programmed to do so. True, but then so do human beings, who have to be taught the rules and strategies of chess just as do computers. If the next objection is that no computer could ever *devise* a game like chess, then that is less easily rejected, but even that does not hold water on close analysis.

Many people, when faced with these arguments, will admit that machines could probably be made intelligent, and indeed do exhibit some intelligence now, but will resolutely draw the line when it comes to talking about 'thinking' machines. No computer, they argue, no matter how cleverly programmed could possibly be said to be capable of thought. Or could it? Of course if one defines thought as some God-given property of human beings and human beings only, then one necessarily rules it out of court for machines, but this is unrealistic and unhelpful. The real tactic is to ask oneself, how could one know whether a machine was thinking or not? It's obviously not good enough to simply say you will believe a machine is thinking when it tells you that it is – you could easily program a telephone answering machine to say 'I am thinking' when it obviously wasn't. An ingenious solution was first put forward as far back as the 1940s – by the great English mathematician and computer pioneer, Alan Turing. He argued that if, in conversation with a computer by way of something like a typewriter terminal, you could not be sure whether what you were chatting to was a computer or a man on the other end of the line, then the machine would have reached a level where it could be said to be thinking. This does not mean of course that it had emotion, that it could see, that it could hear, that it could have needs or wants, nor that it looked anything other than like an electronic device. But if it could talk with you *in such a way that you couldn't be sure whether you were talking to machine or person*, then for all practical purposes you would be dealing with a thinking machine. The argument is a powerful one, and is based of course on Turing's realization that a major feature of thinking is undoubtedly the power to process linguistic information and communicate in a human-like way. Significantly, Turing was confident that machines would one day pass what has now come to be called 'Turing's Test'. Significantly too, many of the other great pioneers of computing and indeed most of the scientists working in this area today would agree. The only thing that would probably have surprised Turing – he died tragically in 1952 – is how quickly computer intelligence has been advancing, and how quickly the thinking computer is likely to be upon us. Many computer scientists for example believe that we may see it within the medium-term future, but even if that is being optimistic by 10 or even 20 years it doesn't really matter for clearly the arrival of machines that think will have a more revolutionary impact on man than any technical or scientific development in history. For the first time man will have on this planet, a being – the fact that he has created it himself is quite beside the point – which exhibits intelligence of a kind which he has always believed is unique to himself. And this being, or beings, will be able to communicate with him on a par, exchanging information and developing ideas. The consequences will be various, but the first and perhaps the foremost will be that man will need to reappraise himself, and see his own personality and identity within a totally new context. It is not clear at this stage just how he will respond and adjust. One possibility indeed might be that he will be so horrified at what he has done that he will immediately set out to destroy the thinking machines, but that of course will not prevent them being generated again at some other time and place. And what are the ethical and moral problems associated with the destruction, let alone the creation of a machine with a self-consciousness and self-awareness – both properties which would seem to spring inevitably from giving the machine the power to think? The question marks are, as you can see, numerous and large. But they are negligible compared to those which will arise as we move from the medium-term into the distant future. For to date we have only been talking about machines with an *equivalent* intelligence to man. We now need to consider, and face up to the staggering scenario which will begin to unfold as we move a step beyond, to the date when man is no longer the most intelligent creature on the planet earth.

Beyond 2000

The permutations of possibilities caused by the creation of artificial intelligence in a computer roughly equivalent to that of a human being are almost unthinkable, but that such a step will be achieved in due course is considered to be not only likely *but inevitable*. The main point at issue facing scientists in this field is *when* it will come about. Objections can of course be made. Some people will argue that it is intrinsically impossible for a computer to rival man in intelligence, but if pressed to give reasons for their belief will rarely have anything to go on other than to say, 'It couldn't be.' But the brain is a device for receiving, storing and processing information and while it is a magnificently efficient one, unmatched at present in the known universe, there is no shred of evidence to suggest that it could not be matched by some other device developed from scratch, or that it necessarily represents the end-point of such information processing devices. The fact that it is biological, and not mechanical is sometimes raised, presumably implying that in some way its biological qualities give it a magical 'something else' which no machine could exhibit. Again there is no evidence that this is so, and, as we've pointed out many computers are already exhibiting types of intellectual processing which are very brain-like in the kind of jobs they perform, whether or not they perform them in the way that the brain does. Another superficially powerful but basically weak argument is that today's computers are relatively stupid despite all the effort that has been put into developing them. In fact, computers have reached their present – admittedly still sub-human standards – in a little over a quarter of a century, a pace of evolution which has a lightning-like quality, whereas the brain has reached its present stage of wondrous power only after an immense period of evolution – hundreds of millions of years.

Another argument frequently advanced is that a computer

can only do what it is programmed to do in the first place. This argument is rendered quite invalid when one replies that the same is undoubtedly true of people. Humans come into the world with an inherited bank of computer programs or software which we know as instincts. The rest of an individual's psychological makeup, including his personality, his knowledge, his approach to social and personal relations, his language, his artistic and creative skills are all programmed into him by the process known as life and learning. In other words, we are all a product of our heredity and our environment, just as are computers. To the further objection that we are self-programming in many respects, the answer is again that so are many computers, if only in a very small way at present. But there is no reason to suppose that they will not become totally self-programming in due course – provided that we find it useful for them to be so. And here comes a new twist to the argument, which boils down to the phrase, 'Why do it, anyway?' Why go to such elaborate lengths to develop a computer, which will mimic a human when there are literally thousands of millions of these 'natural' humans around already? This appears to be a compelling argument but it is by no means a clinching one. In the first place, by learning how to create brains synthetically, we may well learn enough about *natural* brains to be able to uncover some important secrets which will lead to significant improvements in our own brain powers. The second argument is that man badly needs extra intellectual power to help him gain control over his ever more complicated world; much as he needed machinery to assist the limited power of his muscles, now he needs computers to amplify the limited power of his brain. But the third argument is the most significant and within it lies the key to understanding what may well take place with computers in the long-term future.

Computers master over the human race?

The argument is simply this: whereas there is probably a theoretical limit to the power of the human brain, and that any major development in man's intelligence would be a process stretching over millions of years – there appears to be no such theoretical limit to the development of machine intelligence. Furthermore this evolution, instead of taking millions of years might well be encompassed in decades. The chilling argument has perhaps best been put by the distinguished mathematician and philosopher, Professor I. J. Good, himself one of the pioneers of computers. Good argues that the course of current Artificial Intelligence research will lead to the creation of what he calls an 'ultra intelligent machine' or UIM for short. The UIM is a machine which can perform any intellectual task as well as any human. It could have emotions, and it would not need a physical body, only the means to communicate with the outside world. And the UIMs, according to Professor Good are likely to be with us at just about the turn of the century, give or take a decade, and probably within the lifespan of most people reading this book. So when the UIMs come, what then? The argument proceeds as follows. One of the first tasks one will set the UIMs will be to solve the problem of improving their own intelligence. How they will do this does not have to be specified – they will simply be given the job of improving their own software and what other aspects of their 'brains' it is necessary for them to evolve. Progress will at first be slow but because there are a lot of them, and because they can go ahead single-mindedly they will inevitably progress faster than man would have been able to. So, shortly there arrives the next generation of UIMs, slightly, but measurably more intelligent than man. And now, these new UIMs set out to continue the task of developing their own or their successors' IQs, but because they are brighter than the first generation they will be even more successful and the process will proceed more rapidly. Generation three of UIMs now comes into being raising its own IQ so much that generation 4 is perhaps twice as intelligent as the most intelligent human. Barely have these massively intelligent machines come along than they are spawning generation 5, and with each generation the IQ jump gets greater and the pace of evolution gets hotter, until suddenly the machines we have created have left us far behind. And still the process continues remorselessly, for computers are now in control of their own evolution – and their intellectual power soars upwards out of sight. And what now happens to man whose curiosity, drive and intellectual brilliance set this awesome advance in motion? How will the computers treat their creators, those creators whose relative stupidity becomes exponentially greater with each passing year? Will they care about us at all? Perhaps not if they have not had emotions built into them. And what will the computers themselves do with their incredible new freedoms? Presumably, they too will attempt to explore and understand the universe, but what they will find is beyond the scope of this article.

To many people the idea that computers could evolve in this spectacular way is horrifying, and indeed there is something more than unsettling in the idea that work now in progress in laboratories and research centers across the world might be sowing the seed for the eventual replacement of mankind by another species – the super-intelligent computer. The species, which will be capable of reproducing itself, will probably not be biological but there is no reason why it should have to be. We ourselves are machines, if made of living cells rather than minute electronic components, and we, like computers, exist to handle and process information. The sole rule of evolution seems to be that what is most fit to survive does so – there are no built-in subclauses which say that the game can only be played by biological beings. Nor is there any law which says that man is himself the pinnacle of evolution and the ultimate master of the universe.

Perhaps in the vastness of outer space there exist other civilizations with beings stupendously in advance intellectually of our own, and perhaps in due course the super-intelligent machines that we have created will go out and meet them. If they do they will find a friendly, rather than a hostile reception. For it seems to me to be inevitable that the creatures who dominate other solar systems in other galaxies will be like those who will come to dominate our own. They too will be machines.

APPENDIX 1

The following conversation took place between a computer and a research worker using Winograd's System. (Lines in capitals are generated by the computer.)

Pick up a big red block.

OK.

Grasp the pyramid.

I DON'T UNDERSTAND WHICH PYRAMID YOU MEAN.

Find a block which is taller than the one you are holding and put it into the box.

BY 'IT', I ASSUME YOU MEAN THE BLOCK WHICH IS TALLER THAN THE ONE I AM HOLDING.

OK.

What does the box contain?

THE BLUE PYRAMID AND THE BLUE BLOCK.

What is the pyramid supported by ?

THE BOX.

How many blocks are not in the box?

FOUR OF THEM.

Is at least one of them narrower than the one which I told you to pick up?

YES, THE RED CUBE.

Will you please stack up both of the red blocks and either a green cube or a pyramid?

OK.

Which cube is sitting on the table?

THE LARGE GREEN ONE WHICH SUPPORTS THE RED PYRAMID.

Had you touched any pyramid before you put the green one on the little cube?

YES, THE GREEN ONE.

When did you pick it up?

WHILE I WAS STACKING UP THE RED CUBE, A LARGE RED BLOCK AND A LARGE GREEN CUBE.

Why?

TO GET RID OF IT.

Why did you do that?

TO CLEAR OFF THE RED CUBE.

Why did you clear off that cube?

TO PUT IT ON A LARGE GREEN CUBE.

Why did you do that?

TO STACK UP THE RED CUBE, A LARGE RED BLOCK AND A LARGE GREEN CUBE.

Why did you do that?

BECAUSE YOU ASKED ME TO.

ELECTRONICS

Brian J. Jennings

Editor of *Electronic Engineering* – Morgan-Grampian Publishers.

Nearly every human effort depends on electronics, to a degree to which those involved do not fully appreciate. Crystal gazing into the next century, it is impossible to imagine where electronics will not touch human lives. Yet, electronics is a science younger than this century, and even fifteen years ago – just before the breakthrough into micro-electronics – nobody could have predicted that technology would have reached the point where it stands today. Perhaps the biggest phenomenon is the amazing rate of change which it is causing – not only in the technology itself, but in every aspect of its application. In fact, it's beginning to present a major challenge to all concerned.

Nobody is immune from this technological 'octopus', as its tentacles reach out to an ever-increasing range of activities. The manager, the scientist, the engineer, the academic, the doctor, the airline pilot and even the housewife, are all affected. So are cars, machine tools, aircraft, ships, shops, factories, offices and homes.

Electronics works as a slave on the sea-bed and in space; in marketing and on the moon; in hospitals and on the race track. And yet this revolution only began with the thermionic valve, developed after the Wright brothers first proved that heavier than air machines could fly. Radio was the first practical and commercial development; with the advent of two world wars to accelerate development, the second leap forward came in 1947 with the more mature semiconductor technology. Then came a new world of solid state devices on a miniature scale, offering unparalleled benefits of speed and cost.

Pre-transistor equipment is now viewed as crude and clumsy, yet it managed to produce such far-reaching techniques as radar. Even computers were developed before the transistor, albeit elementary, enormous and slow by modern day standards. The revolution, in hindsight, reached its peak in the late 1950s, with the successful breakthrough into the diminutive world of micro-electronics – a new series of micro-miniature electronic devices in the solid state, offering almost limitless scope in designing and producing complete circuits on a tiny chip of silicon, not much bigger than a pin-head. However, the conventional technology of the day – based on the transistor – had been developed to quite a sophisticated level. Thus it was possible to fit essential electronics to earth satellites and space probes, to take the computer to a more advanced stage and to start an entirely new industry, that more than any other changed the post-war industrial scene of Japan – the transistor radio. Micro-electronics, the science and technology of fabricating complex electronic circuits to infinitesimal dimensions, has led to electronic equipments of higher performance, lower cost, enhanced reliability and smaller size than was previously either possible or envisaged. Today micro-electronics stands as the foundation of the industry's total future, offering a potential of almost incalculable proportions. The scene in the next century cannot be precisely predicted, but clearly the efforts of this industry – the first ever to have such a universal range of applications in other industries and endeavors – will be aimed at making life easier. Advanced electronic control techniques will take the drudgery out of most work; the factory and office will largely be the arena of automation and even housework will be more a question of efficient programming, rather than of tedious chores and synonymous with this increased leisure time, itself, probably assisted by electronics.

This outlook poses some of the largest problems that this relatively young industry has had to face. Until recently, the electronics industry has been more concerned with developing the technology, now many of the applications dictate the technology. Yesterday it was hard to imagine another breakthrough in the same order of magnitude as the silicon integrated circuit – today we are at the watershed of a new world – the world of the microprocessor, just fifteen years after the industry came to adolescence.

Although modern electronics is relatively young it has already made a great impact on our lives, often in ways of which we are not aware. Much of the growth of electronics has relied upon telecommunications on which modern society depends. In the relatively short time since the Second World War, electronics has taken giant strides. The results show that we are now capable of putting men on the moon and putting satellites in space to send messages vast distances over the earth's surface. It has also been responsible for significant advances in the medical field, traffic engineering, chemical analysis and a host of other areas.

When the father of electronics – electricity – first appeared people could see immediate benefits in better lighting and heating for homes and a cleaner method of providing power to machinery to turn the wheels of industry. However, electronics is more subtle in its effects. It has shortened the distances between continents by our ability to communicate via satellites. It has relieved us of many of the tedious tasks of working life in offices. Most businesses use computers to store information such as bank account details, in a much smaller space than endless reams of paper could ever do.

Electronics has made travel much easier and safer. For example, nowadays the electronics on board an aircraft accounts for a third of the total cost of the aeroplane, equal in fact to the value of the engines. Most of the equipment comprises communications, navigation and control systems but all are interdependent since the aircraft navigates by receiving pulses transmitted from ground beacons. The flight controller monitors many parameters within the aircraft and ensures that the aircraft is responding correctly to the pilot's manual controls. Also the flight controller can take over the pilot's function and perform all the controls to manoeuvre the aeroplane. By tying in the flight controller into the navigational equipment automatic landings can be made which are important in poor weather conditions.

Back on the ground, we see electronics in the street at every traffic light. Some traffic light sets are monitored by a central unit so that if the lights malfunction at a particular junction the total traffic sequence can be altered if necessary to compensate. In the future cars will themselves contain more electronic bits and pieces and we can see electronic ignition and fuel injection systems in some vehicles already with the promise of electronic dashboard instrumentation to come.

On the railways, there are many stretches of track which are monitored and operated electronically. This means that the position of any particular train is known and adjustments can be made if any failures occur. Where several trains are using the same stretch of track, priorities can be given to the fast or through train, rather than the local train, which has to make frequent stops anyway. This means that electronics must be extremely flexible and have the ability to function correctly in many difficult and hazardous environments.

As we shall see, it is not only transport and communications where electronics has made its mark – there are few areas which have remained free from its influence and as we will explain later, the remaining aspects will not escape for much longer. Unfortunately electronics has its destructive role, and as with many other technologies, much of the breakthroughs have come from the desire to maim or destroy but ultimately it brings some benefit in relieving pain and suffering.

Today sophisticated missiles guided by electronics can unerringly find their targets. Military aircraft fly with special computers used solely for weapon-aiming purposes. However, as with the horrifying prospect of atomic warfare the increased sophistication of weapons makes us more wary of using them.

Medical electronics has been a great compensation for the games of destruction in which the human race has a tendency to indulge. Doctors now work with tools such as lasers and ultrasonic techniques. Lasers have a special use in eye surgery for welding the retina to the back of the eye in cases where it has become detached. Detached retinas cause blind spots at the points of detachment. The laser (which means light amplification by stimulated emission of radiation) produces a thin beam of highly concentrated light which can weld

tissue together. It can also be adjusted to destroy any unhealthy tissue which is likely to cause cancer.

Ultrasonics too has several medical uses. One is to destroy
tumors because in some instances the high frequency sound
when directed at the unhealthy tissue causes it to heat up
and be destroyed while unharming the healthy tissue. This
treatment is often used in conjunction with X-rays. It can
in fact be used in a similar manner to X-rays to produce
photographs of internal organs and is a much safer technique. In gynecology doctors can obtain a complete picture
of a child in its mother's womb which is so clear that the sex
of the baby can be determined before birth, although the
most important use is to ensure that the child is normal and
healthy.

In the intensive care unit of hospitals, electronics can
monitor all the body parameters such as blood pressure, heart
and brain activity of several patients at once. Heart pace-
makers stimulate the heart by an electrical signal so that if
it contracts it can pump the blood around the body. Surgical
techniques and pacemaker design have become sufficiently
advanced to allow the device to be fitted internally. Also
the development of the battery or power cells will soon allow
the pacemaker to be nuclear powered although we are still
a long way from the bionic age that television would have
us believe.

Our leisure time is even becoming more dependent on
electronics. Television was one of the first electronic devices
to captivate us, although the internal components have almost changed beyond recognition. However, their influence
is becoming stronger with the emergence of electronic games.
By plugging a small block box into the aerial socket of the
television, the screen is transformed into a sports arena.
Games such as football, tennis, squash and badminton can
all be played. Late 1978 games will be available to be played
on the TV screen such as crosswords and Scrabble. Hi-fi
equipment has been placed within the range of most people's
pockets and radios became transistorized and flooded the
world. Musical instruments can now be made electronic to

simulate the sound of instruments, to amplify ordinary ones or to enhance or create new sounds such as the moog-synthesizer and electronic organs.

Electronics has changed the face of time for suddenly electronic watches with their novel way of displaying time digitally have appeared on the scene. It has had a tremendous impact on the mechanical watch industry and the Swiss have had to adapt their ability to produce precision mechanical watches to the electronic accuracy required in digital watches. In an electronic watch a small slice of quartz crystal is caused to vibrate over 56,000 times a second by applying a small voltage. This forms the accurate heart of the system and the electronic circuitry counts the number of vibrations to make the seconds, hours and minutes which are displayed on the watch in digital form. Digital watches are far more accurate than our familiar mechanical counterparts – for example a good electronics watch is accurate to 3 minutes a year, while even a well designed mechanical watch may only be accurate to 3 minutes a month!

However to most people electronics appears to be a mysterious art, veiled in strange practices and, those who are uninitiated, are bewildered by the strange jargon of the electronic engineer. Although this confusion and sometimes mistrust of electronics is not the doing of the engineer who would seek to keep his profession secret, but is the result of the speed with which the technology develops. This means, that what may be up to date one month may be obsolete the next. If the experts have problems keeping up with the changing face of electronics, the layman can justifiably sit down and shake his head especially when he hears words such as integrated circuit and solid state. These terms are often used by salesmen when attempting to sell such items as Hi-Fi's, pocket calculators and even washing machines but usually mean little to the layman, except to impress and more often confuse!

To understand how electronics stands today we must take a short excursion into the past. For indeed, the fundamental principles on which electronics relies today were first explained in the nineteenth century. Electronics relates to a special branch of the electrical sciences and derives its name from the fundamental particle of electricity, the electron. Unfortunately since all matter is electronic in nature, the name is really misapplied to such a specialized subject.

It was J. J. Thompson who first proved the existence of the electron in 1897 but electronics as a separate entity did not arrive until much later – in fact the next century. Although even if we compare electronics today with forty years ago we will find that it has changed to be almost unrecognizable.

Valves were the first step along the electronics trail and appeared as a direct result of radio communications developments. In order to send signals further a need to produce a system of amplifying weak signals which had travelled large distances through the earth's atmosphere arose.

Valves are of course quite large devices made of glass and metal and are still to be found in old television sets and in some radio equipment for special applications such as high power and radar. Although the First World War pushed the development of valves forward it was another invention which was responsible for electronics as it is today – the transistor.

The transistor opened the way for miniature electronics

As we have already mentioned the foundations for all electronics was laid in the last century and the coming of the transistor was no exception. It was 1951 when Shockley patented his design of a transistor, having done work with Bardeen and Brattain in 1947. However, the groundwork comes from Planck's quantum theory developed in 1901, followed by papers from Schroedinger and Heisenberg in 1926–7, and, in 1930 Professor Lilienfield patented a type of transistor design but the device did not work because of the poor types of materials the professor used.

The principle which changed electronics was in the use of

semiconductors which are materials such as germanium and silicon, in which effects analogous to valves could be obtained but this time in one solid piece of material, hence 'solid state' rather than having to encapsulate metal electrodes in an airtight glass. Although the manufacture of transistors from semi-conductor materials are fraught with their own special problems it did mean that equipment could be made more compact by the elimination of bulky valves which also required high voltages with which to operate. The main advantages of the transistor apart from its smaller size was the fact that it could run off extremely small voltages compared with the valve (in the order of volts as against thousands of volts) and it could be more easily adapted to perform many other functions.

But the development of the transistor did not stop there. The transistor is made by taking thin slices of silicon or wafers as they are known, and etching out patterns in the silicon to produce the transistor shape. In order to make the transistor operate, certain areas within the transistor have to have small quantities of other materials which are known as dopants, such as arsenic and phosphorus added to the very pure silicon.

This is done on an extremely small scale – so small that the transistor is barely visible and special photographic techniques have been devised to make the small patterns or masks which are laid over the silicon to determine the transistor shapes. These masks can be produced to make hundreds and even thousands of transistors on one single wafer of silicon. The common diameter for a silicon slice is now three inches. The next move was to make complete circuits containing not only transistors but all the other components such as capacitors, diodes and resistors which go into a completed electronics circuit. This was achieved in the early sixties and the blending of separate components onto one single part of a silicon wafer (since it was possible to make several circuits onto one silicon wafer) the term integrated circuit grew. Designers became more ambitious with their circuit designs and transistors have become smaller and smaller with the advancing years and technology. Even now there is the promise of smaller circuits to come.

This extreme miniaturization which is used in industrial systems, in military applications and for computers sometimes comes to light in more identifiable areas in our everyday life. Computers were the first electronic machines which inveigled their way into our lives – in fact, what bill would be complete without them? The computer has relieved us of much of the mundane work in filing and accounting, although it has many important uses in scientific research and in industry for monitoring and controlling many types of machines and processes.

When computers first appeared they were very large beasts. In fact the first practical calculating machine was built just after the Second World War. It was built in America and was called ENIAC (electronic numerical integrator and calculator). It was made entirely with valves (transistors were not available then) – about 18,000 of them – and occupied 1500 square feet and weighed 30 tons. The temperature was so hot that the atmosphere in which the programmers had to work was more akin to a ship's engine room rather than the clinical, comfortable environment modern programmers are used to. Also there was a great problem of reliability since every time they switched on some of the valves were guaranteed to fail!

As we have seen, the capabilities of that early computer can now be performed by computers which are no larger than a typewriter, fitting snugly on a desk top.

Fear to familiarity

While the amount of information possessed by man is estimated to double every seven years, the human capacity for retaining it remains more or less constant. Thus the student of today possesses a smaller share of available knowledge than his counterpart of 50 years ago – not in spite of educational and technical advances, but because of them. How therefore is the ever-increasing collection of information to be reconciled with the demand for education, given the inevitable restrictions of the human mind? The answer is through more efficient education; shifting the emphasis towards developing in students, not the capacity to retain more information, but the ability to use it. Using just human resources to reach this ideal situation would require almost as many teachers as students, plus a never ending expenditure on aids and information material. However, computers can gather, assemble and store data faster, more accurately and more completely than any other known means.

Simple programming languages, such as BASIC, enable the student to communicate with computers in a conventional way by typing statements on video displays or teleprinter terminals and receiving immediate responses. Time-sharing computer systems have added another dimension, in that several terminals can inter-communicate simultaneously with a single, central system.

Taking the USA as an example, to accommodate 55 million students only 2000 computer systems are in active instructional use; within ten years computer access will extend to nearly all children at every grade level. Today, most of the mistrust of computers as sinister instruments of depersonalization resides in the adult populace, not amongst school children. Today's students will be among tomorrow's teachers and professional leaders. Their attitudes, not of fear but of familiarity, will set the example for their own and successive generations.

However, computers form a very small part of the wide field of electronics. The ability of semiconductors to be made into large circuits in a very small area has meant significant changes in many fields. Companies who need large numbers of special circuits can have them made economically since the greater number of circuits required the more dramatically prices drop.

Since most of the activity of electronics takes place over a small area it appears to perform most of its marvels unseen, and to many people it is impossible to visualize all those electrons whizzing around bits of silicon and in fact it is of little use without the phosphor screen to provide the picture;

an expensive Hi-Fi would be a folly without a loudspeaker to convert the electrical signal into sound. Many such devices do exist in the half world between electronics and human interpretation. Some of the most interesting devices are those involving light-displays. Semiconductors can be made to have several other properties to make it perform several functions. By mixing the right semiconductor materials, the semiconductor can be made to flow or become luminescent as current travels through the circuit. This has lots of uses in many display applications as they can be designed in many shapes. Some of the more obvious examples are for pocket calculator display and as displays for digital watches. The reason why digital watches using light emitting diodes are not continually glowing is that they consume so much power compared with the rest of the circuitry that the battery would soon be flat if the display were continually on. But on the whole the battery should last for several months. Other types of display which are used in watches are known as liquid crystal display. These are not made of semiconductor materials and the effects of liquid crystals were first noted in 1888. On application of a voltage the liquid crystals align themselves into a predetermined set of shapes. These do not glow or emit light as light-emitting diodes do, rather they reflect it. This means they require little power compared with light-emitting diodes and can provide a constant display. Their disadvantage is that the display is naturally not as bright and the ability to read the display is dependent on the angle at which the display is read.

The optical communication revolution

Light in fact will have an important role to play in the electronics of the future. Since as our need for more and more communications grows, the more congested our telephone networks become since conventional telephone cables can only carry a limited number of telephones in the order of 500 calls down one cable. Each telephone conversation is separated so that callers cannot hear other calls by a method called multiplexing. However, light gives us the answer. A method has been devised whereby speech and other similar types of signals for television and computer information can be superimposed in light. The light wave which is generated by a tiny laser carrying speech guided down a fine glass tube known as an optical fibre. This glass fibre has a very special property since it has been designed so that any light launched into it at one end will travel down the tube without escaping through the sides no matter how many bends or loops there are in the fibre. Since we are taught that light only travels in straight lines the optic fibre appears to bend light travelling through it. An amazing feature of the fibre is that the thickness of the fibre is only that of a strand of hair (about 50 millionths of a metre!). Not only that, but this tiny fibre can carry almost ten times as many telephone calls as the large cables currently being used. Its smaller size will give many advantages since it can be used in places which were once inaccessible.

The future for electronics is an extremely exciting one if only we can keep the rapid progress technology makes in perspective and let our enthusiasm be guided by the light of our reason!

Electronics, like any other field of human activity, has its landmarks, events which radically alter the path that the technology follows. Three such landmarks have appeared in the last twenty years – the transistor, the integrated circuit and the microprocessor. The transistor dominated the electronics industry in the 1950s and made mass production of products such as portable radios and computers possible. The 1960s were the age of the integrated circuit and saw the arrival of pocket calculators, cheap electronic organs and many other products that could not be economically made with discrete transistors. We are only just entering the age of the microprocessor but already it is clear that this new device will play an important role in our future. In the 1980s, microprocessors will be everywhere – in cars, washing machines, traffic lights, shops, offices, toys and even in our homes.

The incredible microprocesser — a mini home computer for $150

To understand the importance of the microprocessor, it is necessary to go back to the integrated circuit (ic) and the computer, both of which can be thought of as the parents of the microprocessor. The integrated circuit, it will be remembered, is a vast array of unencapsulated transistors formed on a small slice of silicon. The individual transistors are interconnected so that the final device performs a specific function. A peculiarity of the semiconductor world is, that it costs almost as much to make one ic as it does to make 100,000 ics. This is because most of the cost is incurred in the initial design stage and the tooling-up stage. Once the design is complete and the masks made, the individual ics can be churned out in great quantities at a comparatively low cost. Semiconductor manufacturers, therefore, only make new ics when they are confident that there will be a large market for the products. This means that there are two basic kinds of ic known as standard and custom. Standard ics are possible because most electronic systems can be partitioned into smaller subsystems which can be considered as basic 'building blocks', in the same way that a painter can produce any color by mixing the correct amounts of primary colors. Since any color can be produced by mixing various combinations of red, blue, yellow, white and black paints, paint manufacturers could produce only these colors and leave the creation of other colors to the painter. However, thousands of painters would be constantly mixing, for example, blue and yellow to obtain green so the paint manufacturer produces popular shades of green, brown, pink, etc., to save the painter a considerable amount of time. Similarly, most electronic systems can be built from just four kinds of transistor but the manufacturers allow designers to save time and space by producing ics which carry out commonly required functions. These are the standard ics — oscillators, amplifiers, adders, decoders, etc. — which can be connected together to carry out millions of different functions. If a semiconductor manufacturer discovers that a particular combination of standard ics is used by a large number of different customers, he will probably integrate the whole combination into a single new ic. There comes a point, however, when there is little value to be gained from increasing the level of integration, because the versatility of the ic diminishes as its complexity is increased.

Beyond this point, a customer wanting a higher level of integration can order a custom ic. This means that the semiconductor manufacturer will design an ic specifically for the customer's requirements and, in most cases, will supply the ic only to the customer. The drawback of course, is cost – because of the high cost of design and tooling the customer will need to order at least 100,000 ics. We have here two conflicting requirements : in order to minimize the number of ics needed and to cut down on system design time, the customer wants an ic which performs all his own required functions. The manufacturer, by contrast, wants to avoid the expensive design and tooling stages and simply churn out a standard product that everyone can use. To return to the paint analogy, it would be very convenient if a paint manufacturer could produce a single paint which could be given any desired color by, for example, stirring for the correct number of minutes. Such a 'programmable' paint would be profitable for the paint manufacturer since he could devote all his resources to producing a single standard product. It would also be beneficial to the painter, since he would only have to buy one kind of paint and 'program' it to his exact requirements. Although no such paint exists, its counterpart in the semiconductor world does, for this is exactly what a microprocessor is.

Because a microprocessor can be programmed to carry out particular tasks it is essentially a computer, although the difference is more than just a matter of size. Like all computers, the microprocessor carries out complex tasks by performing simple operations one at a time, although the familiar 'large' computers are much faster and much more powerful. The importance of the microprocessor is that it is

much smaller and much cheaper than a computer, occupy-
ing only a few square inches and costing from about $10.
This means that the computer concept can be applied in
situations where previously, because of restrictions of space
or cost, it would have been impossible. Products using microprocessors are already available, with many more on the
way. One of the first microprocessor based products to appear is the Singer Athena 2000 sewing machine. This
machine has 24 control buttons, each of which selects a
different stitching pattern. Some are straightforward functional patterns, others are decorative patterns but in every
case the user simply touches the control button and sits back
while the microprocessor controls the movement of the
needle. The information that the microprocessor needs to
control the stitching process is held in a semiconductor
memory. Many thousands of transistors are needed to store
this information and, while this is no problem with modern
ic technology, it would be impossible to incorporate the computing power into a sewing machine without microprocessors.
Apart from the convenience to the user, the use of microprocessors benefits the manufacturer – about half of the
usual 700 cams and gears are eliminated, making maintenance much easier. In addition, if the manufacturer
wanted at a later stage to change the stitching patterns, he
has only got to alter the instructions which tell the microprocessor what to do, a much easier task than designing a
completely new electronic system. Other household products
that now use microprocessors are washing machines, electric
cookers and TV games. Although, at the moment, these
products are rare outside of the USA, they will soon be
available everywhere. In these, and other non-household
applications such as the new breed of 'intelligent' gas
pumps and cash registers, the use of microprocessors allows
the manufacturer to incorporate attractive features that save
the user time or make it more difficult for mistakes to occur.
Because microprocessors can carry out individual operations
at the rate of over 100,000 each second, a microprocessor
can appear to perform several different tasks simultaneously.

For example, a single microprocessor can be programmed to simultaneously monitor and control a dozen gasoline pumps, keeping track of the amount of gasoline delivered by each pump, calculating the cost, working out cash discounts and the number of trading stamps to be given out, adding tax and warning the garage owner when his gasoline stocks are low.

The real job threat is only just beginning

Although the domestic applications of the microprocessor will be the first to attract public attention, the graver implications of the 'microprocessor revolution' are to be found in the industrial world. The majority of jobs in industry, whether manual or clerical, are essentially repetitive, requiring little human judgement. This means that they can be automated, i.e. the human worker can be replaced by a computer which costs less to run, doesn't need training, doesn't strike and rarely makes mistakes. Replacing human employees by computers has other advantages to the employer since computers do not need toilet and canteen facilities, they will work round-the-clock shifts and they do not need a payroll department to make weekly payments. In short, they are, in many cases, ideal 'employees'. Automation using conventional computers has been possible for many years and has been the cause of many disputes between unions and management. There is a general suspicion of computers today, fuelled partly by the underlying threat to jobs for many different kinds of worker and partly by the occasional gas bill of $0·00 (which is invariably caused by a human error, not a computer error). But while many people shy away from large computers, it is difficult to see an 'intelligent' washing machine as a threat. The microprocessor provides the means for the computer to pervade our daily lives, not as a vaguely sinister machine but as an unobtrusive component in all kinds of products. Whether we will regret this or not will depend on the uses to which the microprocessor is put and the way in which we adjust to them. What is certain is that microprocessors are here to stay and are now playing an increasing role in our lives. Much of it will be beneficial, reducing the amount of time we spend on tedious tasks, reducing some kinds of crime and providing new means of entertainment. Some of the ways in which microprocessors will be used in the 1980s are already evident.

Cook, play, protect your home, order your groceries – all with your TV and telephone!

First of all, the home computer will be a reality. The computer will be a microprocessor which will be linked to the television, the telephone and a small keyboard. The facilities that such a system could provide are virtually limitless but among the most likely possibilities are security control, entertainment, shopping, information retrieval and domestic environment control. As a security system the microprocessor will monitor door and window sensors to detect intruders and heat and gas sensors to detect fire and warn of gas leaks. It will automatically sound an alarm if burglar, fire or leaking gas is detected and will also call the appropriate authority via the telephone. The TV screen can be used to show a plan of the house, pinpointing the location of the intruder, fire, etc. The computer will also contain programs for video games, not merely the simple games currently available but also complex games which may also be educational. In one existing programme a child can 'answer' back to the TV set using a simple keyboard, and if his spelling is correct a picture of the object appears! It will no longer be necessary to shop in person. By requesting the information using the keyboard, the user will be able to display on his TV screen the current prices of what he wants to buy and to ORDER THE GOODS FROM his own home. Another microprocessor in the shop will process the order and send out regular bills. As well as shopping information, the new CEEFAX and ORACLE information services will be built into the system so that by pressing the appropriate buttons the TV screen will be made to display weather forecasts, entertainment guides, stock exchange prices, etc. Finally, the computer will allow the householder to program his own domestic environment by instructing the computer to switch on or off his lights and central heating, to maintain specified temperatures or even to sound a buzzer at the start of a particular television program. By using the telephone to link his own computer to that of his bank, the owner will be able to call up his bank balances and statements at any time. There are many other possibilities which will arise if there is a sufficient demand. All these will come about solely because the microprocessor enables simple but cheap computers to be built – none of the facilities so far mentioned are beyond the capabilities of today's microprocessors and the whole system would cost no more than a good stereo system.

It can immediately be seen that a home computer of this kind has several beneficial features. Housebreaking would become a much more hazardous occupation and as an added deterrent, the house owner would be able to instruct the computer to turn individual lights on and off at realistic times when the house was empty. Elderly people, particularly those who are housebound, would be able to maintain a stronger contact with the outside world and if they lived on their own they would have the security of knowing that if, for example, they fell downstairs and could not move, the computer would automatically summon assistance. Not all the ramifications are good, however. By encouraging people to stay at home and interact with the outside world by the telephone, TV screen and computer, the arrival of the home computer may result in greater personal isolation. It is not at all clear today what the long term social effects of the home computer will be – the disadvantages could well outweigh the benefits.

The electronic (automatic) driver

Outside the home, microprocessors will play a prominent part in all forms of transport. Here the effects of the microprocessor will be most beneficial, leading to greater safety, fewer car thefts, and less pollution. The railways are already using microprocessors for in-cab signalling equipment, allowing the train drivers to be warned of any hazards further along the line. Because rail traffic follows a pre-determined time-table, it is relatively easy to allow thousands of microprocessors to take over part of the running of the system. Road traffic is less predictable, but even so, micro-processors will be used to reduce traffic jams, prevent dangerous driving and locate stolen cars. Microprocessors will be used in two ways in this area – as part of the traffic control system and as an on-board unit in every car.

Computers are used today to control lights at important road junctions in some cities but the high cost of the computers has prevented this approach from being used except in the most important cases. Most traffic lights operate according to a fixed pattern, regardless of the flow of traffic. The next generation of traffic lights will have a microprocessor at each junction which continuously monitors the traffic density at every road leading to the junction and adjusts the sequence of lights to get the fastest traffic flow. Each microprocessor will be able to communicate with the microprocessors at other nearby junctions so that each set of traffic lights will be able to take into consideration the traffic conditions at all nearby junctions.

A small number of cars incorporate microprocessors today but on-board computers will be a standard feature by 1981. At first, microprocessors will be used to control the car's engine, allowing a very accurate control of ignition leading to reduce pollution, greater mileage and overall better performance. At a later step, microprocessors may be used inside cars in an entirely different way as navigational and location aids. By picking up signals from wires buried underneath the road surfaces, the microprocessor in the car will be able to communicate with a control computer located in a local traffic control center. Each car would then be able to respond to or transmit a unique identifying code. The central computer would be able to send instructions to the individual cars and the microprocessor in the car would interpret these instructions and inform the driver. For example, a driver who was unfamiliar with the roads of London would be able to tell the central computer where he wanted to go and would then be told which way to go at every road junction. The central computer would be able to separate local traffic from through traffic, the driver would be instructed to follow routes which avoided crowded roads and all traffic could be automatically diverted away from the scenes of accidents. In the same way, the central computer would be able to locate and track any specified car, making life difficult for car thieves and drivers who had had their licenses revoked. Since stolen cars are used in the majority of robberies, the incidence of their crimes could be considerably reduced.

A third way in which microprocessors could be used in cars is to detect and prevent dangerous driving, whether due to alcohol, fatigue or just recklessness. By monitoring the car's speed, acceleration and steering parameters and the distance from other vehicles, the microprocessor would be able to decide whether the car was being properly driven. Signals from the buried wires would inform the microprocessor of the appropriate speed limit. If the microprocessor found that the car was being improperly driven it would be able to alert the driver, inform the police via the central computer or even automatically pull over and stop the car. Drivers who wandered on to the wrong side of the road, drove too close to the car in front or overtook at dangerous points would therefore face immediate detection and would not be able to avoid punishment. Many drivers will find these prospects outrageous and the introduction of this kind of supervision would certainly be resented and resisted by many. The effect of these measures can only be beneficial, however, in virtually eliminating hit-and-run accidents, drunken driving and car thefts. There is, of course, a sinister side of the coin concerned with the increased monitoring and surveillance of individuals. Unfortunately, like all inventions that can benefit the human race, microprocessors can be put to less welcome uses. Chief among these is surveillance.

Electronic crime prevention

Access control, i.e. allowing access to buildings or parts of buildings only to individuals presenting valid cards, is already an important part of the security arrangements for important establishments such as power stations, military camps and research centers. With the availability of cheap microprocessors, this concept can be applied almost everywhere – in hospitals, libraries, stations, banks and airports. The simplest way to do this is to issue everybody with an identity card bearing a unique secret code number and to have card readers at the entrances to all public buildings, buses, trains, etc. This does have some benefits in making travel, for example, much simpler. There would be no need to buy tickets for train journeys, for example, since it would be a simple matter for the microprocessor in the train's card reader to check the credit rating for the card presented to it, refuse entry if the credit rating was too low and bill later on if the card is accepted. A single identity card would take the place of all credit cards, driving licenses, library tickets etc., that we carry around today and eventually it may even replace cash. The more sinister implication here is that the police would be able to track the movements of any individual – it would be extremely difficult to obtain medical treatment, draw money from a bank, board a train or drive a car without the police being notified of the exact time and place of the event. The argument in favor of these measures will be that the only people with cause to complain will be criminals and that law-abiding citizens will have nothing to fear. It must be remembered, however, that once introduced, it is almost impossible to remove such a system – it becomes part of the framework of everyday life.

It is impossible, at this stage, to predict what overall effect the microprocessor will have on our daily lives by the mid-1980's because so much depends on the political and social climates in which the technology develops. We may find that many of the frustrations of contemporary urban life have been reduced or eliminated, that no one need spend their time doing monotonous jobs and that the crime rate has been sharply cut. All these are possible, but it is equally possible that we may find ourselves living in a world where personal contact is rare and superficial, where underlying frustrations find expression in crimes of violence and where our every action is recorded by the authorities. What we can be sure of is that the microprocessor is here to stay and will profoundly influence our future. Whether we will find it a blessing or a catastrophe will depend not on the technology but on the uses that we make of it.

MEDICINE AND SURGERY: TO 2001...

John Newell

Took his degree in Zoology and Botany at Cambridge. Joined the BBC in 1959 and has since worked full time as science correspondent, producer, interviewer and narrator of science programs on radio and earlier on TV. Now Assistant Editor of Science, Industry, Agriculture and Export unit of the BBC Overseas Services which produces regular programs about science in English and talks for broadcasting in many other languages.

Introduction

Remember Interferon? It was going to be the all-purpose miracle anti-virus drug of the 1960s. It still hasn't really got off the ground – though Interferon is just now beginning to creep into limited use. I've no doubt that some of these forecasts, cautious and conditional though they are, will follow the same pattern. Some will fizzle out, some will be overtaken by events. I expect to look pretty silly if I'm still around in 2001 – but, to be honest, I do expect to have scored some bull's-eyes too.

The least predictable factor is money. It will be in a very real sense tragic if the British research projects I've tried to describe in this chapter have to be slowed up or put in cold storage through shortage of funds, yet for some of them it's more than possible as Government spending cuts continue. If this chapter of this book makes a modest contribution to making people aware of the potential benefits of medical research today then it will have been very well worthwhile.

Perhaps the most important problem in medicine we have to solve before 2001 – well before then – is, how to bring the world's resources of research and money to bear more effectively on the medical problems confronting the third world. This is not only a humane duty, it is a sensible precaution. The population explosion *must* be brought to an end, malaria *must* be eradicated, sleeping sickness *must* be controlled if the economics and standards of living of the developing world are to prosper, if the gulf between rich and poor is not to widen to the point where it provokes world war. Medical science can find – is finding the answers; what is lacking is not brains and ideas but money and effort.

The other big problem, or rather set of linked problems, is that of medicine in the West. Doctors are becoming further and further removed from their patients. Expensive 'high technology' is no substitute for personal interest and care. This impersonal nature of so much modern medical care is a symptom of society's sicknesses, as also are the diseases of affluence and stress, lung cancer, alcoholism, ulcers, above all coronary heart disease. The limited resources of the planet mean that, by 2001, we in the richer nations shall have had to become less conspicuous consumers, on a planetary scale and in our domestic lives as well. Medicine has already shown that the move to a simpler and more austere way of life will do our health nothing but good. Here doctors' findings march along with the environmental movement, and the dictates of dwindling resources.

From every point of view, if humanity is rational we can expect the richer nations of the world to move towards a simpler, plainer, more small-community-minded life-style. If they/we don't do it voluntarily they/we will be forced to it anyhow.

Targeted drugs – the mighty micropill

Drug delivery systems as far removed from today's pills and 'jabs' as guided missiles are from grenades are already being tested in London hospitals and research centers in Britain and the United States. Among these are the *liposomes* (nicknamed micropills). These are tiny droplets of oil, only visible under the microscope, yet each made up of several layers like miniature onions. By mixing the right oily compounds with drugs and subjecting the mixture to ultrasonic vibrations, the oily onion can be made to encapsulate several different drugs between its layers. These tiny liposomes can then be introduced into the body in solution, in an injection with an ordinary hypodermic.

Once in the bloodstream, the drugs inside are protected from breakdown until the liposomes reach a target organ which requires treatment. Drugs have already been given via liposomes to cancer patients. It has been shown that drugs delivered in this way concentrate in the tumors. Deficient enzymes can be supplied in the same way to patients suffering from enzyme deficiency diseases. Vaccines administered in liposomes are more strongly protective than normal vaccines There are other uses.

Guided missiles

The simple oily droplet is only the beginning. The next stage, already well advanced experimentally, is to coat the liposomes with *antibodies* – protein compounds which react specifically with, and only with, particular types of organs or cells. In this way, liposomes can be targeted, like guided missiles, to deliver their drug warheads only to specific organs where they are required.

Cancer cells have surface proteins, *antigens*, not found on normal cells. So the targeting of liposomes containing anti-cancer drugs could be relatively simple. Some organs have special *receptor sites* for chemical messengers (hormones) into which the hormones fit, on arrival, like keys into locks. It is possible precisely to mimic these 'keys' with molecular groupings, attached to the liposomes, and so to trick cells and organs into accepting the liposomes' loads of drugs, like miniature Trojan horses. In this way, missing enzymes can and surely will be supplied to organs, like the liver for example, deficient in them,

Over the next twenty-five years drugs targeted in this way by linking them to antibodies which react only with specific organs, or cancer tumors, will come into use on such a large scale as to constitute a revolution in drug therapy. And, because they will allow much higher doses to be delivered selectively to specific organs or tumors, they will dramatically improve survival rates after cancer, speed up the rate of cure of infectious diseases, and greatly improve therapy for deficiency diseases due to genetic defects. Liposomes will be used to deliver, not just conventional drugs including antibiotics, but also *antibodies*, the human body's own specific weapons against infections. These will be mass-produced by living 'production lines', of genetically engineered bacteria and cell cultures. And targeted liposomes will begin to be used to deliver correct versions of missing or deficient genes, to organs in need of the guiding blueprint. In this way, presently incurable genetic diseases will begin to become curable.

Great progress is being made today in identifying the specific antigens of various types of cancer cell (the protein compounds on their surfaces which are unique to them and are not found on normal cells). At present, this knowledge is only of value in diagnosing cancer and following the progress of treatment. But, in a few years, it will begin to become possible to use antibodies to the cancer antigens – antibodies mass-produced by the genetically-engineered cultures mentioned earlier – as targeting mechanisms for liposomes filled with cell-killing drugs. Radioactive elements, as well as all kinds of cell poisons, will be delivered to tumours by these 'guided missiles'.

Hounds in the blood

That's one way in which the marvellous effectiveness of the human body in identifying and tracking down foreign cells and organisms will be improved upon and put to work. And I am convinced that it is in this direction, in the further progress of immunology (as the study of our natural defences against disease is called) that the greatest hope of new breakthroughs against disease now lies. The immune system has at least two arms. We are defended by living cells, white cells in the blood, as well as by the antibodies they produce. The white cells home on to infections, and other sites of illness, as effectively as bloodhounds once they pick up the scent of the foreign proteins. At the Hammersmith Hospital, in West London, doctors are already using white cells which have been removed from the body and made harmlessly radioactive so their movements can be followed, as a means of locating deep, hidden foci of infections, deep abscesses. Here the breakthrough – during the last months of 1977 – has been in the development of a unique technique for making the cells radioactive without in any way interfering with their tracking abilities. Once replaced in the bloodstream, white cells treated in this way can be followed around with standard equipment for detecting low-level radioactivity.

Doctors are already beginning to use this technique to locate tumors, as well as infections. A different kind of blood cell, the *macrophage*, locates and clusters around tumors. Areas of rheumatic inflammation can be located in the same way. Other blood cells are to be pressed into service; the tiny dish-shaped cells called *platelets* can be used to locate incipient blood clots, in time to dispel them with clot-dissolving drugs.

These 'hounds in the blood' can be used to destroy disease, as well as to locate it. Blood cells, taken from the body, will be loaded with drugs, given the 'scent' of the infection or tumor they are to attack (by exposing them to characteristic antigens) made harmlessly radioactive, and then released back into the body and followed all the way to their prey. This technique, in its infancy today but already in clinical use, has tremendous potential as a means of following what goes on in the body with no need to probe about inside.

Non-addictive opiates

The new ideas about body and brain chemistry which are now emerging will be dealt with more fully in the section of this book devoted to biochemistry. But two developments, now at an early stage are of such enormous importance for the next quarter century that they must be mentioned here. One of these is the discovery – about two years ago – of a naturally occurring opiate in the brain, the substance *enkephalin*. In a breakneck research program enkephalin has been isolated, purified, analyzed and synthesized. Now medical scientists in several countries are racing to find a way to make this brain chemical, or a derivative from it, into a pain-killer which could be ten times as potent as morphine, yet safely non-addictive.

Happiness pills?

The significance of enkephalin may extend even beyond that. The latest research suggests this compound helps to control mood as well as pain. It looks as though enkephalin and its relatives – more than one compound seems to be involved – are able to induce euphoria, when produced in the brain. Morphine, which attaches itself to the same sites in brain cells as enkephalin, induces sleepy euphoria. Is a sleep-and-happiness pill on the way? It's more than possible. Much remains to be done to make a drug out of enkephalin – or out of some compound which will cause it to be released in the brain, an alternative approach. But the next quarter century should provide ample time for the required research and development.

Selective aspirins

Another completely new approach to pain killing, and indeed to drug therapy overall, is emerging from the new concepts now being shaped concerning the respective roles of nerves and chemical messengers, hormones, in carrying messages around the body. Over the last few years, a whole new family of locally-acting chemicals have been discovered which are released from nerve endings and controlling events in small regions of the body. The old distinction between nervous and hormonal controls is breaking down.

One such local hormone is histamine. In 1976 the first drug came on the market which prevents or cures ulcers by blocking the action of histamine on the stomach and preventing the excessive secretion of acid. Cimetidine, as the compound is called, is far more effective than any previous treatment. This will almost certainly be only the first of a whole new family of drugs which act by selectively and specifically blocking the action of local hormones. Aspirin, it's been shown, acts by modifying the action of another complete family of local hormones, the prostaglandins. Drugs which act like aspirins, but only on single specific prostaglandins, and so only on specific parts of the body are now being developed fast.

This explosion of new knowledge, about local hormones and their release by nerves, has come about very largely in just the last year; it's too new really to evaluate yet. But it is certain that the drugs of today will seem very crude affairs compared to the precise chemical tools which will be developed in the next twenty years or so.

Anti-clotting – anti-heart disease hormone

Towards the end of 1976, scientists at the Wellcome Foundation's laboratories in Kent announced their discovery of what may become the most important drug to be derived from work on prostaglandins. The compound they have discovered is itself a prostaglandin, a local hormone, the particular job of which is to prevent the formation of blood clots. *PGX* (the only name given to the compound so far) is released by the walls of arteries when blood cells, platelets, bump into them. PGX stops the platelets sticking to each other and to the walls of the artery, and so prevent the formation of clots. So PGX is a *natural* clot stopper – and therefore likely to prove a more effective one than the compounds used today.

PGX may be the body's chief means of preventing hardening and thickening of the walls of arteries. So, as well as an anti-clotting drug, PGX may become a means of preventing, or of treating, coronary heart disease. Though far from even the earliest experimental trials, the idea of developing PGX for these purposes is a good deal more already than just a gleam in the researchers' eyes.

Contraceptive pills for men

The control of fertility – both ways – will see some big changes. The contraceptive pill for men is definitely on the way. The chance discovery of a chemical ten times as potent as the male sex hormone testosterone could accelerate the development of the male pill since, paradoxically, the male system can react to such compounds by shutting down the production of sperm.

The super sex hormone was discovered when some female chicks on a Rhodesian chicken farm began to behave aggressively and to grow combs like young cockerels. The change was traced to chemicals in the litter on the poultry house floor, from which the active sex-hormone-like substance has now been isolated.

For women, the once-a-year contraceptive injection should become a popular alternative to the pill. This technique is already being tested in India. It works by making a woman's body react against the chemical signal produced by any fertilized egg. So the signal is neutralized and the woman's body is never stimulated to start the chemical sequence involved in pregnancy.

Fertility without multiple births

At least one drug which induces fertility for some infertile women without causing multiple births is now in use. This is *bromocriptine*, now used in St. Bartholomew's Hospital London and elsewhere. Others are being developed. Surgical as well as chemical techniques for dealing with infertility are being developed. Fallopian tubes can be transplanted, from someone newly dead, so as to bypass a blockage in these tubes, which normally carry egg cells down from the ovaries to be fertilized.

So-called test tube techniques for external fertilization, already used successfully, will be perfected. In these, blocked Fallopian tubes are bypassed by surgically removing eggs and fertilizing them, outside the body, in a laboratory dish after which the fertilized eggs are replaced to continue development normally.

Deep-frozen embryos

This technique, already in use but with a relatively low success rate, can potentially be made much more effective by keeping the embryos at a very early stage of development – a microscopic ball of cells – in deep freeze until the mother is at just the right point for reimplantation of the embryo. Mouse embryos have already been cooled to $-79°C.$, stored in liquid nitrogen over long periods, and then replaced in mice where they developed and were born normally. These and other techniques will make it possible for the majority of women who are infertile to have their own babies if the sophisticated resources required are available.

Nerve freezing rivals drugs

A technique in which short stretches of pain-carrying nerves are located and then frozen, so as temporarily to cut the nerve and stop the sensation of pain, has already proved so successful – at the Pain Relief Unit at Abingdon Hospital near Oxford – that it is now being used to relieve temporary pain after operations, as well as long-term intractable pain due to nervous disease, severe arthritis or forms of cancer. Nerve freezing has to be repeated at intervals, to keep the patient free from long-term pain – though this doesn't apply to short-term, post-operative pain. But the first results are so encouraging that it's likely that nerve freezing will come into wide use, over the next twenty-odd years.

The technique is relatively simple. The surface of the body is anaesthetized, at the point where the probe is to be inserted. Then the long, slim probe is passed in and moved about while a mild electric current is switched on. This locates the pain-carrying nerve fibres. The tip of the probe (about the size of a knitting needle) is then applied to these fibres and the tip is briefly cooled to well below freezing twice over.

The effect is not to kill the nerve cells involved but only a short stretch of the long extension (axon) – often feet long – which carries the pain message back from the nerve endings to the brain. But it does effectively cut the nerve without, apparently, any of the harmful side-effects which can follow cutting by scalpel, and with no need for a general anaesthetic. After a few weeks, the nerve regenerates. But often the complete relief of pain persists for several months. If the pain is only temporary, then no harm is done because the nerve grows again. If the pain is intractable then, when it begins to return, a repeat chilling can be performed.

Potential advantages over the use of drugs are important. Nerve freezing gets rid of the problem of addiction, and of ever-increasing doses. It will have to be made more precise and selective to reap the full benefit, so as to avoid any damage to the motor fibres controlling movement, which run alongside the sensory ones carrying pain. This should certainly be possible. Surgeons at the Pain Relief Unit have already discovered that, during chest surgery, it's quite a simple matter towards the end of the operation while the chest is open to apply the cold probe to the principal nerve responsible for post-operative pain, which is actually visible inside the body cavity.

In twenty years or less shall we wake up after major chest surgery miraculously free from pain, with no need for mind-numbing imperfect opiates? Many of us certainly will, perhaps very many. Minor operations, like hernias, could be carried out with no need for a general anaesthetic.

Syringe replaces scalpel

Surgery, as well as anaesthesia, will be almost miraculously simplified, for some common purposes, through techniques, now just starting to come into use, whereby the hypodermic syringe replaces the scalpel. Transplants of ductless glands, such as the testis, pancreas, thymus, adrenal and parathyroid, which have proved very complex and largely unsuccessful, will be outdated, by replacing organ transplants, with all the complex plumbing and reconstruction of ducts and blood vessels involved, with a simple injection of living secretory cells of the type required, made into a solution in a hypodermic syringe.

It sounds too good to be true. But in fact the technique *is already being used*, on a small and semi-experimental scale. Ductless glands do their job by producing their vital chemicals, hormones like thyroxin, adrenalin and insulin, directly into the bloodstream. Only parts of each gland are involved in this vital task. Techniques have now been developed for separating the vital secretory cells from the rest of living glands, removed from cadavers. These cells can be kept alive, stored in deep freeze, perhaps even multiplied in tissue culture. Then, when required, a solution of them can be injected into a patient's bloodstream, or into some suitable part of the body where the cells will continue to produce their hormone, just as though they were still in their original home.

All the major technical problems involved have already been solved. Secretory cells can be separated, kept alive chilled in suspended animation, taken out when required, tested to see they are working properly, injected into sites, in the body where they will not be rejected or will be minimally rejected by the immune system of the recipient, and there checked to see if they are working properly. All this has been done.

Even if the cells injected are recognized as foreign and so killed off, by the immune response of the recipient, all that is required is a topping up injection a few months later. While, no doubt, long-term problems will arise, this will certainly be an improvement on having to subject the patient to a second transplant operation, with repeats whenever immune rejection occurs.

In Britain children who might otherwise have died have already been successfully treated for several enzyme deficiency diseases, using the technique to supply them with secretory cells provided the missing enzyme. In the United States, injections of insulin-secreting cells have been used as an alternative to the, very difficult, transplantation of a complete pancreas. Injections of cells from the thymus gland or liver are being tested as an alternative to the complex and painful bone marrow grafts used today in attempts to transplant the ability to resist infectious disease to those born without it. Other such experiments are being performed. Some of these will fail. But enough will, it now seems certain, succeed to bring about a revolution in many forms of treatment over the next quarter century.

Pill replaces syringe?

Some drugs which now have to be given by hypodermic will be given as pills instead. The identification, in recent months, of the active part of the insulin molecule will allow the development of insulin substitutes which can be taken by mouth. Other such drugs will be protected by slow-dissolving coatings to allow them to pass right through the digestive canal and into the bloodstream without being broken down.

Wafer replaces pill!

The last step in this chain of simplification will be the replacement, for many purposes, of pills to be swallowed by
wafers simply stuck to the skin for a few hours. These are
already under test in a London laboratory and, because they
deliver drugs slowly and evenly, direct into the bloodstream
through the skin, they have some clear advantage over pills,
for many purposes.

Bacteria to prevent infection . . .

And speaking of skin and the prevention of disease, research
at a London hospital has revealed that among the human
skin's principal natural guardians against harmful bacteria
are – other bacteria. Only about one in five people appears
to be naturally gifted with these unlikely guardian angels.
But doctors at the Westminster Hospital have now developed
a cream containing the protective bacteria which can transfer them to others with no natural defences of this kind. The
cream has already been successfully tested on animals and
should become widely used over the next ten to twenty years.
And, by the way, it may lead to rather different ideas about
the desirability of over-frequent washing.

Senile dementia curable?

Hope for finding new treatment for progressive nervous and
muscular diseases, such as multiple sclerosis, must rest in a
better understanding of how they are caused. It would be
very wrong to give the impression that such work is far
enough advanced to begin to put any timetable to this. But at
least there is now solid evidence that some such diseases are
due, in part at least, to a virus acting in an unusual way, a so-
called *slow* virus. Unexpectedly, research on slow viruses has
suddenly given new hope, though fairly distant hope, of understanding and ultimately treating or preventing a brain
disorder previously thought to be accelerated ageing.

Psychiatrists and medical research workers, over the past
year or so, have acquired several separate strands of evidence

which together strongly suggest that *senile dementia*, the sad and, apparently, unstoppable deterioration of mental powers which often accompanies old age, is in reality due to an infective organism and so is, in theory, potentially preventable or curable. This can explain why the symptoms of senile dementia is quite often found in relatively young people, who are perfectly healthy in other ways. If this is proven, then all the massive research effort now going into slow viruses as a cause of multiple sclerosis and allied diseases, could, as 'spin off', benefit huge numbers of elderly and other people threatened with mental decline.

Mind-power shapes muscles

British and American scientists have discovered in the last few months that muscles can be made completely to alter their shapes and functions, merely by submitting them to patterns of very mild electric stimuli, too slight to be called shocks. The discovery is already being used to help patients with immobilized limbs and doctors plan to use it to help to rehabilitate victims of strokes, and to help to prevent the formation of clots after surgery. It also begins to explain mysterious muscle-wasting diseases. Strangest and most intriguing, however, is the light this throws on athletic success.

Some skiers, for example, undoubtedly succeed and excel because their legs have an unusual quantity of the type of muscle fibre responsible for maintaining posture rigidly. It begins to look as though determination – mind power – may be able to alter muscle structure through alterations in the patterns of electrical stimuli fed to the muscles.

I doubt if this line of research will lead to anything as startling as some means of directly controlling the patterns of muscle-development by mental discipline, at least not in the next twenty-five years, though it could happen eventually. What it is leading on to already are means of helping to redevelop muscles which have become wasted, through disease or injury, by the implantation of painless electrical stimulators, which feed regular patterns of 'shocks' to muscles, at a level too low to be detected consciously. Incidentally, electrical stimulation, of a parallel kind, is now, after years of uncertain experimentation, being developed fast also as a means of stimulating the regrowth of damaged bone.

The story behind the discovery of the power of electrical stimulation over muscles deserves telling in full. In the human leg, there are two kinds of muscles, the slow and the fast. The fast muscles, as their name implies, are used for movement; running and walking. The slow muscles are used when we stand still to stop us crumpling up and collapsing; they maintain posture. You might suppose that, like most other things in our bodies, the natures of these muscles are irrevocably laid down from the moment we were born, indeed before that, by the genetic blueprints for them. But it isn't so. Slow muscles can change into fast ones, and vice versa, in only a few weeks.

This transformation was first demonstrated a few years ago by British scientists in England and America who also showed that it's the *nerves* supplying muscles which determine the structure of the muscles. They did this by anaesthetizing experimental animals, cutting the nerves supplying the fast and slow muscles in one leg, and crossing them over – cross innervating – so that the slow muscles were now innervated by fast-type nerves and vice versa. Later, they took samples of muscle tissue under anaesthetic and found that, after about a year, the slow muscles had very nearly changed into fast ones, and vice versa.

The next question was, *how* does a nerve influence the structure of the muscle it supplies? There were three possibilities; either it's the pattern of electrical stimulus supplied from nerve to muscle, or it's some mysterious chemical (a trophic substance) trickling from nerve to muscle, or a combination of the two. Scientists of Birmingham University and the Boston Biochemical Institute set out to discover what it was and the answer is, it's an electrical stimulus which controls muscle structure.

What they did was to bury a tiny electrical stimulator in the abdomen of a rabbit, with leads to a fast muscle in the rabbit's leg, so that it gave the muscle a non-stop stimulus of about ten tiny shocks a second – at much too low a level to be detected consciously. In fact, this was the nearest possible intimation of the *natural* electrical stimulus which the nerves supplying slow muscles give to the muscles they innervate. Slow muscles receive a more-or-less non-stop low frequency stimulus of this type, to keep them at work maintaining posture, whereas fast muscles only receive brief, occasional bursts, of much higher frequency – fifty to one hundred stimuli per second – when they are to perform a movement. So a fast muscle was being given a slow muscle type of electrical stimulus, but, because this was coming from a piece of apparatus and not from a nerve, there was no possibility of any trophic chemical also being involved.

The results showed that the slow-muscle-type electrical pattern was quite enough on its own to change a fast muscle into a slow one. Within only *four weeks* the change was well advanced. And in twelve weeks it was practically complete – a much quicker transformation than that brought about by cross-innervation.

Now the researchers plan to build another stimulator, which can produce different patterns of electrical pulses to order – receiving its orders by radio link from a transmitter in the laboratory. Using this, they plan to discover more about the detailed ways in which electrical stimuli control muscle structure, and how to bring this process under artificial control. It could be very valuable indeed. Electrical stimulation based on these discoveries is already being used clinically, in California, to help to straighten twisted legs. It will soon be used to improve blood supply to immobilized limbs, and so to help to prevent dangerous clotting, and to help to restore health and movement to muscles paralyzed and wasted after strokes. It provides new clues about the causes of muscle-wasting diseases, like muscular dystrophy.

How far will the combination of new knowledge about the effects of electrical stimulation from nerves, plus new knowledge about local hormones released from or controlled by nerves, take us in the next twenty-five years? One hesitates to be optimistic for fear of raising false hopes among sufferers from neuro-muscular disease. And indeed, unfortunately, even when the causes of such diseases are understood, there will be a long way to go to put them right. Two things, however, are certain. First, that it is only through research along these lines that better treatments, cures and means of prevention will ever emerge. There are no short cuts. Second, that the subject is moving fast now, and a lot of really valuable new knowledge is emerging, in a field which had been relatively barren for a long time.

Restoring communications in the brain

Understanding of the brain, at a level which might lead to new treatments for mental diseases, including diseases affecting the muscles as well as the mind, is a long step behind understanding of diseases affecting just nerves and muscles, a step which reflects the relative complexity of the two systems. In 25 years' time the brain will be the most exciting area of medical research. What goes on beneath the neck will be wholly understood. Human physiology will be completely worked out. But knowledge of the brain is unlikely to be much more advanced than that of the body is today.

New drugs for mental illness

Nevertheless, we can hope for much more effective treatments of common mental illnesses by AD 2001 or thereabouts, just as antibiotics have been used to attack bacteria, and vaccines to protect against viruses, long before a full knowledge of either type of organism was achieved. And, along with such drugs, we shall certainly have the power to change moods and desires and motivations, for good or ill. And we shall develop far more sophisticated means of passing on instructions from the brain to the prostheses of a paralyzed person.

Around the end of 1975, London doctors first reported results described as dramatic in treating *myoclonus*, a brain condition resulting in uncontrolled muscular tremors and jerkings, using a derivative of one of the chemicals responsible for communications in the brain, a neurotransmitter. Myoclonus can have several different causes, attempted suicide with drugs, clots in arteries supplying nerves, accidental overdoses of anaesthetics, concussion and other physical damage and so on. Treatment with a chemical very closely related to *serotonin*, one of the four or five chemicals now known to carry out or to control communications in the brain, proved remarkably effective in suppressing the jerks.

Myoclonus is the second mind-muscle disorder to be treated in this way. The first was Parkinson's disease, in which brain damage also leads to muscular jerks and tremors. Parkinsonism can now very often be effectively treated with L-dopa, a chemical out of which the brain makes the neurotransmitter *dopamine* – just as it makes serotonin out of 5HT.

The fact that two different mind-muscle disorders have now been successfully treated with two different neurotransmitters makes it now appear much more likely that breakdowns in chemical communications, which may be at least partially reparable from outside, are the principal causes of other common mind-muscle disorders. These too may, in the future, become treatable.

Schizophrenia

Purely mental as compared to mind-muscle disorders present a worse problem. But biochemists at Cambridge University believe they are on the track of the basic cause of severe schizophrenia. A disorder in the neurotransmitter, *dopamine,* seems to be involved. Research is presently delayed, through the unwillingness of relatives of newly-dead schizophrenic patients to offer brains for research and chemical analysis. But there is a very good chance of this and other work leading, over the next twenty-five years, to drug treatments, able to restore even severely mentally-ill patients to completely normal life. It will take much longer to find ways of putting right the underlying causes, so that a cure can be effected. Until that time comes, the continual use of drugs by severely-ill mental patients will have to be accepted.

Memory pills?

Research into schizophrenia at the University College of Wales in Cardiff led last year to the first tests of drugs which significantly improve memory and learning performance. The pills are of a drug now known as *Nootropyl* ('Towards Mind') first developed as an experimental anti-seasickness compound in Belgium. While it was being tested there several subjects remarked upon how nootropyl appeared to improve their memories and learning abilities.

These were only subjective impressions, but sufficient to interest a research worker at University College, Cardiff, in view of animal experiments which showed him what the true effect of Nootropyl is. It improves communications and increases 'nervous traffic' between the two hemispheres of the brain, through the tract of nervous tissue, the corpus callosum, which alone links the hemispheres.

Scientists were already deeply interested in interhemispheric communications, through studies of schizophrenia. They had already shown (in experiments reported in 1973), that communications through the corpus callosum are impaired in schizophrenic patients, by means of a technique in which two pictures are shown simultaneously one to each eye, in such a way that one picture is seen only by one eye, and the other only by the other. Because each eye is connected only to the opposite hemisphere of the brain, left eye to right hemisphere and vice versa, this means that the ability of the subject to compare the two pictures depended entirely upon the link between the hemispheres through the corpus callosum. Schizophrenic patients, almost without exception, did considerably worse than normal people in such tests, when they were asked to compare the two pictures shown to their two eyes and to say if the pictures were the same, or different. So, perhaps, the literal meaning of schizophrenia, split mind, is a realistic definition. This sort of defect appears to be unique to schizophrenia, because when the same tests were performed on people suffering from several other mental disorders or illnesses, *their* performances were on average no worse than normal. Several other lines of evidence also suggested that the corpus callosum becomes damaged in schizoprenics.

The team had been following up the idea that drugs which improve communications between the two halves of the brain might ameliorate schizophrenic symptoms. But their trials of nootropyl, the results from which were recently reported, were carried out on normal students. A group of sixteen students were given either nootropyl pills or identical-looking but inert placebo pills, every day for a fortnight. They were also given tests of their abilities to learn and remember – at the start of the experiment; every day during the two weeks; and at the end of the experiment. The tests involved memorizing sequences of letters glimpsed on a fast-rotating drum, and memorizing lists of words spoken rapidly into ear phones.

At the start of the fortnight, there was no difference between the abilities of the eight students who were to be given nootropyl, and those who were to receive the placebo. But by the end of the experiment, the group who had received the nootropyl were all doing fifteen to twenty per çent better than the placebo group – and fifteen to twenty per cent better than they had been at the start.

How long-lasting such effects will be, and how far they can be taken, remain to be seen. But they are not the only experiments to suggest that memory-enhancing drugs can be developed. Experiments an animals, suggesting that specific learnt skills are laid down in chemical traces in the brain, which can be transferred from one brain to another, have remained tantalizingly unconfirmed.* Memory and learning aren't

---

*One way of searching is actually to try to transfer memories between animals. The most sensational experiments here were done a few years ago by two American scientists, James McConnell and Allan Jacobson. They trained tiny pond animals called flatworms (planaria) to respond to simple stimuli: for example, to stretch out towards a light or curl up in a ball when given a mild shock. They then minced up the brains of these trained flatworms and fed them to untrained worms. They claimed that on this cerebral,

just chemical traces. The search for memory molecules seems fruitless, unless accompanied by a search for some electrical-cum-chemical pattern linking them, which is probably un-transplantable.

(Bits of) brain transplants?

On the other hand, chunks of brain matter may one day be transplanted, from dead brains kept in deep freeze, to replace brain tissue lost in injury. Whether connections between the transplant and the rest would be made is an open question. But brain tissue is not rejected, because the cells and mole-cules responsible for rejection are kept out of the brain. And recently biologists have come, cautiously, to the conclusion that, contrary to previous dogma, brain tissue can be per-suaded to regenerate, on a limited scale. Brain cells do not normally divide and grow in adulthood. But new connections, to a limited extent, can be made between existing cells. My bet is that the development of drugs to encourage and extend such regeneration – itself by then well understood – will be at least a major field of research in 2001.

Reading machines for the blind

It seems certain that by 2001, as a result of work now in pro-gress, numbers of suitable blind people will have been fitted with reading prostheses, in which miniaturized TV cameras, linked to the visual centers in the brain, will allow them to read books and get a general picture of the world about them – though such prostheses will still be exceedingly primitive compared to even the simplest true eyes.

This will come about as a result of experiments which have been going on now for several years, in which neuro-surgeons in Britain and the USA have implanted minute electrodes

cannibalistic diet the untrained animals now rapidly learned to respond to the same stimuli. Since then Jacobson and others have claimed the same kind of results with higher animals (and even between species) fed with RNA and proteins extracted from the brains of trained animals.' Quotation from *The Biocrats* by Gerald Leach published by Pelican Books 1972. But other scientists have completely failed to repeat these results.

into the brains of blind people and, by passing tiny electric currents through the electrodes, have caused the people to see simple dots and patterns of light. Animal experiments have shown how cells in the visual cortex of the brain are assembled in patterns in which one set of cells responds to points, another to boundaries, lines, at one particular angle to the head, and so on. The linking of miniaturized TV cameras to such prostheses is now in progress.

New limbs for old

As for the blind, for those paralyzed below the neck, medical science is unlikely to offer anything more than the crudest replacement of normal function within the next twenty-five years. But if – and this is a great 'if' throughout medical research today – a way can be found to overcome the immune reaction against grafted tissue, whilst leaving unimpaired the body's defences against disease, which are dependent upon the same system, then amputees might benefit from spare-part arms and legs. Certainly the plumbing and grafting required should present no great further problems to surgeons who already replace virtually severed hands with some regaining of control and sensation. Replacing nervous communications which have been cut, as in paraplegics, raises greater problems because, once again, of the greater complexity of the nervous as opposed to the muscular system. Yet experiments with artificial metal nerve substitutes suggest that crude nerve substitutes could be in experimental clinical use within twenty-five years.

Growing your own spare parts

Biologists working in this area already talk about the real possibility of growing our own spare parts, limbs and organs, instead of waiting for transplants. Animals as relatively close to us, in evolutionary terms, as the amphibia do it already. Within the next twenty-five years research will have shown just *how* they do it, and what the difference between us and the amphibia is, the factor that, somewhere along the evolutionary line, has lost us the ability to do the same thing. Work will have begun on techniques – probably a combination of drugs and electrical stimulation – to promote the regeneration of limbs and organs. I doubt whether it will have got much further than the stage of promoting complete regeneration in experimental animals more closely related to us than newts. But I could be over cautious. In the USA an opossum, a mammal with a body structurally and functionally very like our own, has already regenerated a completely recognizable foot and the greater parts of three toes. Certainly, finding ways to grow our own spare parts seems ethically preferable to stealing them from animals.

New ethical problems

This is a convenient moment to mention some ethical problems which medical progress will bring about, some well within the next twenty-five years. Amputating the limbs of dogs, cats and monkeys, to see if they can be made to grow new ones, may be accepted by the majority as a step towards the long-term aim of permitting amputees to grow their own new limbs. But, if rejection problems are overcome, then a very different prospect is opened up, that of maintaining huge numbers of animals, for the foreseeable future, as a source of organ transplants for humans. Vivisection is accepted by many largely because only a few years of animal research, which then need never be repeated, can bring about permanent benefits for many generations to come. Whether animals should be bred for their organs as commonly as they are bred for food today, to replace human organs so as artifificially to extend the natural span of life, as well as to replace diseased organs, is a more controversial question.

The trend of thought today, especially among the young who will lead the world in twenty-five years time, is to think of man no longer as a unique and deservedly dominant animal; but as one among many equal and inter-related species who, through his short-sighted view of himself as all-important, is upsetting every kind of natural balance, threatening nuclear destruction and using up natural resources at a rate which cannot last more than another quarter-century.

How such an attitude will view the further exploitation of
the animal kingdom to avoid the natural process of ageing,
remains to be seen.

Fingers from foetuses?

An alternative to obtaining spare parts from animals, and a
more realistic one, so far as appendages are concerned (after
all, who really wants a paw or tail?) is to obtain them from
aborted foetuses. The supply of foetuses will lessen in most
countries, as contraceptive measures improve. But enough
will remain to allow, if it is considered ethical, a plentiful
supply of spare parts.

Research workers in a London heart hospital believe that
parts of diseased hearts in humans could be most effectively
repaired by grafting in healthy heart tissue taken from
aborted foetuses. They are more compatible, less likely to be
rejected than spare parts from animals. At a rheumatic diseases research centre in Britain, doctors have removed toes
and fingers from three- or four-month human embryos and
kept the digits alive and growing in tissue culture for periods
of several days, during which the digits not only grew half as
long again, but also continued to develop normally internally.

One immediate application of such a ghoulish sounding
experiment is that of watching the development of joints and
bones through all the critical stages, another is to study reactions which cause arthritis. However, it's almost certain that
such a technique could provide banks of appendages, and
perhaps of organs for transplant, which could be stored in
deep-freeze to allow the best possible matching of transplant
to recipient. Will society accept the idea? What is certain, is
that now is the time to begin to decide.

The thought-controlled wheelchair

Reverting to prostheses for the disabled, developments already completed show the way, well within twenty-five years,
to allowing paraplegics to lead much fuller and more varied
lives. All such patients retain control over some muscles, in
the face, or neck or shoulders. Tiny sensors can already be

implanted in such muscles, to pick up the minute electric currents which pass through them when a message is sent to the muscles from the brain. These currents can be picked up and amplified many times over with ease, transmitted over a radio beam to a small antenna and from there used to control any required function. The muscle movements involved are all under conscious controls. So an assortment of such devices, known as *Emgors* (Electro-Myogram Sensors), implanted in different muscles, could enable a disabled person, simply by transmitting a mental command to a particular muscle, to switch on lights, change TV channels, turn a page of a book, activate or steer a wheelchair and much more. It's certainly not too far-fetched to think of such devices allowing a paraplegic to drive a car or to operate a walking machine.

Artificial lungs

Major heart surgery today would be impossible without the aid of the artificial heart-lung machine, which takes over both the pumping action of the heart and the oxygenation of the blood while heart and lungs are stilled for open heart surgery. Oxygen can be fed into the blood, and the waste gas carbon dioxide extracted from it, through a membrane with the oxygen supply one side and the flow of blood on the other. The problem is, how to get enough oxygen into and enough carbon dioxide out of the blood. The answer – recently developed by a research team at the Department of Engineering at Oxford – is to crinkle the membrane separating the blood from the oxygen into furrows, which enlarge the surface area and stir up the blood.

Implantable lungs?

The new lung machine has already been tested on animals, and has proved so successful that the research team think it will soon be used to develop a portable heart-lung machine, to replace or extend the range of the cumbersome machines now in use in hospitals. And – says the project's Director – by the year 2000 an implantable artificial lung may have been developed, for patients whose own lungs have been irreparably damaged or diseased. The first plans for implantable lungs are already sketched out.

Wearing artificial kidneys

Wearable artificial kidneys should certainly be available within the next twenty-five years, perhaps undergoing their first human trials within the next decade. Today's artificial kidneys, required for the vast majority of patients who, through lack of donors or of surgical facilities cannot have a transplant, are large, cumbersome, expensive affairs, due to the large quantities of liquid required to cleanse the blood. So as to simplify and speed up the process, what's required is some absorbent material which will absorb all the toxins in the blood more effectively than today's slow process of extracting them through a membrane. No such substance yet exists. But, in recent years, a combination of work, on new absorptive compounds and on enzymes, which could be used to break down the toxins to harmless compounds, has got to the point where wearable artificial kidneys with everything required to clean the blood without need to go to a dialysis unit can easily be envisaged.

Spare gut into artificial kidney

There is an alternative approach to the undoubted need to make artificial kidneys simpler and easier to use; one which sounds strange and even gruesome at first hearing, but which potentially has big advantages, and which has, very recently, been shown to be highly practicable. A London surgeon has developed a technique for making a loop of large intestine, severed from the rest and connected to new openings in the body wall, into a natural artificial kidney, utilizing the excellent properties of the gut membrane as the separation and filtration system required.

The basic need in an artificial kidney is a membrane with pores of just the right size; big enough to allow impurities in the blood to be washed out of it but small enough to retain vital blood cells and big protein molecules. Just such a membrane exists, naturally, in the shape of the wall of the gut, through which all sorts of relatively small molecules are absorbed into the bloodstream, while none of the blood's vital constituents move the other way out into the gut. And we all have a good deal more gut than we really need, for moderately good health at least. So, why not convert some of the surplus gut into a conveniently-portable naturally-occurring artificial kidney?

It's not, in fact, an original idea. Over the past thirty years, about thirty patients with kidney failure have been treated as follows. Under anaesthetic, a loop of the patient's small intestine has been cut away from the rest at each end, and the two cut ends of the main part have been joined together, leaving a length of gut floating loose, and a substantially shortened small intestine. Two neat round holes have then been cut in the side of the body wall, each about two or three centimetres in diameter, the diameter of the small intestine, and the cut ends of the loose length have been sewn to these. So, what is revealed is a tube which enters the body from outside, loops about inside for two or three feet, and re-emerges to the outside. It still retains its rich blood supply into which food was absorbed from the intestine.

When the new structure has healed up, which takes a surprisingly short time – indeed the whole procedure is surprisingly simple and easy – the next step is to use the tube as an artificial kidney. This is done, simply by passing a strong solution of sugars and salts through the tube from one end to the other. It's easy to do and quite painless. The strong solution exerts an osmotic pressure across the gut membrane in the loop. That's to say, it draws liquid containing relatively small dissolved molecules out of the blood supply to the loop, and through the gut into the loop, leaving blood cells and proteins behind. In this way, as the blood circulation passes through the arteries to the loop of gut, the blood is gradually all purified.

This technique had some advantages over the conventional artificial kidney; it didn't involve the use of expensive apparatus and it was pleasanter for the patient, once the operation was over. But the snag was, that the small intestine is the most vital part of the gut and removing a loop of it has unpleasant and sometimes dangerous side-effects. So a surgeon at the Westminster and Kingston hospitals in London, decided to experiment instead with using the *large* intestine, the last part of the gut, as an artificial kidney.

Tests on sheep have now shown that the *colon*, as the large intestine is called, is nearly as good for use as an artificial kidney as the small intestine – not *quite* as good, because its blood supply isn't so rich and so it takes longer to clean up blood. But the great advantage of using the colon is that sheep and, as is known from other surgical operations, ourselves, can do very well with half our colon bypassed and made into an artificial kidney. Some experiments have already been tried, using volunteers who have had part of their large intestines removed and a new entry made to the remainder, and the results have been most encouraging. It may not now be long before the first trials start on human patients.

The surgeon sees the technique as a likely treatment for patients whose kidney failure is slow, or only partial, rather than for those with rapid or total kidney failure. It could be used to help people awaiting transplants. And it might become the treatment of choice for parts of the world where there are no facilities either for dialysis or for transplants. The surgery is simple, risks during or after operation are minimal, and patients can easily be taught to perfuse – to pass the salt and sugar solution through – their own built-in artificial kidney.

Back to heart transplants

Heart transplants have appeared rather discredited in the past few years, owing to the very low survival rates for most patients after the operation. But American surgeons, at the Stanford Medical Center in California, who have been working quietly to improve techniques are now, without much publicity, getting results about as good as those obtained on average with cadaver kidney translants. If – once again that great if – the problems of immune rejection can be overcome, then heart transplants, with the technical improvements which are bound to come in the next ten to twenty years, could become as common as kidney grafts are today.

Animal hearts, or sections of hearts, could be used, though their different anatomies would make such a transplant imperfect. So could parts of hearts taken from foetuses. Wholly artificial hearts will certainly be usable well within twenty-five years. They have been experimentally implanted into animals for several years. But size, weight, damage to blood in contact with surfaces of man-made materials, and other factors present big problems. The ultimate heart transplant unlikely to be perfected, but still probably in widespread use by 2001, is going to be part-man-made and part-living, perhaps with pump action and basic frame made of man-made material, while every surface which comes in contact with blood is coated with sheets of living tissue grown from samples of the patient's own tissues so as to raise no rejection problems.

Overcoming rejection?

Overcoming the rejection problem – selectively eliminating just one reaction against one set of foreign proteins, those of the needed graft, while leaving intact all other reactions against all other foreign compounds, so as not to diminish the body's defence against disease; this is the greatest challenge facing medical research laboratories today. If the problem can be solved then along with it, in parallel, must come ways of selectively stimulating specific parts of the immune response, so as to make the body's own attack on an infection, or a tumor, more effective.

This approach has proved disappointing in recent years, after earlier bright hopes. But, undoubtedly, a more detailed understanding of the immune response, and a better knowledge of how to manipulate it, could bring *immunotherapy*, as it's called, into the forefront of medicine.

There will be no such advance here, no miraculous 'breakthrough' – that much abused word. The complexity of the immune response will be mastered step by step, as is the way of complex subjects. And, a long step behind, will come first hypotheses, then laboratory tests and finally cautious clinical trials of more specific ways to intensify or abolish immunity.

Three brief examples of work already done, or going on, shows how different approaches, ranging from the simple to the sophisticated, will be used to prevent rejection, or to stimulate it. Skin grafts are already well on the way to being rendered unrejectable, thanks to work at Dundee University. Medical research workers there have shown that skin treated with enzymes to dissolve away its distinguishing proteins, can be rendered so anonymous that when such skin has been implanted into rats, after 280 days there are still no signs whatever of rejection. The enzyme treatment leaves nothing by which the skin can be recognized as foreign to start rejection. The first clinical trials of skin treated in this way have begun. Soon the technique should make the plastic surgeon's job much easier.

'Blinding' the immune system

Organs like kidneys couldn't be treated in that way without destroying their function. But here a selective 'blinding' technique, developed at Uppsala University in Sweden, could be the answer, as recent experiments have shown. It's possible to identify white blood cells – the agents of immunity – which have become or can become sensitized to a grafted organ. It's then possible to immunize the patient against these white cells, so that the rest of his immune system turns upon these white cells and eliminates them. In this way the patient's own seek-and-destroy system is turned against just that part of his defences which is – from his point of view mistakenly – preparing to attack a friendly ally, the graft. So far it's only been tried on rats. But it could well be in use for humans, on a large scale, before 2001.

Starch fights cancer

Basic research into the immune system is helping to stimulate it as well as to switch it off. At the Clincial Research Center,

London, medical workers have worked out the sequence through which one 'arm' of the immune system, that involving so-called *macrophages*, scavenging cells in the body and blood, is stimulated to attack cancer tumors. They have gone on to develop, based upon their research, a completely new chemical approach to stimulate macrophages to attack cancer tumors more effectively. This involves the use of a starchy compound, known as COAM, which can be administered in larger doses, and is otherwise apparently more effective, than existing stimulants. This work is at too early a stage, really to measure its value. But it's a good example of how basic research is beginning to lead to a much finer and so more effective control over the immune system, which is bound to blossom into new treatments before 2001.

Blood swaps stop heart attacks

Today, patients with overmuch cholesterol – the fatty substance which is among the causes of arterial hardening – in their bloodstreams are advised to change their way of life. In twenty-five years' time those who are most in need – and, no doubt, those who can afford it will instead be told to change their blood. A complete blood exchange, London doctors have already shown, while it doesn't remove the cause of high blood cholesterol can, if repeated every few weeks, reduce the risk of heart attack to normal or even sub-normal.

The technique has been pioneered over the past two years by a London doctor, who has now begun to introduce it to the United States. Five or six patients have now been treated for up to two years with very good results. All of them are people suffering from a common condition, *familial hypercholesterolaemia*, in which a genetic defect prevents cholesterol being taken out of the blood in the normal way, when it's present in excess. There are about 100,000 people with this condition in Great Britain alone. A few of these suffer much more seriously than the rest, because they carry a double version of the genetic defect.

Diet can do little to help any of these people. The doubly-afflicted usually die from a heart attack before the age of twenty. These suffering from the far more common single genetic defect have a much higher-than-average risk of a heart attack in their forties and very often die of a heart attack before the age of sixty. The new treatment brings the cholesterol levels of such people down, straight away from two or three times normal, on average, to one third normal.

The treatment is no cure. It has to be repeated every two or three weeks, by which time the cholesterol level has built up to near its old concentrations. But, used at this frequency, it reduces the risk of arterial hardening and coronary blockage to normal.

The question so far unanswered is whether the treatment also *reverses* symptoms already present, whether it can do anything to reduce fatty deposits already laid down on the walls of the coronary artery. There is some evidence that this may happen.

The patients' blood is simply shunted painlessly – while the patient watches TV – out of a vein through a centrifuge in which the blood plasma is separated from the blood cells. The plasma, or most of it, is removed for treatment to remove the cholesterol. Meanwhile the cells, coming off in a sort of sludge, are diluted with fresh blood of the right group from which cholesterol has been removed, taken from a blood bank. Nothing is wasted. The patient's blood, with cholesterol removed, is put into the bank to await recycling into another high-cholesterol patient of the right group.

The blood swap is quick and simple and can be done on an outpatient basis. The centrifuge used at Hammersmith Hospital, London, where the technique has been pioneered cost about $44,000 but much simpler and cheaper models could be developed, given the demand. The doctors who have developed the treatment see it becoming easily as common as dialysis (the use of the artificial kidney) is today, and coming into use for a number of other purposes as well as the removal of blood cholesterol. More useful would be the removal from the blood of rheumatic patients of the so-called *immune complexes*, huge molecules which clog the blood and provoke inflammation and pain. Many other uses for blood swaps are envisaged and they could all happen within much less than twenty-five years.

Washing cells to fight cancer

Undoubtedly blood will be taken outside the body for more and more reasons, including the treatment of leukaemia and perhaps of other cancers. Techniques which allow the recognition of cancer cells in blood could be coupled with techniques for their elimination, if the blood could be exposed for long enough outside the body. In Dublin doctors have for some time circulated blood from some cancer patients outside the body and literally washed the white cells responsible for immunity with ordinary washing-powder enzymes. They found the washing cleansed the surface of the white cells principally responsible for identifying cancer cells. It removed a coating of chemicals which appears to form as a result of cancer cells' action, which covers and partially blinds the *T cells*, as the white cells involved are called.

Doctors have shown that the technique substantially strengthens the reaction against tumours and leukaemic cells, and believe that it will prove valuable in combating infectious disease as well as cancers.

Removing organs for clean-ups

Surgery outside the body is only a step on from dialysis and blood swaps. More than ten years ago, a surgical team at the University of Minnesota showed that dogs' stomachs could be removed, kept in cold storage for several hours, treated surgically and sewn back into place to continue working normally. The removal of organs for surgery would allow surgeons much more freedom to manoeuvre than they have inside the body cavity. Radiation and drug treatment of cancer could be made much more intensive if applied to an organ isolated from the rest of the body's healthy tissue.

The only real problem is that such removals involve cutting the nervous supply to an organ. But the nerves could be persuaded to regenerate.

Eliminating infectious diseases

Several new vaccines will certainly emerge over the next quarter-century providing between them the ability to put an end to other scourges as smallpox is being dismissed from the earth. Among them malaria, sleeping sickness, influenza, some forms of cancer and the bulk of tooth decay. Whether the full potential of these and other new vaccines is realized, will depend upon whether national funds and international agencies can provide the required backing for nationwide immunization programs.

The first human trials of a vaccine to protect against malaria will, probably, begin within five years, certainly within ten. This can be said with confidence, following the announcement, made this year, that two separate American research teams have found ways to grow the parasites which cause malaria on a large scale in the laboratory, and have shown that vaccines made from cultured parasites can protect monkeys from malaria.

The vaccine is apparently able to protect against all guises of a malarial parasite, in spite of the fact that, inside the human body, the parasite is able to dodge the immune system by repeatedly changing its surface proteins (antigens). So is that other parasite the trypanosome which causes that other scourge of the tropics, sleeping sickness. No vaccine against this has yet been developed. But there are signs of an imminent breakthrough, from the Molteno Institute at Cambridge where scientists have worked out how the parasite changes its coat of antigens and are well on the way to preparing a vaccine containing *all* the key antigens which the parasite is capable of making. They are also developing drugs to prevent the parasite altering its antigens, so the immune system can annihilate it at leisure.

'Jab' prevents tooth decay

Like the first trials of a vaccine against malaria, human trials of a vaccine against dental caries, tooth decay, are only five or less years away. This vaccine, too, developed at Guys Hospital, London, has already been thoroughly tested on monkeys and shown to reduce the incidence of tooth decay by about 80 per cent. But dental care will still be required, not only to deal with the other 20 per cent but to prevent gum disease which causes the loss of more teeth than tooth decay.

Vaccine for burns

Infected burns pose a big and growing problem today, because the bacteria involved, especially in hospitals, have in some places evolved resistance to all the drugs used against them. A revolutionary alternative, announced in 1976, is a vaccine made from killed bacteria, given *after* someone has been severely burnt. This vaccine acts so quickly – within 24 hours – that any infecting bacteria are killed off by the patient's immune system, stimulated by the vaccine, before infection can become established.

Other twenty-four-hour vaccines against other infections, will replace antibiotics for other purposes, providing a convenient way around the growing problem of the evolution of resistance to drugs on the part of bacteria.

Anti-cancer vaccines

French doctors have already produced and tested a vaccine to give protection against hepatitis B, a highly infectious virus disease of the liver. It is clear, from African studies, that infection with this virus greatly increases the risk of a person contracting liver cancer. So the vaccine can greatly reduce the risk of liver cancer, as well as protecting against hepatitis B. As it comes into use, over the next few years, this will effectively become the world's first anti-cancer vaccine.

Another common African cancer, Burkitt's tumour of the lymph glands in the neck, is caused by another type of virus infection. West German doctors, at Göttingen medical center, have shown that monkeys – the nearest thing to man – can be protected against a similar cancer by a vaccine made from the same type of virus. This is only one step away from a vaccine to protect humans. This should be in wide use by 2001. And other human cancers, very possibly including forms of leukaemia, will certainly prove to be caused by viruses, and preventable by vaccination, over the next ten or twenty years.

Screening out cancers

Some cancers will be combated with protective vaccines. Others can be screened out by removing their causes. In 1976 an international conference on cancer and the environment concluded that very roughly 80 per cent of the total incidence of cancer throughout the world was due to carcinogenic chemicals in the environment, in foodstuffs, in the air, in water, coming into contact with human skin and so on. New sensitive tests, developed in the past two years, now enable the great majority of such carcinogens, new and old, to be identified. More such tests will be developed over the next few years and their sensitivity and ease of performance will be greatly improved.

This will allow, given sufficient funding and sufficiently strict legislation, the incidence of cancer through the world to be reduced by perhaps 50 per cent. (It is probably unrealistic to hope for much more, since some existing carcinogens, like asbestos, have already become so widely disseminated, and because the less scrupulous nations and concerns will not enforce vigorous testing and screening-out of industrially promising new organic compounds.)

Still, between them, vaccinations and screening tests will reduce the incidence of cancer greatly. And the proportions of patients who are cured after treatment for cancer and suffer no recurrence will improve quite dramatically for some cancers, as the drug-targeting techniques described earlier come into use. Earlier diagnosis, will be possible through X-ray machines like EMI's whole body scanner, able to make a slice through the body as visible as if it had been sliced like salami.

Ultrasonic body scanners

Within the next few years, an ultrasonic version of this machine will follow the X-ray scanner into production. Between them, these machines will put an end to a vast amount of painful, frightening and time-wasting exploratory surgery and other primitive procedures for taking a look into the deeper recesses of the body.

Looking into the living brain

A few years later – still well before 2001 – revolutionary scanners using the principle of *nuclear magnetic resonance* will enable doctors and scientists to look deep into the living brain and other organs while they are working, without in any way harming or affecting the workings. This device, which depends upon measuring the changes in radio wave absorption by living tissue subjected to a magnetic field, can produce a picture, if its data is processed through a computer. Crude pictures of the insides of organs have already been produced using NMR as the technique is known for short.

Because it does not involve subjecting the patient under examination to any form of harmful radiation, and because of the wealth of additional data it can provide, imaging by NMR promises to bring about as big a revolution in diagnosis as the EMI scanner has already done.

Screening out doomed children

Techniques already exist for taking samples from foetuses at an early stage of pregnancy, in women at special risk of giving birth to a seriously abnormal child; so that, if requested, the foetus can then be aborted. We can expect a steady improvement in these and, in some cases, the achievement of the ability to deduce an abnormal foetus from a sample of a pregnant woman's blood, a much simpler and safer procedure than sampling the foetus. This year, a technique has been described which allows a foetus suffering from *thalassaemia*, a very severe form of anaemia which usually leads to death around the age of twenty, to be identified in the womb. At present such tests carry a serious risk to the foetus, which must be taken into consideration by the pregnant woman, who may find she has caused the inadvertent death of a healthy foetus. The improvements in safety and accuracy confidently to be expected in such tests over the next twenty years or so will mean that medical services will have the power – given co-operation by parents and religious bodies – to eliminate a good many genetically-caused abnormalities from the population at large.

ACKNOWLEDGEMENT

My thanks are due above all to the staff of the Medical Research Council's research units and laboratories, who have spent so many hours over the years patiently explaining their ideas and techniques to me. Their ultimate contribution to world medicine will, I am very certain, add up to many times their total cost. In particular, I'd like to mention the staff of the Clinical Research Centre at Harrow, the MRC staff at the Hammersmith Hospital, and of the National Institute for Medical Research at Mill Hill.

FURTHER READING

I'd heartily recommend Gordon Rattray Taylor's *The Biological Time Bomb* first published by New American Library in 1968 and since in paper back, which gives an excellent account of the problems and ethical dilemmas with which modern biological research is now beginning to confront the world.

GENETIC ENGINEERING AND BIOCHEMISTRY

Susan Goodman

After receiving an honours degree in physics from the University of Oxford, Susan Goodman joined a research team at the Royal Dental Hospital Dental School in the University of London and investigated the physico-chemical nature of teeth. She subsequently worked in medical physics and is now a full-time writer and broadcaster on medical science.

What is life?

The aim of the molecular biologist, or genetic engineer as he is more popularly called, is to discover how genes function. The biochemist wants to understand the chemical processes which are found in living organisms. Basically, both the molecular biologist and the biochemist are searching for an answer to the same question – What is life?

In this chapter I shall attempt to identify some of the more important areas of recent research which have provided scientists with considerable insight into the life processes. The growth of knowledge in this field will inevitably enable the scientists of the future to regulate and manipulate the fundamental processes of living organisms. These methods of control can be used to benefit mankind by, for example, curing the malfunctions of the body which cause disease and disability; but these same methods can also turn the horror stories of science fiction into reality, with headlines like 'Escaped Bug Kills Millions', and 'Factory Starts Manufacturing Babies' appearing in our newspapers.

Mankind has some difficult ethical decisions facing it in the years ahead. We shall have to decide to what extent we should allow doctors and scientists to interfere in the life processes. What will happen depends on us, and the decisions we make. What can happen is the subject of this chapter.

Life in the cell

A study of life processes is really an investigation into the function of cells. The human body is a community of cells, all of them functioning in such a way as to maintain the life processes within themselves and in each other. The cells can be studied in isolation from the body, but for them to survive independently they must be provided with exactly the right sort of environment in a 'culture' which contains a supply of all the essential chemical nutrients required by the cell. In the body the cells work co-operatively to make sure they each have all they need to survive.

A single-cell organism like a bacterium is much better equipped for life alone and therefore does not require as much pampering as does a human body cell in order to grow in a culture. For this reason and also because they reproduce very quickly bacteria are widely used for studying certain life processes, especially those which interest the molecular biologist. Extraordinary though it may seem, bacteria and human cells have a great deal in common when it comes to gene function. In fact bacteria are used to test whether certain chemicals will cause cancer or other changes in the gene structure of man.

Let us look at the human cell in more detail and see what actually goes on inside it. A cell is basically a blob of jelly, or cytoplasm, contained within a cell membrane; at the center of the cell is a small packet filled with proteins and large, complex molecules of so-called nucleic acids, namely ribonucleic acid, RNA, and deoxyribonucleic acid, DNA. It is this packet, or nucleus, with its nucleic acids which directs the activity of the rest of the cell.

The DNA is divided into units known as genes, and each gene carries instructions for the production of one particular protein – most of these proteins are special chemicals called enzymes, each of which promotes a particular chemical process in the body. The instructions for making the proteins are passed by the DNA in the nucleus to the surrounding cytoplasm, where the proteins are formed, by means of the nucleic acid 'messenger' RNA.

Genes are also able to pass their protein-making instructions from one generation to the next; in effect they are responsible for transmitting all hereditary information to new cells and new individuals. This is achieved through DNA's unique ability to make an exact copy of itself. In the 1950s Watson and Crick recognized that the DNA in the nucleus is actually a double helix, with each of the DNA spirals containing the same genetic information. Cells multiply by dividing, and preceding division the DNA is organized into separate sections, known as chromosomes. Each chromosome contains a very large number of genes, and is basically two strands of DNA joined together which separate as the nucleus and cell divides with each strand of DNA duplicating itself. This means the new cells each contain the full complement of genetic information contained in the original cell.

The molecular biologist is interested both in the control the gene exercises on cell function and, conversely, the control which chemicals in the cell exercise on the function of the gene. His work tends to look at the cell in isolation from the body; but the biochemist is generally much more interested in the organization of cells in the body as a whole. He wants to know how different cells communicate with each other, and how the messages they receive actually modify the chemical processes going on inside them.

Communication between cells

The quickest way in which two cells can communicate with each other is through the nervous system. For example, if you touch something very hot the cells in your finger tips promptly send a message off to the muscles in your arm, they contract and you move your hand away. The message is sent along a nerve fibre and when it reaches the end a chemical substance, known as a neurotransmitter, is released and this then triggers off a series of chemical reactions which result in the contraction of the cells in the muscle.

Another way in which cells in one part of the body pass information to distant cells is by the release of chemicals known as hormones. Certain special glands produce these hormones and none of them acts wholly independently of all the others. There are complex feedback mechanisms which ensures that the blood is carrying around the body just the right amount of all the hormones. If you start running and therefore need your cells to generate more energy the thyroid gland will send out more of its thyroxine hormone, and this hormone, through an unknown mechanism, stimulates the release of energy by cells. If there is not enough thyroxine in the blood, the pituitary gland will send out a thyroid-stimulating hormone.

Recent research has revealed that in some instances these two methods of cell communication are very closely linked. In fact the chemicals which are released at nerve junctions, the neurotransmitters, are also capable of acting as local hormones. This work has already led to the development of a new type of drug which doctors claim will cure ulcers by acting on neurotransmitters located in the gut wall as well as controlling related areas of the brain.

A number of different giant molecules have been found both in the nerve cells of the brain and gut and also in gastrointestinal cells which behave rather like hormone-producing cells. These chemicals carry impressive names like vasoactive intestinal peptide, somatostatin, and substance P. The biological significance of the occurrence of the same chemicals in nerve and hormone-producing cells remains a mystery, and yet month-by-month more of these chemicals are being discovered. It is suspected that research into the function of these substances will provide new insight into many diseases; the anti-ulcer drug is probably the first of many new developments in drug therapy which will come from this field of research.

Brain kills pain

There has been over the past year or so, a sudden surge in the discovery of new giant molecules in the brain. The function of some of these substances, or peptides, has been broadly identified, and one that has promoted an explosion in concentrated research effort is the peptide named enkephalin. At the end of 1975 enkephalin was identified as being the brain's own morphine; the excitement was tremendous, for here, the scientists thought, was a powerful, natural pain killer, which in the purified form could replace morphine and would be non-addictive. It is a doctor's dream to have a pain killer as powerful as morphine with none of its side-effects. In the early months of 1976, it soon became clear that enkephalin was not going to fulfil its original promise. For one thing it was very rapidly destroyed in the blood stream and had to be injected directly into the brain of a rat to have any effect. But worse was to come : experiments showed that experimental animals could become addicted to the brain's own enkephalin! However, research into enkephalin turned up a whole host of related peptides and the picture of the brain's own pain control mechanism has become very complicated; but many biochemists working in this area believe that once the chemical processes concerned with pain control are identified then it may be possible to stimulate them with certain enzymes; this would mean effective pain control with no risks.

Enkephalin has been found in the areas of the brain associated with mood, and some preliminary research indicates that our state of happiness or misery could depend on the concentration of enkephalin and other chemicals in the brain. In the next twenty years these chemicals, and the way in which they function, could be sufficiently well understood for doctors to be able to maintain us in a perpetual state of euphoria with doses of happiness-inducing drugs.

Getting the message

The discovery of enkephalin caused much excitement, but this substance had not suddenly appeared quite unexpectedly during the course of research by some unsuspecting scientist. Far from it; in fact scientists had spent a lot of effort looking for enkephalin and its relatives. The existence of these peptides was first postulated when some outstanding research, a few years ago, revealed how morphine interacts with cells in the brain. It seems that there are special sites on the surface of some brain cells which are chemically adapted so that morphine molecules can easily attach themselves. These sites on the surface of cell membranes are called receptors; and they serve a very specific function, and present theories maintain that one type of receptor can only receive one specific type of chemical. Clearly if there were brain cells with receptors for morphine, there was probably a chemical naturally produced in the brain which looked very much like morphine; and that is why enkephalin was discovered.

This theory of receptors sited on the surface of cells has been devised to explain how certain chemicals can influence the internal behaviour of the cell. Recent research has shown that cell membranes are not the rigid structures they were once thought to be, but are very mobile and fluid with considerable movement of proteins throughout the structure. There are suggestions that chemicals interact with the cell by immobilizing some of these proteins, and of course the proximity of chemically active substances sited on the cell membrane actually alters the cell environment.

This study of the cell surface is a comparatively recent science, and it has already revealed the existence of receptors and much information relating to the human body's immune process.[1] There is essentially one technique which has made all this work possible. It utilizes antibodies which bind to the cell surface, and it is now a highly refined method but its origins go back to 1900 when it was used to identify the existence of different blood groups and thereby made possible blood transfusion between individuals with the same blood group. This technique has also enabled doctors to identify tissue types so that they can carry out successful organ transplants by using donor organs that have very similar tissues to those of the patient – this avoids 'rejection' of the transplant.

Study of the cell surface will improve our understanding of disease, for it is here that chemical messages are passed from outside the cell to inside it, and of course vice versa. Recently a new mechanism has been identified which controls the number of receptors on the cell surface, and this might considerably improve our understanding of diabetes with the result of effective long-term cure. The experiment was designed to observe the binding of insulin to specific insulin receptors on cell surfaces; the presence of insulin actually caused some of the receptors to disappear. Exactly the same process has been noticed with other hormones and with some neurotransmitters. It is suggested that this mechanism is very important in controlling the number of receptors on the cell-surface, and this in turn acts as a control on the sensitivity of the cell to that particular hormone or transmitter.

What happens outside the cell clearly affects what is going to happen in the cytoplasm surrounding the nucleus, and this must influence the response of the genes in the nucleus. The biochemist will in the future, probably be able to control certain cell response mechanisms by regulating the enzymes of specific chemical processes. Where does the genetic engineer fit in? He looks to the central nucleus of the cell, where manipulation of the genes can alter specific life processes. But this work has stirred up considerable controversy.

Mankind threatened?

The main fear raised by genetic engineering experiments is that someone might intentionally, or accidentally, introduce into bacteria the ability to produce poisons they couldn't make before, or even cause cancer. If these bacteria leaked out of the laboratory the effect on mankind could be devastating. This is why in 1974, American molecular biologist, Dr. Paul Berg, called for a voluntary world-wide moratorium on dangerous genetic engineering experiments. His appeal was heeded and scientists have generally refrained from those experiments which are regarded as hazardous. The US and British Government are now formulating codes of practice which will contain recommendations of how genetic engineering experiments should be carried out. The codes, already in draft form, suggest that all research in this field should be conducted under the same conditions as those used when handling dangerous viruses and bacteria. They also suggest that, wherever possible, special 'disabled' bacteria should be used; these bacteria can only survive in a highly artificial environment created in a container in a laboratory, so that even if they did happen to escape into the atmosphere they would die immediately.

There are no immediate plans to make the code law. But of course, the responsible scientist will always take every precaution to ensure that his experiments are as safe as possible.

Can we take the chance that all scientists, in every country of the world, will always act responsibly? Is the threat to mankind too great a risk to take, even though genetic engineering does offer us considerable benefits? I hope the rest of this chapter will help you answer these important questions.

Killer diseases

One of the real threats to mankind's continued existence on Earth is the possibility that a new killer disease will suddenly appear and rapidly infect the world population. I'm not thinking about a disease escaping from a genetic engineering laboratory, because it is very much more likely that a strange virus will be passed to us from an animal, insect or plant. To survive we shall have to find faster and better ways of fighting disease – genetic engineering can provide them!

If the likelihood of disease devastating mankind doesn't sound very plausible to you, just remember that the outbreak of swine influenza following the First World War actually killed more people than were slaughtered in the four years of bitter trench-warfare. And only last year, in 1976, it was feared that this same influenza virus had re-emerged, killing an American soldier in Fort Dix, New Jersey. President Ford immediately planned to vaccinate the whole American population, but it took more than six months to prepare enough vaccine to protect only a section of the population. Recent research in genetic engineering could provide a much quicker method of preparing vaccines, but more of that later.

1976 brought accounts of many strange diseases suddenly appearing and claiming the lives of hundreds of victims. We read of large numbers of people dying in Africa from a little known infection, usually called Green Monkey disease but more accurately 'Marburg disease'. There were also cases in Britain and Canada of another poorly understood disease, originating in Africa, called Lassa fever. And do you remember the outbreak of a strange disease in Philadelphia which killed dozens of ex-servicemen attending a convention? It now seems unlikely that we shall ever know what caused their deaths, but eminent virologists suspect that it was a new, hitherto unknown virus.

Curing patients with these diseases was almost impossible; none of our antibiotics have any effect on these types of viral diseases. The only defence is the body's own immunological system. But this immune system has a very difficult battle in conquering diseases it has never encountered before. Normally as soon as a foreign substance, such as a virus or bacterium, invades our bodies we produce 'antibodies'. These complex chemicals are capable of inactivating or killing the foreign organism. In fact in recent cases of Lassa fever and Marburg disease, antibodies from the blood of people who have had the disease have been injected into other victims. A boost of the right antibodies can help a patient fight a disease very effectively. Genetic engineering experiments have indicated that they will be able to provide large amounts of any antibody whenever it is needed (see section 'Breeding Antibodies').

• How great is the threat of a new disease appearing quite unexpectedly? All animals, plants and insects have their own diseases and it is possible at any time for one of these viruses to mutate and adopt man as its host, making him very ill indeed. Take for example the diseases already mentioned: swine influenza, as its name implies, is normally found in pigs; Marburg disease is found in the African green monkey but experts believe that this disease really belongs to insects or perhaps plants; and Lassa fever is a rat infection.

Our chances of encountering new viruses have considerably increased over recent years as we clear away jungles in Africa and South America to make way for new agricultural and industrial developments. These projects bring large numbers of people very close to a new animal, insect and plant environment which might harbour many unknown deadly viruses. We have rapid movement of people, inside one country and between different countries; this means disease is carried further and faster, and the problems of containing an outbreak have become much more difficult.

This picture is frightening enough and yet we must also

add the depressing fact that many bacterial infections have become resistant to the action of available antibiotics and flourish in their presence. There is considerable fear that the more we use antibiotics, the more likely it is for resistant bacteria to develop. It seems that our most powerful weapon against bacterial disease is slowly being inactivated. But here again, as the next section will reveal, it is genetic engineering that comes to the rescue by increasing our understanding of antibiotic resistance in bacteria and by providing the means to develop new antibiotics.

Understanding antibiotics

The first antibiotic, penicillin, was accidentally discovered by Sir Alexander Fleming in 1928, when he noticed that bacteria would not grow in a culture contaminated with a particular mould. Clearly the mould produced a chemical which killed the bacteria. Antibiotics are not only made by moulds, they are also made by most living things. The majority of antibiotics used in medicine such as tetracyclines and streptomycin are actually produced by a group of bacteria called streptomycetes.

It is only in the past few years that scientists have discovered how streptomycetes make antibiotics and this has led to a whole new branch of genetic engineering. It has been found that the information necessary for a cell to produce an antibiotic is not found in the chromosome in the nucleus of the cell but it is carried in a gene which exists quite independently in the outer tissues. These independent genes or 'plasmids', are transferred between bacteria when they mate, allowing genetic characteristics to be exchanged between different individuals.[2] It has recently been found that these same plasmids also carry information relating to the drug resistance of a particular bacterium, and the free movement of plasmids between neighboring bacteria means that resistance to antibiotics can be rapidly passed from resistant to non-resistant bacteria.

Genetic engineering experiments with plasmids have tremendous possibilities. The rate of production of an antibiotic can be considerably increased in one of two ways. Firstly, plasmids can be transferred into bacteria which are especially easy to grow, and in fact plasmids carrying antibiotic codes have already been transferred from streptomycetes into *Escherichia coli* (*E. coli*). Secondly, bacteria which have plasmids carrying information for making a useful antibiotic can be crossed with a strain of mutant bacteria which reproduce at ten times the normal rate and will therefore produce antibiotic-making plasmids at this elevated rate.

The transfer of plasmids from streptomycetes to *E. coli* opens up another field of research. The biochemistry and genetics of *E. coli* are much better understood than those of streptomycetes, and so it may be possible to find out the series of chemical reactions, or so-called 'biochemical pathway', which leads to the production of the antibiotic. This pathway could then be stopped, with a suitable chemical, at a certain point before the antibiotic was completed. By attaching different chemicals to the molecules of incomplete antibiotics, new synthetic antibiotics could be made and these would be able to kill bacteria resistant to the usual antibiotics.

The main risk associated with this work is that this new strain of *E. coli* which can make large quantities of antibiotics will escape from the laboratory. This may not sound very important, but *E. coli* normally lives in our intestines and if the antibiotic-producing *E. coli* reached us it would make us very ill by killing all the beneficial micro-organisms which normally live inside us. However, this risk can be avoided by using a disabled strain of *E. coli* which cannot live anywhere except in the highly artificial conditions created for it in a laboratory.

Here then is a safe method with the potential of providing us with new antibiotics to fight diseases caused by bacteria resistant to existing drugs. But antibiotics are not able to help us in our fight against diseases carried by viruses; here our immune system must generate its own antibodies to inactivate the virus, but genetic engineering will provide vital help in the future.

Breeding antibodies

Antibodies, or so-called 'immunoglobulins', are formed in the spleen and lymph nodes in response to the presence in the body of foreign micro-organisms, called 'antigens'. It takes several days for the body to produce antibodies when an unknown antigen enters the body and by this time the infection is well established. But if you catch this same infection again, the body can promptly produce antibodies, and so once you have had an infection you are 'immune' to that disease. This is the principle associated with vaccination – a virus or bacteria causing a particular disease is slightly modified so it is not as dangerous as its disease-causing relatives, this is then injected into your bloodstream and stimulates an antibody response, thereby preparing your body in case you should ever encounter the disease itself.

Although our bodies do produce antibodies when faced with a new virus, sometimes we are unable to produce sufficient of them to totally eliminate the viruses, or bacteria, which are making us ill. Ideally doctors would like to have supplies of different antibodies available so that they could boost the body's response to the disease. At present about the best we can do is to stimulate antibody production in an animal and then use these antibodies in patients. But animal antibodies are not quite the same as human antibodies, and anyway after a short while the animal stops making antibodies.

Recent research at the Molecular Biology Laboratory in the University of Cambridge looks like providing the solution with a technique capable of producing human antibodies very quickly.[3] The original work was done with mice but it was not anticipated that there would be any difficulty in developing it for human use. In the experiments at Cambridge, scientists stimulated a mouse to produce antibodies by injecting an antigen into it. Then they took a sample of spleen cells and identified which specific cells were responsible for making these particular antibodies. These cells were then fused with cancer cells, which have the peculiar pro-

perty of reproducing very quickly. These hybrid cells reproduced passing on to successive generations of cells all the
genetic information contained in the two original separate
cells. In the case of these hybrid mouse cells, they were able
to continue making antibodies indefinitely – in the mouse
they would have stopped after a short while.

For the production of human antibodies, doctors would
not have to extract spleen cells, they could use white blood
cells which are also responsible for making antibodies and
these of course can be obtained in any blood sample from
the patient. They wouldn't have to fuse these antibody-
producing cells with cancer cells, any fast-growing cell could
be used.

Even more recent research at the Cambridge laboratories
has shown that bacteria could perhaps be used to produce
unlimited numbers of antibody copies.[4] Dr. Terry Rabbitts
has already used bacteria plasmids to make millions of copies
of a gene from a rabbit cell. The gene, which was incorporated into the bacteria, wasn't complete and so the bacteria
could not actually use the gene to make the protein which the
gene was usually responsible for producing. The next stage
in this work is to provide bacteria with a complete gene, so
that the bacteria can make millions of copies not only of the
gene but also of the protein associated with it. Supposing the
gene were responsible for the production of a particular antibody, then this technique could be used to produce large
quantities of antibodies very quickly.

But these experiments can be dangerous, because it is very
difficult using conventional methods, to be sure that the
bacteria is being given only one particular gene. Usually a
gene is obtained from an animal by chemically removing it
from a chromosome, but it is impossible to separate out an
uncontaminated gene. The scientist might well give the
plasmid some unwanted and unexpected characteristics and
these will be reproduced along with those belonging to the
gene of interest.

Dr. Rabbitts got round this problem by taking the gene
out of certain cells which do not have the genes contained

in complex chromosomes; these cells only have bits of genetic material which can be easily purified to provide uncontaminated genes.

Vaccines and drugs from bacteria

Techniques for transferring genetic information from one organism to an unrelated one are being developed and refined in molecular biology laboratories all over the world.[5] Research is being supported by chemical and pharmaceutical manufacturers because they hope that these methods will lead to the development of production lines with bacteria producing drugs and vaccines. The secret is to get the right bit of genetic material incorporated into bacteria in such a way that the bacteria's synthesizing machinery can use it as if it belonged to the cell – in other words, produce a limitless number of copies of the chemical compound required, whether it be a drug or a vaccine.

Rabbit blood from frog eggs

We have seen in the preceding sections that gene manipulation with bacteria and other micro-organisms can achieve amazing results;[6] but perhaps the most astonishing work in genetic engineering has been done with animals. And it is perhaps this field of research which offers some of the most terrifying glimpses into the possible future accomplishments of genetic engineering.

Work with animal cells really began back in 1962 when Dr. John B. Gurdon of the University of Oxford extracted nuclei from cells found in the tissue-lining of the intestines of a frog; he transplanted these nuclei into frog eggs from which the nuclei had been removed. A number of these eggs developed into normal, perfectly healthy adult frogs, which proved that cells in the intestine contain all the genetic information required for normal development. This method of transplanting nuclei into frog eggs has formed the basis of a whole range of important experiments designed to investigate the control mechanisms which are exerted on genes, for example : how they are switched on and off. This research has, as we shall see, produced some bizarre and rather disturbing results, but it is also providing scientists with essential information on what actually goes wrong in a cell when it becomes cancerous, or produces birth defects and metabolic disorders of genetic origin. Understanding the control process might be the first step towards putting it right when it does go wrong.

The frog egg has been widely used for this work because it is such a large cell, more than 10,000 times larger than a frog liver cell; this makes it much easier to inject substances into it. In the early 1970s a sample of messenger RNA was extracted from rabbit blood cells and injected into frog eggs; this particular messenger RNA was responsible for directing the synthesis of the protein carried by red blood cells in the rabbit, and in the frog egg it did exactly the same thing. The result was that the frog egg produced proteins found in rabbit blood-cells. The experiments were repeated using messenger RNA from duck and mouse red blood cells; the frog eggs produced duck and mouse blood components. The experiments went a stage further : messenger RNA from the venom glands of the honeybee was injected into frog eggs, and sure enough, the frog eggs produced honeybee venom.[7]

Breeding identical animals

I have so far only touched upon the significance of Gurdon's work with frog eggs. In his original experiments he was actually able to stimulate a frog egg to produce a tadpole without any fertilization by a sperm; all that was needed was a nucleus from any cell to be transplanted into an egg from which the nucleus had been removed. Supposing the transplanted nuclei all come from the cells of one particular frog; all the eggs will develop into tadpoles containing exactly the same genetic information as the one original frog and these tadpoles will grow into absolutely identical frogs; such *genetically identical individuals* are described as members of a 'clone'. In fact this is just what Dr. Gurdon did, and by doing so he brought us one step nearer to turning the fiction of Aldous Huxley's *Brave New World* into reality.

... and identical humans?

How close are scientists to producing the identical humans who people the 'Brave New World'? Recent genetic engineering experiments have brought us much closer to this reality than the early work of Dr. Gurdon with his frog clones.

In late 1975 at the Department of Zoology in the University of Oxford, Dr. Derek Bromhall fertilized an egg cell of a rabbit, not with a sperm, but with the nucleus of an ordinary non-reproductive cell. He had succeeded in doing the same with a rabbit egg as Dr. Gurdon had achieved 13 years earlier with a frog egg. Fertilizing a rabbit egg in this way is much more difficult than doing it with a frog egg. For one thing, a frog egg is about one thousand times larger than a rabbit egg and this makes the micro-injection of a transplant nucleus into a rabbit egg a difficult and delicate procedure. Also, a frog egg naturally develops outside the body, whereas a rabbit egg needs special treatment if it is going to survive outside the rabbit's womb.

In his first experiments Dr. Bromhall fertilized the rabbit egg with the nucleus of a non-reproductive cell taken from another rabbit; this 'sperm substitute' cell was injected into the egg cell and the two nuclei fused. In subsequent work he showed that if the sperm substitute received a sufficiently strong stimulus to enter the egg then its nucleus doesn't fuse with the egg nucleus, instead it totally replaces it and the egg develops according to the genetic information contained in the sperm substitute nucleus. It is not difficult to imagine how this technique could be used to produce rabbit clones, in the same sort of way as frog clones have already been produced. So far, no attempt has been made to grow these fertilized rabbit eggs into animals, so we don't yet know what these creatures would look like if they were encouraged to develop, but it is highly probable that they would grow up looking like ordinary rabbits. In fact they would be a set of absolutely identical, ordinary rabbits.

If it can be done with rabbits, then certainly there is the strong possibility it can be done with human beings. Scientists in this field of research have expressed complete abhorrence at this idea, but I suppose there is a chance that at sometime in the future, somewhere in the world, a scientist with the right sort of knowledge might want to play at creation. He could select a particular man or woman who he thought worth duplicating and then produce dozens, or hundreds, of members of the clone. Perhaps he finds a very strong, tall man – someone exceptionally powerful, who could almost be regarded as a freak of nature because he has such strength; the scientist could use cells from this man to produce a whole army of identical superstrong individuals. The imagination can conjure all sorts of frightening prospects, but it is always important to remember that genetic engineering has a very positive side to it. We have already discussed the negative and positive aspects of work with bacteria. The experiments with animals clones also have a positive contribution to make.

Dr. Bromhall's work is likely to find application in cancer research, where it is essential to discover the role of the nucleus when a cell becomes malignant. If the nucleus of a cancer cell were used to fertilize an egg cell, then valuable information could be obtained on the relative importance of the nucleus and the surrounding cytoplasm in determining the cancerous nature of the cell.

This technique could also be used to rapidly multiply the numbers of a particular breed of pedigree cattle. The nucleus of the egg from an ordinary cow would be replaced by a nucleus from a cell of a pedigree cow; the fertilized egg would then be implanted in any cow and allowed to develop naturally. A stock of these fertilized eggs could be kept in suspended animation at very low temperatures, and then when required they would be implanted in a cow.

Human-plant hybrid

In 1976, American scientists at the Brookhaven National Laboratory announced that they had fused tobacco plant cells with human cells and produced a living organism.[8] In this experiment the actual nuclei of the cells have not themselves fused. However, similar experiments in Hungary seem to have achieved fusion of the plant and human nuclei. It is hoped that animal–plant hybrids will develop into fully-grown organisms. The scientists do not envisage that the resulting hybrid will turn out as a monster, half plant and half human. They expect the technique will be used to pass on to plants certain animal qualities, for example : the flavor and nutritional value of animal flesh could be introduced into certain plants.

In Britain there already has been success in fusing animal cells, specifically red blood cells, with yeasts. As yeasts grow much faster than plants it is more likely that they will be used to incorporate animal flavors and proteins. Here then is the food for the future; it is nutritious, tasty, cheap and abundant.

This cell fusion technique has already produced remarkable results with plant hybrids and there are fully grown 'fusion hybrids' to testify to the success of this method.[9] The original work was done in the 1960s with two wild tobacco species; the leaf cells of the plants were suitably treated so that they fused, and when a small plant developed in a culture medium it was grafted on to other tobacco roots and grew into a mature tobacco plant with all the genetic characteristics of the two original plants. Moreover, it produced seeds which carried this genetic information and developed into identical plants, showing that these fusion hybrids breed true to type. The plant breeder will be able in the future to use this method to produce vigorous new species which combine the most desirable features of each parent plant, such as increased resistance to diseases, increased yield, or higher nutritional value.

Cell fusion: Cure for genetic disease?

Cell fusion techniques are probably going to provide the world with not only its future food needs, but also a cure for genetic diseases. These diseases are due to a defective gene which fails to make a particular enzyme, without this enzyme a whole series of chemical reactions are unable to take place in the body; the result is usually a progressive deterioration in a baby's condition and death in early childhood. These babies would grow up into healthy adults if doctors could find a way of giving the body a constant supply of the required enzyme; cell fusion methods could provide that vital enzyme supply.

Recent research with hybrid cells produced by fusing mouse with chick cells have demonstrated that if the mouse cell has a specific enzyme deficiency, then it can acquire the ability to produce that enzyme after fusion with a chick cell carrying the necessary genetic information. This work implies that it might be possible to perform the same trick with human cells which have a defective gene. Cells from the organ normally responsible for the production of the particular enzyme which is lacking, could be fused with normal cells from somebody else. The fused cells containing the essential genetic information for producing the deficient enzyme can then be injected into the patient, and they will migrate back to the organ from which they were originally taken.

There have already been reports of success in treating children with enzyme-deficiency diseases. They have not actually used cell fusion, but a rather similar and in fact simpler technique. The doctors involved in this work, all based at London hospitals, have injected healthy human cells into patients whose own cells can't make the necessary enzyme. The injected cells are carefully selected so that they match the patient's tissue type and are therefore unlikely to be rejected by the body as foreign intruders. So far doctors have used this method to treat children suffering from the disease mucopolysaccharidoses and they are hopeful that this technique will find application in many other enzyme deficiency diseases.

Cell fusion and cancer

The cause of cancer remains unknown despite intensive research all over the world. But there is mounting evidence that the disease is linked with gene damage. Cell fusion studies have provided supportive data for this argument. There are reports that after fusion between normal and cancer cells, some of the cells produced are not cancerous. But this normal hybrid can give rise to cancer cells in subsequent generations when a loss of chromosomes is also observed. Even more extraordinary are reports of normal cells arising after the fusion of two cancer cells and it is suggested that this occurs because the combined cells contain a full complement of genetic information. If cancer is, as these experiments imply, due to a loss of certain genetic information it might be possible to replace the lost or mutant genes with future genetic engineering techniques. In fact molecular geneticists have already taken several important steps along the road towards gene replacement surgery.

Gene replacement surgery

One of the essential ingredients in this work is obviously the gene itself. The scientist must have available a pure supply of the genes he requires; these genes can be obtained either by isolating them from their neighbours in a DNA strand or by making them artificially. The actual isolation of a specific gene found in *E. coli* has been achieved and a picture of it obtained with an electronmicroscope. Once isolated a gene can be investigated to discover the precise sequence of nucleotide molecules which will produce this particular gene; this is exactly what has been done with the *E. coli* gene and at the end of August 1976, scientists at the Massachusetts Institute of Technology announced that they had made the first complete, functioning artificial gene. In order to make this gene they had to link up 126 nucleotides in precisely the right sequence so that an *E. coli* cell treated this artificial gene just as if it were a natural part of the cell's heredity.

The prospect for making an artificial human gene is at present rather remote, because a single human gene contains millions of nucleotides and this is obviously much more difficult to put together than the 126 of the *E. coli* gene. It is perhaps more likely that single pure genes will be isolated from strands of human DNA.

The next problem facing the genetic engineer is how he is going to introduce these genes into the cells with defective genes. He cannot consider directly inserting the replacement gene directly into each cell, the enormous number of cells involved makes this an impossible task. Instead he can use certain types of harmless virus which willingly pick up bits of genetic material and transfer them to other cells. These viruses could be encouraged to incorporate the specific gene of interest and then patients with the defective gene could be infected with the virus. There are already reports of an attempt to use this technique with two human patients suffering from an enzyme deficiency disease; there has as yet been no success but it is early days and the technique is regarded as having considerable potential.

Getting genes turned on

Certain genetic diseases are thought to be associated with the mechanism which turns genes on and off. In the past few years certain proteins associated with chromosomes have been identified as responsible for regulating the activity of genes. Scientists working in this field believe that it may not be long before proteins that control the activity of specific genes are isolated; this would introduce the possibility of a certain kind of genetic engineering whereby proteins might be inserted into cells in order to modify the action of a gene and this technique could be used to correct genetic abnormalities which cause cancer and many other diseases.

The whole mechanism of gene regulation is fundamental to our understanding of the processes of life and research in this field has already produced significant results. In earlier sections I have described Dr. John Gurdon's important work with frog eggs fertilized with a cell from intestinal tissue. This and subsequent work has demonstrated the constancy of DNA in all cells. This effectively means that all cells potentially have the ability to produce all the proteins that the body needs, and yet it is obvious that different cells do in fact perform different functions, therefore only a limited number of genes are expressing themselves; the rest of the genes remain unexpressed and do not contribute to the function of the cell.

For a gene to function, 'express itself' as the molecular biologists say, there must be a chemical which triggers the action of a particular gene, leaving the others dormant. It has been the task of the biochemist to identify these chemicals. His interest was aroused by the presence in higher organisms of certain proteins which were clearly linked with the chromosomes but didn't seem to contain any genetic information. These chromosomal proteins have been divided into two groups : the histones, which share many chemical characteristics, and the non-histones.

Are the histones responsible for the control of the genes? The first important biochemical studies, in the 1960s, demonstrated that histones are very closely linked with the synthesis of RNA. Experiments at the California Institute of Technology in 1962, showed that the production of RNA by DNA could be inhibited if histones were added and this inhibitory effect could be reversed if more DNA were added. At about the same time work at Rockefeller University proved that removal of histones from nuclei increased the rate of RNA synthesis. The chemical properties of certain histones have been identified as having some sort of influence over the rate at which cells divide. The histones appear to be the switching off mechanism for genes, but some recent work with insects suggests that histones cannot influence specific genes. In these insect studies it has been observed that the areas on the chromosome where a gene is actively synthesizing RNA are extended, forming what are called 'chromosome puffs'. Insect chromosomes are so large compared with animal chromosomes that these puffs are clearly visible under an electron microscope. The amount of histones in these puffs is no more or less than the amount associated with the inactive genes on the chromosome.

The search goes on, for there must be certain chemicals which can switch on particular genes. What about those other chromosomal proteins which are not histones, that is, the non-histones. Investigations into non-histones has produced a very complicated picture. With histones they were chemically very similar but the non-histones show amazing diversity; some of them have a lifetime of only a few minutes while others are as stable and long-living as DNA itself, and it is estimated that there actually are an enormous number of different types of non-histone proteins. The one thing they all share is the fact that they are proteins closely associated with the chromosome. Perhaps their diversity is what should be expected if they are each capable of exercising control over one specific gene.

The experimental work done in this field by the biochemist and molecular biologist is complex and ingenious. It is not an easy task to encourage a cell to yield such fundamental secrets of life. But there is mounting evidence supporting the idea that the non-histones have the ability to switch specific genes into action; if the action of the genes is then switched off by histones. If scientists can discover which proteins switch on which genes then they would clearly be much closer to understanding, and perhaps controlling, such fundamental processes as cell development, hormone action, as well as metabolic diseases, cancer and genetic diseases of all types.

How to stay healthy

The genetic diseases referred to in the preceding sections cause gross abnormality in specific body processes. These diseases are well known to doctors and can be readily diagnosed. For most of them, cure lies in the future work of genetic engineers and biochemists. But there is a whole class of genetic diseases which I haven't even touched upon; these diseases include brain haemorrhages, heart diseases, malaria, bilharzia, thrombosis, diabetes and stomach ulcers! You are probably surprised that I should have categorized these conditions as genetic diseases, but in a sense they are. An individual's susceptibility to contracting any one of these diseases is probably genetically determined; in fact there are many distinguished scientists working in this field who believe that our susceptibility to all diseases is in some measure genetically determined.

With so many doctors telling us what to eat and how to live in order to stay healthy, you are probably wondering whether it is worth bothering with a good, wholesome life if everything is genetically determined anyway. But this ignores one very important factor – the effect of the environment on your genes. We have already seen in preceding sections how certain chemicals are responsible for controlling the function of genes; these chemicals form part of the cell environment and this may be influenced by the external environment. This cell environment certainly seems to be affected by what we eat, drink and inhale. The scientists are suggesting that the external environment can encourage certain genes to operate which would normally have remained inactive throughout our lifetime. Let's suppose that a group of people all carry genes which make them susceptible to developing lung cancer; if any of the people in that group smoke then they will alter the cellular environment in such a way that the gene responsible for susceptibility to lung cancer will immediately make that cell cancerous. In effect, smoking tips the balance; and this is possibly why some heavy smokers never develop cancer – their genes do not carry the information which makes them susceptible.

There are many substances which are capable of converting, or mutating, perfectly healthy genes into specimens which are unable to carry out their proper function. The list of these so-called 'mutagens' includes substances, like benzene and carbon tetrachloride, which are mainly found in chemistry laboratories. But today human populations are continuously exposed to an ever-increasing number of chemicals, most of whose potential for damaging genes remains unknown. These chemicals are being pumped into the air we breathe by modern industry, and they are being added to our foodstuffs as preservatives, coloring and flavoring. Also drugs are being extensively used in medicine with little comprehension of the complete effect they have on our bodies or the changes they are inducing in our genes. It is the change in our genetic material that is most frightening because this is what we pass on to all subsequent generations.

In the next decade we can look forward to the development of better methods for testing whether chemicals cause gene-damage in humans. At present tests for chemical mutagens use bacteria as the test-cells, but methods using human cells will emerge with our increased understanding of the mechanisms controlling human–gene function. For the meantime the best way to stay healthy is to be kind to your genes, and that means keeping your intake of unnecessary chemical additives and drugs to a minimum.

... and live longer

Suppose, through the combined efforts of the genetic engineer and the biochemist, all your genetic defects could be eliminated so that every cell in your body worked smoothly and efficiently, free from all disease. The big question is, would you then live longer?

An intriguing series of experiments has indicated that, unfortunately, immortality is not a property of human cells, but it might be possible to delay the ageing process. In these experiments Dr. L. Hayflick employed cell cultures of a particular group of cells called fibroblasts; these cells are responsible for making collagen, a substance which forms the packaging material of the human body. One might expect these fibroblasts to live forever in a tissue culture; the cells would divide and produce new cells which could subsequently divide to form another generation, and so on. But this isn't what happened. In fact fibroblast cells, taken from a four-month-old embryo, divided over and over again for approximately 50 cell generations, and then the population of cells died. The experiment was repeated with the cell division process interrupted by placing the cells in cold storage for a few years; the cells were allowed to divide for 30 generations before being placed in suspended animation in cold storage; when they were thawed out a few years later, they began to divide again but after a further 20 generations the cell population died! The cell population insisted on living for just 50 generations and then dying, and the cells did not forget, during the years of suspended animation, how much more life was due to them. This 'remembering' by the cells implies that the whole process is somehow genetically controlled.

Further research has revealed that chromosomes develop more structural faults the older they are. This has so far only been demonstrated in cells grown in culture but scientists do think that the ageing process does involve mistakes during the replication of DNA or in translation of genetic messages. These errors could eventually cause a deterioration in normal cell function. If this is the case, as seems possible, then the ageing process could be delayed by using enzymes to repair damage to DNA. Scientists already know something about the way animals and plants are able to repair DNA and Dr Hayflick has suggested that perhaps animals with longer life spans have evolved repair mechanisms that are more effective than those of animals with shorter life spans.

If future research does support present evidence then ageing will be catalogued as yet another genetic disease. Here the loss of genetic information results in the deterioration of chemical processes in the cells. But the genetic engineer is developing a whole range of different techniques for introducing genetic information into cells, for example: cell fusion and gene replacement surgery, both of which have been fully discussed in earlier sections. The ageing process could be delayed, but for how long? That we just don't know.

The secret of immortality

Man has always longed for immortality : for really there is something very unsatisfactory about the way in which life processes suddenly terminate at death. But today, molecular biologists maintain that we can achieve immortality through the survival of our genes. In fact Richard Dawkins argues in his book *The Selfish Gene* that the whole purpose and function of the human body is to promote the survival of the gene – it is this part of us which lives on in our offspring and in all subsequent generations. The genes we possess belonged to our distant ancestors, and each gene may be tens of thousands of years old or perhaps a great deal older than that !

In Professor Norman Rothwell's fascinating book *Understanding Genetics*, he suggests that we could ensure immortality for ourselves by breeding identical human copies. We have already seen in earlier sections of this chapter that scientists are on the threshold of being able to achieve this; it is possible that in the next twenty years we could each of us achieve immortality by providing genetic material for egg cells which will develop into people who are exact copies of ourselves.

It is curious to think of immortality in these terms. And yet the ancient philosophers, Plato and Aristotle, came to similar conclusions when considering the transience of an individual's life.[10, 11] For them the maintenance of the species by reproduction is the way in which living forms achieve immortality. For us the species has been replaced by the gene.

REFERENCES

1. M. C. Raff, *Scientific American* (May 1976, pp. 30–39).
2. R. Sager, *Scientific American*, 212 (No. 1), pp. 70–79, 1965, Genes outside chromosomes.
3. *Nature*, August 7th, 1975.
4. *Nature*, March 18th, 1976.

5. S. N, Cohen, *Scientific American*, July 1975, pp. 25–33, Manipulation of Genes.
6. W. Hayes, *The Genetics of Bacteria and Their Viruses*, 2nd ed. (John Wiley & Sons, Inc., New York, 1968).
7. C. Lane, *Scientific American*, August 1976, pp. 61–71, Rabbit Hemoglobin from Frog Eggs.
8. *Science*, July 30th, 1976, p. 401.
9. P. S. Carlson, et al., *Proc. Natl. Acad. Sci.*, 69, pp. 2292–4, 1972.
10. Plato, *Symposium* (207a–208b).
11. Aristotle, *Generation of Animals*, Book IIi.

PSYCHOLOGY AND MIND CONTROL

Martyn Partridge

Martyn Partridge is a freelance journalist specializing in the interrelationship between science and society. He read Politics, Philosophy and Economics at Oxford and formerly worked at the Science Museum in London. He is a member of the editorial group of *Undercurrents* magazine.

Human life is carried on in a sea of influences. Advertisements seduce and cajole from every direction, making us want goods and services whose existence we had not even imagined. Writers and film-makers, artists and musicians, teachers and journalists, all present to us a chosen facet of their own experience and invite us to call it the world. Representatives of authority, governments, administrators, policemen, enforce certain modes of behavior on the population at large using gentle persuasion, sometimes, or otherwise threats and sanctions. Doctors and psychiatrists try to inculcate particular concepts of health and sanity. Politicians, pressure groups and evangelists cry out their faith from banners, newspapers, walls and soapboxes. No inhabitant of twentieth-century urban society can avoid any of these things.

Indeed, we are all part of it. If this list of not-so-hidden persuaders were extended it would eventually include everybody. As well as being the recipients of such influences each of us, to a greater or lesser degree, does likewise. Whether at home or at work, in social or in public life, we habitually, and for the most part unconsciously, try to affect the way that other people think and behave. We give orders or beg favors, choose clothes and hairstyles, use etiquette of one sort or another, make signals when driving, argue, quarrel, and so on.

You may doubt the value of this all-embracing insight : it is reminiscent of the man who spoke prose all his life without realizing it. If influence and persuasion are just two more words meaning ordinary, everyday human interaction, then how do they assist our understanding of what's going on? The answer to this perfectly sensible question lies in the above list of professions, organizations and agencies all of whom *specialize* in mass persuasion. The rest of the human race, the majority, are just amateurs who dabble on a very small scale. This is a distinction worth taking seriously because at no time in history have the means of communication and persuasion been so powerful, so circumscribed with trade secrets and esoteric language, and so concentrated, as they are in the present day.

Behavioral modification

One of the most influential modern psychologists is the American, B. F. Skinner, who is a leading member of the school of thought known as *behaviorism*. This, like many other branches of psychology, can be regarded as 'folklore' given a scientific treatment through experiment and numerical assessment. People have known for centuries that it is possible to train animals to behave in a predetermined way using simple techniques of reward and punishment – not only working animals such as dogs, horses, camels and elephants, but more exotic creatures such as doves, lions and dolphins too. Skinner takes this process one stage further and applies it to human beings. He argues that it is possible to analyze traditional types of rewards, such as money, esteem, approval and status, and discover which are most efficient in which circumstances. Such knowledge produces a powerful tool, a truly scientific method of social control.

Instead of talking about punishment and reward, behaviorists prefer the words *aversion* and *reinforcement*. The value of this distinction is that, for example, the idea of reward in the popular imagination does not encompass all the things which are, in practice, rewarding. A smile, a pat on the back, a few words of gratitude, all constitute, in most circumstances, reinforcement : when one responds in this way to someone's action it becomes associated in their minds with a favorable response, and they are more likely to do the same thing again. Similarly, aversion means more than just punishment; a scowl or an icy retort will reduce the probability of an action being repeated. Controlling a person's activities by conscious manipulation of aversion and reinforcement until new, automatic modes of activity become ingrained, is known in the jargon as *behavior modification.*

The essence of this process is that the new behaviour pattern should become automatic. It is not sufficient that people should consciously assess the probabilities of reward and punishment and act out of judicious self-interest; this is a recipe for inner conflict and instability. The object of behavior modification is that people should eventually want to adopt the approved mode of action, and feel gratified in so doing. As a form of *conditioning*, therefore, it has to go deeper than the conscious mind. The notion that people's behavior is influenced by parts of the mind of which they are not directly aware, loosely speaking the *subconscious* or *unconscious*, is of crucial importance in psychology and will be encountered frequently in the following pages.

Skinner argues that every human action is determined by an inescapable network of aversion and reinforcement, so we might as well put this network on a rational basis and use it efficiently in order to achieve desired social goals.[1] The function of government should be to decree a pervasive and comprehensive system of rewards which will gently divert people's activities away from antisocial habits such as crime, aggression or drug abuse, in the direction of virtues such as hard work and consideration for others. The problem, of course, is that not all social goals are generally agreed upon, and the incoherence of existing reward systems reflects the glittering diversity of human ambitions. In an open, pluralistic society it is difficult to imagine mass behavior modification making significant changes in people's lives, although, as we shall see, that does not rule out the use of the technique in specific circumstances.

The vestiges of free will

There are many sensibilities which the clinical calculations of behaviorism offend, and heated argument surrounds the work of Skinner and his followers both in theory and in practical application. The idea of human beings as virtual robots helpless before the insidious goading of behavior modification is not an inspiring one. We have inherited a tradition which regards each individual as a free agent capable of making decisions by an effort of will, a concept deeply rooted in the idea of an independent 'personality'. What becomes of such hallowed notions in the light of a science which can not only 'explain' effects which had been previously attributed to 'personality' and 'free will', but can also short-circuit such old-fashioned ideas by causing people to behave in a predetermined way?

Such objections are similar to the ones which greeted Pavlov's discovery of the conditioned reflex over sixty years ago. Pavlov showed that it was possible to make a dog salivate by ringing a bell which had previously been rung at meal times so that the dog had come to associate the ringing with food. Again, when possible human applications were mooted similar protests filled the air. Free will and the inviolability of the personality were invoked in faithful opposition to a science of behavior.

The answer to all such objections is the same. We have come to regard the human mind as a deterministic system, which is to say, a system whose nature is caused by discoverable external stimuli, rather than by any innate properties of its own.[2] This is in principle a giant leap from the traditional concept of the mind as an independent moral being capable of free will, but in practice it does not make very much difference.

We still think in terms of free will, but now think of the mind as the activity of brain and the human nervous system. The nervous system being composed of an astronomical number of combinations of the 10^{11} (100,000,000,000) neurones that compose it.

But although psychology and neurology have not abolished free will, they have affected the balance of forces within which it is exercised. Where people have no strong predilection for any particular course of action the science of persuasion can make a lot of difference. Although you cannot modify the behavior of all of the people all of the time, you can exercise a significant influence over what most of them do at least some of the time. A more penetrating criticism of behavior modification than the philosophical problem of free will is the political problem of who uses it and when.

Mind control versus crime

A form of behavioral control common to most societies is criminal law; for whatever reasons certain actions are proscribed and steps are taken to dissuade people from committing them. Those who do not respond to the anticipation of punishment, stigma or ridicule are made to experience it in actuality in the hope that they will refrain from crime in the future. The law is a purely aversive system of behavior modification: penalties are ordained tor transgressors, but there is no explicit system of reinforcement for those who obey. Psychologists consider that an aversive system is in principle the least effective method of influencing a person's behavior, for not only does it cause resentment and trauma, but it lacks a continuing supportive structure to reward correct actions spontaneously.

There have been several recent experiments in the United States to investigate the potential for behavior modification in the case of prisoners. The philosophy underlying these experiments was expressed by Dr Edward Schein,[3] an associate professor at the Massachusetts Institute of Technology, who argues '... in order to produce marked change of behavior and/or attitude, it is necessary to weaken, undermine or remove the supports of the old patterns of behavior and the old attitudes'. This may be done '... either by removing the individual physically and preventing any communication with those whom he cares about, or by proving to him that those whom he respects are not worthy of it'. In other words, the structure which reinforces undesirable behavior must be replaced by one which reinforces desirable behavior.

One such experiment, known as *Special Training and Rehabilitative Therapy*, was carried out at Springfield, Missouri by the US Bureau of Prisons. A number of prisoners were assigned to the START program and, in the first instance, stripped of such amenities of prison life as showers, exercise, visitors, reading matter and personal possessions. As they began to conform to appropriate behavior patterns they were rewarded by the restoration of these comforts. Such appropriate behavior included the avoidance of abusive language, avoidance of anger and irritability, and willingness to perform chores.

In other experiments efforts were made to dissuade prisoners from committing crimes after release. One such method was to administer an *emetic* (a drug which makes the recipient violently sick) while showing the subject a film of a bank robbery, the object being to condition the prisoner into feeling nauseous whenever he contemplated violent crime. This technique was foreshadowed by Anthony Burgess as long ago as 1962 in his novel *A Clockwork Orange*, more recently the subject of a well-known film. Professor James McConnell,[4] of the Department of Mental Health Research at the University of Michigan, has stated : 'The day has come when we can combine sensory deprivation with the use of drugs, hypnosis and the astute manipulation of reward and punishment to gain almost complete control over an individual's behavior.' We will encounter these other techniques below.

The behavior modification experiments in American jails have provoked furious controversy among prisoners, psychologists and civil rights campaigners. In the first place the technique has been criticized on the grounds of ineffectiveness : 'It demands a conformity to prison without any consideration of developing behavior which would be adaptive outside prison, thus crippling the prisoner even more,'[5] in the words of one of the critics. It has also been condemned as an infringement of the human rights of the subjects, and unfavorably compared with other methods used to break down the willpower of prisoners, such as beatings and starvation. The use of chemicals has been termed 'drug assault'. But such methods are still in their infancy.

It must be stressed that the kind of prison regime described here is by no means an *inevitable* consequence of the assumptions and techniques of behaviorist psychology. Rewards and reinforcement of a more benevolent sort could equally be used in order to give criminals a sense of security, a feeling of belonging, and an appreciation of the advantages of living within the law, the main obstacle being our social and political traditions which strongly resist the idea of 'curing crime with kindness'. The Biblical injunction of 'an eye for an eye' still casts a shadow over efforts to deal with crime rationally and scientifically, but there are certain exceptions to this pessimistic tendency. Grendon psychiatric prison, in England, is an institution specializing in the psychological problems of persistent offenders, with a better than average record of keeping recidivists out of further trouble.[6] The aim of the prison is to help inmates regain their self-esteem through group discussion and a sympathetic attention to the individual's personal experiences. This is a rather more complex process than straightforward reinforcement conditioning, but it serves as an example of how behavioral problems can be approached in a non-punitive way.

It is an open question which of these vastly different regimes will characterize the prison system of the future. Both schools of thought reflect the turmoil of changing social values and will persist for many years. In fact, a final decision between the two might never be made, for recent neurological research has opened up a third possibility – the abolition of prisons altogether. . . .

The electronic conscience

When, in *Nineteen Eighty-Four*, George Orwell wanted to portray a plausible society whose inhabitants were the objects of constant scrutiny, two-way television and hidden microphones were just about the limit of his imagination. Subsequent discoveries in electronics and neurosurgery have combined to make his nightmare obsolete, however. As a result of ten years' experimentation with laboratory animals scientists believe that it will soon be technically possible to implant tiny radio transmitter-receivers inside a person's skull in order to measure the electrical activity of the brain and convey the signal to a central control many miles away. This monitoring station would be aware of unusually agitated brain activity which could indicate the contemplation of a crime, and would be able to respond with further radio signals which could alter the subject's brain patterns. The ultimate development of this technique would be an implanted microcomputer capable of assessing brain patterns and reacting automatically to quell antisocial behavior at the ideas stage. Installed in the heads of potential criminals (however defined), or even of the population at large, such devices would mark the final redundancy of prisons because people would be unable to sustain the willpower to commit crime.[7]

How much of this is fantasy, and how much is fact? Wilder Penfield, a surgeon at the Montreal Neurological Institute, was able to perform a remarkable series of investigations over twenty years ago in the course of operations for the removal of brain tumors. Because this only required a local anaesthetic the patients remained conscious throughout these operations, so he took the opportunity to stimulate the surface of the brain with minute electrical currents and asked the patients to describe their responses. The result was that the stimulation of specific areas of the cerebral cortex was associated with the tingling sensations in specific parts of the body, and Penfield drew the tentative conclusion that these areas of the brain were uniquely associated with, maybe, in a certain sense, 'responsible for', corresponding areas of the body. Subsequently there has been a vast range of experiments carried out all over the world with the object of further defining the roles of specialist areas of the brain in the processing of thought and behavior, and there has emerged a pattern of *centers* concerned with speech, memory, vision and so on.[8]

Since our awareness of the world consists of the millions of tiny electrical currents and chemical reactions which are taking place all the time inside our brains it should, in principle, be possible to short-circuit reality (that is to say, the world of which we are aware through the normal exercise of our five senses) by introducing electrical stimulation directly into the brain. Experiments have shown, for example, that when an electrode touches a speech center the patient will emit a quite involuntary, inarticulate grunt. And when the auditory or visual centers are activated the patient will report a sensation of sound or color. Experiments with laboratory animals, carried out by Dr Olds, have located 'pleasure centers' in the brain which, when stimulated, can throw an animal into uncontrolled ecstasy. Rats were presented with a simple lever which enabled them to trigger their own pleasure centers, and it was found that they would cling grimly to the lever for hours, or even days, until they eventually died. The possible human applications are awesome.

According to the same principle it should also be possible to observe and analyze the electrical activity of the brain and thus draw some broad conclusions about what the subject is experiencing. General states of mind such as rest, sleep, meditation, hypnosis, excitement, agitation and aggression are detectable in this way, In recent years systems such as toposcopes, have been developed in order to translate 'states of mind' into brain waves, so that by looking at the brain waves on the oscilloscope we can tell much of what people are thinking.

But there is a world of difference between detecting such general states of mind on the one hand, and actually being able to monitor it completely, although it is likely that many of telemetric monitoring as a means of behavior control now confined to 'broad spectrum' interventions such as de of telemetric monitoring as a means of behaviour control now confined to 'broad spectrum' interventions such as de tecting agitated brain patterns and responding with a diver sionary stimulus such as the activation of a pain center, or perhaps, with dubious benevolence, a pleasure center, will soon be replaced by more powerful methods. This will, of course, represent a significant change in the relationship between the powers of authority and the population at large.

Another prospect seemingly out of the realm of fantasy is the manipulation of brain activity using microwaves, which is currently being investigated in the USSR.[9] Microwaves are radio signals of very short wavelengths which at the moment have an innocuous domestic application in high-speed ovens. As a means of thought control they are potentially more insidious than the system just described because they do not require any sort of receiving equipment installed in the subject's head.

The existence of this research was made public in a US Defense Intelligence Agency document which was declassified in November 1976. Apparently the Russian experiments show that 'sounds and possibly even words which appear to originate within the head can be induced by signals at very low power densities. Combinations of frequencies and other signal characteristics to produce other neurological effects may be feasible in several years.' Physiological effects have also been demonstrated : heart seizure has been induced in frogs using such methods, for example. The report holds out the possibility that nerve disorders, heart complaints and blood circulation problems could be induced using microwaves. It might be noted in this connection that there was recently a full-scale diplomatic row between the USA and the USSR over the alleged beaming of microwaves at the US embassy in Moscow by the Russian authorities. A State Department administrative order of November 1976 declared the Moscow embassy to be an 'unhealthy place'.

Is direct communication with the brain using microwaves feasible? Theoretically, very much so. One possibility is that assymetric neurone contacts in the cortex could act like a radio crystal and demodulate the microwaves, setting up a small electric current which would function in much the same way as a current introduced with an electrode. Furthermore it is known that the rate of certain chemical reactions is affected by microwaves of certain frequencies, and determined research would probably elicit ways in which reactions in the brain could be stimulated or retarded in this way. But again, the same limitations apply as with electronic 'bugs'. the brain is immensely complicated and such methods would almost certainly be restricted to quite non-specific stimuli which could perhaps seriously affect a person's mood, but not actually 'program' their thoughts.

It is a matter of speculation how these techniques might actually be applied. It seems unlikely, given the present state of knowledge, that microwaves could be beamed over long distances such as from orbital satellites in order to broadcast propaganda or interfere with the health of military personnel. But over short distances microwaves might even now be part of the panoply of interrogation techniques, used to weaken the resistance of subjects by affecting their moods, or undermining their sanity by introducing disembodied voices into their minds.

Interrogation

The analysis of brain activity has brought psychologists and neurologists into the realm of the most profound, yet the most commonplace, secret in the world – the contents of other people's minds. The essence of our individuality is that we can never directly share our experience with others : we have to communicate it by way of language, which is often a confusing, tenuous link. Most of what we perceive, our memories, our daydreams, the ceaseless tumult of our private thoughts, are forever concealed from the world.

When people wish to keep secrets they can do so with extraordinary tenacity, and few human enterprises could have inspired as much hard-hearted ingenuity as the development of even more efficient interrogation techniques. The ability to decode brainwaves will perhaps put an end to the interrogator's business for good and all, but in the meantime the insights of psychology are increasingly used to pry vital information from the minds of suspects and informants. Whereas in the past physical torture was all too often used for this purpose there has recently been a movement in favor of direct mental pressure, partly arising out of a moral objection to bodily harm, partly because psychological tools often prove more efficient. *Sensory deprivation*, which involves keeping the subject in a dark, noiseless room with a minimum of physical sensation, exploits the fact that when denied external stimulus the brain becomes extremely agitated and eventually deranged. It has often proved very effective in breaking down prisoners' resistance to interrogation; it is used by police and military authorities in many parts of the world, and was at one time used by British security forces in Northern Ireland.

The quest for the mythical 'truth drug' continues, and current research into where and how the brain 'blocks' the verbalization of memories might any day yield results. Such a drug would probably be in the nature of a *euphorient*, and would function by relaxing the subject and generally reducing inhibitions. A euphorient could be something as mundane as alcohol (policemen traditionally ply informants with drink – not necessarily out of gratitude!) or sodium amytal, which was used during the Second World War, or possibly an hallucinogen such as LSD, but no really effective substance has been developed to overcome determined resistance. Where drugs are used in interrogation they more usually have the function of inducing stress and anxiety, example being prolixin (which is a powerful depressant related to largactyl, a drug used in the treatment of schizophrenia), and anectine, a derivative of curare, the South American arrow tip poison. Both these drugs have been used in the United States on convicted criminals, but the practice is extremely controversial.[3]

In Britain public opinion would almost certainly be repelled by the use of sensory deprivation and drugs in the course of normal police work, so rather more subtle methods of persuasion have been developed.[10] One of the most common is a process known as *cognitive shock*, which involved gradually building up the subject's hopes of, say, release, or a meal, and then abruptly disappointing them. Also, there is the renowned 'nice guy/nasty guy' technique, in which the subject is interviewed alternately by sympathetic and unsympathetic people. A third method is the use of what virtually amounts to psychoanalysis in order to discover the subject's deep fears and anxieties, and then to exploit this knowledge in order to secure co-operation. All these procedures rely on producing stress and confusion and have been developed into fine arts through psychological study. They are not particularly pleasant, but in pursuance of a socially necessary function they are, perhaps, preferable to most of the alternatives.

A technique used with spectacular success in certain cases is *hypnosis*, although at the moment this is only effective with willing witnesses who have difficulty in recalling crucial information. Hypnosis makes use of the fact that the brain retains far more details of experience and observation than the conscious mind can gain access to – a process known as *crypto-amnesia*. When a witness consents to being placed in a hypnotic trance events can sometimes be recalled in minute detail. In one murder case a witness was unable to recall anything about the crime, except for having noticed a 'vague figure' at the scene, but under hypnosis was able to provide a 1000-word description of the assailant.

Since it is well known that lights can induce a trance state in some people when they flash in phase with brain activity (a phenomenon discovered in the case of radar screen monitors), it is conceivable that even an unwilling witness could be made to yield information after being placed involuntarily in a trance. This would be a considerable innovation, and contrary to a widely held assumption of hypnosis, that it is not possible to force a person to behave in a way opposed to the wishes of their conscious mind, but it may yet prove feasible. There is a problem about whether evidence given under hypnosis should be admissible in court, however. Is the otherwise mute unconscious mind subject to the moral and legal disciplines of the oath and perjury?

Psychology in court

The involvement of psychologists in the work of the police and the courts is constantly increasing, because many of their traditional practices fall into areas in which the use of scientific observation and experiment proves most fruitful. A recent controversial example has been identification evidence : since perception and memory are two subjects of major interest to psychologists it is now possible to analyze such testimony objectively. In an experiment carried out in 1974 by Professor Laurie Taylor, of York University, the ability of an audience to identify the 'criminal' in a staged incident proved remarkably low – only 50% under the sort of ideal conditions which would rarely obtain in reality. In the light of such investigations, and inspired by a number of sensational court cases in Britain, judges are now reluctant to admit uncorroborated ID evidence.

Meanwhile, other psychologists have been working to improve the reliability of identification procedures. Testing the assumption that the judgement of witnesses at an identification parade is impaired by nervousness, shock, embarrassment, or socially conditioned reluctance to stare at people's faces, Helen Dent and Fiona Gray,[11] at Nottingham University, conducted an experiment using color photographs of 'suspects' instead. The accuracy of identification was considerably improved using this method. There is here a conflict between historically established legal procedure and the conclusions of scientific investigation, but it is by no means clear how strong such conclusions have to be before the latter can prevail over the former. The traditional insistence that an accuser should physically accost the accused is slightly reminiscent of pre-scientific magic, as if physical proximity would somehow cause the truth to manifest itself.

We have already encountered the concept of cryptoamnesia, which is the inability of the conscious mind to retrieve as much information as the unconscious mind can store. It is now recognized that the act of remembering details of an event is not a 'passive' act comparable with rerunning a newsreel, but is rather more a 'creative' process in which people tend, quite unwittingly, to make guesswork assumptions in order to cover gaps in actual recollection. There have been numerous experiments to determine the accuracy of eye-witness evidence, most of which have been discouraging. In one study witnesses were shown a film of cars crashing at various speeds; there was found to be no significant correlation at all between the actual speeds and the speeds estimated by the witnesses.

The 'leading' of witnesses, that is to say, phrasing questions to imply a particular answer, is a device much frowned upon in the courts, but at the same time the subtle practice of this is part of the weaponry of any self-respecting barrister. A recent experiment[12] has shown that it is possible to bring about fundamental changes in the way a witness remembers an event simply by astute phraseology. In one instance subjects saw a film of a car crash, and then a number of them were given a questionnaire which asked, 'How fast were the cars going when they contacted each other?' Other subjects were given different questionnaires in which the word 'contacted' was replaced by 'hit', 'bumped into', 'collided with', or 'smashed into'. Their estimates of the speeds were significantly determined by the choice of question: the more dramatic the verb, the higher the estimated speed. Subsequently witnesses were asked about details such as broken glass at the scene of the crash; again there was a strong correlation between the choice of verb in the question and the witnesses' tendency to recall broken glass. (Actually there hadn't been any.) This strongly indicates that the careful choice of words in phrasing a question can affect the way in which a witness unconsciously fabricates information in order to supplement the memory.

Mass manipulation

Of course, there is nothing new about the careful choice of words: this is not an invention of psychologists, or even of barristers. Ever since language began it has been used to inform and persuade; that is its function. The ancient Greeks had a science of *rhetoric*, which was an attempt to codify the way in which language is used effectively. The advice of the old rhetoricians still has its uses, but psychology promises to supersede its insights in the quest for ever increasing eloquence.

The advantage of psychology over rhetoric is that whereas the latter proceeds from a number of assumptions about the innate quality of language, the former takes a closer look at the way in which the audience responds, and assesses the response quantitatively. For example, the use of fear as a means of persuasion is a well-worn technique and psychologists have, in recent years, discovered a lot about the sort of fears which people are vulnerable to. This applies particularly to unconscious fears which people are not consciously aware of, but which affect their responses nevertheless. But is the exploitation of fear necessarily a powerful means of manipulation? Traditionally the answer is 'Yes', as the determined efforts of military propagandists and hell-fire preachers testify. But investigation shows that reality is not quite so simple.

Persuasion based on fear arouses stress and anxiety, but this does not always cause the recipient to submit to persuasion in order to 'cancel' the anxiety. Other possible responses range from inattentiveness or mishearing, aggression towards the persuader, or subconscious *repression* of the fear-arousing information. In an experiment[13] three groups of American schoolchildren were shown different films about the care and maintenance of teeth, the content ranging from the mildly informational to excruciatingly graphic descriptions of tooth decay. Although the latter evoked the strongest feelings of concern or anxiety when the subjects were questioned immediately afterwards, subsequent investigation showed that when it came to taking regular action as advised in the films the 'strong' appeal had the least effect. It seems that fear is not a good basis for long-term behavior modification because people tend to set up 'blocks' to shield themselves from anxiety.

We have reached an era in which the last major obstacles to general good health are habits over which people have direct control, such as excessive eating, drinking, smoking or lack of exercise, and it seems quite likely that pressure will grow for the public authorities to take action to persuade the populace into better lifestyles. Indeed, this has already happened in the United States in the case of anti-smoking 'commercials' on television. Rather than just trusting to intuition, or to the advice of ancient rhetoricians, public health campaigners will be able to approach their task in an enlightened way and concentrate on the positive rewards of clean living, rather than the dire effects of its absence.

The search for ever more effective means of persuasion is a search for ever-increasing subtlety. Most people's conscious opinions and prejudices are well-rooted in experience and are difficult to change, whereas the subconscious, over which they have no control but which exercises a strong influence on their behavior, is every persuader's prize. An advertiser might have difficulty persuading a housewife at a rational conscious level that one brand of soap powder is any better than another, but if she can be conditioned into unconsciously associating that brand with some image which appeals to her, be it soft towels, towsle-haired children, leisure or elegance, then she is more likely to pick that brand up off the supermarket shelf. There is an immense amount of psychology applied to advertising techniques every year, the problem being that people develop resistance which demands a spiral of increasing sophistication in order to keep them wanting products which they had never wanted before. Advertising experts do not seem to be running out of ideas yet, but if ever they do the effect on the economy will be startling.

If anything the range of weapons available to advertisers is increasing rather than decreasing. There has been a lot of research aimed at discovering how people respond to different colors in order to assist the design of display material and packaging.[14] There are 'warm' colors, such as red, orange and yellow, and 'cool' colors, such as green, blue and violet, and they affect people's reactions independently of the product concerned. Warm colors are associated with movement and tend to make packages seem larger, while cool colors suggest tranquility and smallness. Bright, intense colors are redolent of masculinity and hardness, while soft, muted pastel shades imply softness and feminity. In a study concerned with the marketing of coffee it was found that coffee out of a brown can seemed much stronger to people than coffee out of a yellow one, even though there was no actual difference. Manufacturers argue that products are bought in order to indulge a wide range of psychological satisfactions and that if the seemingly irrelevant color of an item can be chosen in order to increase this satisfaction, then this is a quite justifiable addition to the quality of the product.

Another process which has received a great deal of attention, particularly because of its relevance to propaganda and advertising, is *subliminal perception.* The theory is that an image or a message which is seen but not consciously noticed, either because it is concealed or because it is only instantaneously visible, in some way goes direct to the behavior-initiating part of the mind instead of having to run the gauntlet of conscious censors and blocks.[15] This could be used when designing advertising graphics so that the 'brand image' could be concealed in order to overcome conscious rejection or ridicule. Instead of portraying beautiful women prostrate at the feet of the strong, silent cigar smoker, for example, erotic images could be incorporated into the smoke swirling up above him. The unsuspecting potential customer would not be aware of the symbolic pattern eating its way into the subconscious, but vague intimations of sexual gratification might just determine an otherwise arbitrary choice of brands when next at the tobacconist. Ambiguous images have been a regular preoccupation of psychologists since the rise of the *Gestalt* school in the 1920s. A diagram which can mean either of two quite different things has a disturbing, irresistible fascination about it.

The use of subliminal persuasion in political propaganda has frequently been cited as a potential nightmare of modern life. The idea of an entrenched authority using quite indetectable techniques in order to manipulate the population is not an edifying one, whether used for good or evil. The development of mass media such as television has greatly increased the feasibility of such a practice. Slogans flashed onto the screen for a fraction of a second would evade conscious attention and eventually come to dominate the minds of the population. But a lot of research would be required before this technique could be used effectively – the method is clearly not as straightforward as it seems. During the British general election campaign of 1970 a Labour Party television broadcast had the injunction 'Vote Labour' incorporated into its closing titles in one fiftieth of a second flashes. The Labour Party lost the election!

Mind and media

As a means of assessing 20th-century mass media psychology is a better discipline than aesthetics. The defining property of a mass medium is its capacity to feed the same stimulus into thousands, or even millions, of human brains simultaneously. The niceties of form and content are irrelevant: they are judged by what they achieve. Perhaps it is important to remember that broadcasting, cinema and recorded music have been with us for only half a century or less. They are still very young. They live within the shadow of books, orchestras and canvas, and have not yet even begun to explore all their possibilities.

Apart from dominating the industrial world's supply of information, television also functions as a surrogate for 'real' experience in a restricted and overcrowded urban society. Since the passing of the era of exploration, colonialism and demographic expansion, people's opportunity to indulge adventure and excitement has been curtailed, and the tendency to conquer and to dominate is turned increasingly inward. Although a filling TV diet of crime and war is often cited as a cause of alleged proliferation of aggressive behavior, the connection is far from proven. One alternative hypothesis for which there is a certain amount of support is the notion that TV violence might actually be *cathartic* in effect.[16] In other words, it provides the viewer with the opportunity to entertain violent fantasies in a harmless way, thereby reducing the incidence of actual antisocial aggression.

If this theory is ever justified to general satisfaction then, in the cause of peace and harmony, we might expect TV and other media to become increasingly violent, increasingly repulsive, fully utilizing psychological discoveries about fears and emotional responses to spin a web of vicarious bloodlust. This idea was alluded to in *Nineteen Eighty-Four*, where Orwell describes the use of a daily TV program called *Ten Minutes' Hate*, which functioned as a politically stabilizing release for the population's pent-up aggression. Should all this sound far-fetched, stop and consider the sort of changes which have taken place in the cinema in the last decade or so, the single-minded pursuit of ever more explicit, ever more disturbing horror. William Peter Blatty, who wrote and directed *The Exorcist*, learned his excruciatingly effective skills while working for the Psychological Warfare Division of the US Defense Department at the Pentagon during the Second World War. The connection is, of course, circumstantial; what is being described is a tendency, not necessarily a determined policy.

Madness, sanity and normality

Having read this far you could be forgiven for assuming that modern psychology is a Pandora's Box of sinister techniques designed to enable power-seeking, authoritarian figures to exercise control over the rest of the human race. In fairness to the various sciences of mind some of their not-so-mixed blessings should be mentioned.

To most people psychology is associated first and foremost with the treatment of mental disorder. It is a matter of speculation whether mental illness is any more prevalent in our era than it has been in the past – there are few reliable statistics and, in any case, the definition of what constitutes a mental disorder is constantly changing. One thing which is certain, however, is that every year in Britain one man in fourteen and one woman in seven will consult a doctor about some form of mental disease. The stress and the tensions which have become by-words for modern life claim thousands of victims every year.

The alleviation of mental disorder has been surrounded by furious controversy ever since psychiatry first began to break free of medieval quackery towards the end of the 19th century. The dividing line between madness and sanity is invariably problematic, for to define insanity is to define normality by antithesis, and the definition of normality is a matter of profound political consequence. What are the limits of acceptable behavior, of acceptable modes of perceiving things? In the USSR the practical definition of insanity embraces political dissidence, presumably on the assumption that anyone who questions the wisdom of the Politburo could hardly be *compos mentis*. There are also reports of religious people committed to asylums for manifesting symptoms of 'delusion' and 'hallucination' – practices sanctified elsewhere as prayer. Even so, there are few grounds for complacence about western definitions of madness : to what extent does the label 'insane' simply reflect intolerance of people's eccentricities, and indifference to their needs?

There are, very roughly speaking, two separate types of approach to the problem of mental disorder. One regards human beings as rather like pieces of machinery whose imperfections are 'organic', i.e. caused internally, and treatable in isolation from other people. The other approach emphasizes mental illness as 'social', i.e. proceeding from faults in the intricate pattern of societal relationships with which individuals are surrounded.

The mechanistic school of thought is the one which corresponds most closely to the western medical tradition. It centers on the one-way relationship between a detached, 'objective' therapist, and the subject, or patient. In the course of treatment the therapist assesses the symptoms of the patient and chooses from various techniques to intervene to quell the symptoms, and possibly to eradicate the causes. These techniques include *psychotropic,* or consciousness-altering drugs, *electroconvulsive therapy* (ECT), otherwise known as 'electric shock treatment', and *psychosurgery,* such as lobotomy, which consists of making incisions in the brain, and sometimes the removal of brain tissue.

Of these three categories it is the development of increasingly sophisticated drugs which has marked the greatest progress recently. With a growing knowledge of brain chemistry and the substances associated with different states of mind, it is now possible to make comparatively accurate interventions into the consciousness of a patient. For example, it has been possible to identify the operation of a chemical 'reward system' within the brain, and the stimulation of this system is the basis of most modern antidepressants. The 'subjective' effect of this on the patient is an all-round elevation of mood and increased feelings of motivation. Similarly, *benzodiazepines*, which are the most widely used class of tranquillizers, including librium, valium and nobrium, go to work specifically on the areas of the brain which specialize in punishment, uncertainty and stress, and again, the effect perceived by the patient is that of enhanced relaxation, reduced agitation, diminished neurotic behavior.

The problem with artificially synthesized drugs is that they do not naturally belong in the body, and sometimes give rise to side-effects such as nausea, trembling and confusion whose disadvantages have to be weighed against any benefits. The most promising area of research is the detection and analysis of such substances as the brain manufactures for its own purposes, since these might be expected to have less deleterious side-effects. One recent success has been the discovery of *enkephalin*, the 'natural opiate', an analgesic peptide molecule which is produced within the brain in order to mitigate the sensation of pain, acting like morphine. One possibility is that this substance might be produced in response to acupuncture, therefore accounting for the anaesthetic effects of this traditional Chinese medication.[17] A similar discovery, made by a research team at Harvard Medical School, has been dubbed 'factor S'; it can be separated from the cerebral fluid of sleepy goats and induces drowsiness in rats and rabbits when transfused into their brains. It is likely that enkephalin and 'factor S' will pave the way for a new generation of naturally produced psychotropic drugs happily devoid of side-effects, eventually including sub-

stances capable of inducing relaxation, stamina, fast reactions, perhaps even enhanced perception, sensual pleasure
and so on. Even today there are several candidates for the
role of *soma*, the pleasure drug of Huxley's *Brave New World*,
of which marijuana is probably the most widely used. If brain
substances responsible for sensations of pleasure can be
isolated they would obviously have far-reaching social effects.

The use of ECT in the treatment of depression and schizophrenia is entirely pragmatic : no one is at all sure how it
works, its effects are variable and unpredictable, and it has
been likened to tapping a television set which is 'on the
blink'. The practice was first introduced into psychiatry by
an Italian, Ugo Cerletti, in the late 1930s; the idea occurred
to him after visiting a slaughterhouse in Rome where pigs
were stunned with a 125-volt shock across the skull prior to
being killed. Patients treated with ECT frequently show a
remission of disturbing symptoms, usually accompanied by
partial amnesia, but due to the lack of a theoretical justification the procedure is often branded as a modern form of
witchcraft.

Psychosurgery has been most often used in an effort to
control the symptoms of aggression. It is acknowledged to be
a risky business because it involves the removal of brain tissue, and can therefore lead to permanent personality changes.
As a result of experiments with chimpanzees performed during the 1930s it was discovered that tantrums and violent
behavior could be significantly reduced by the surgical
removal of the frontal lobes of the cerebral hemispheres of
the brain. *Frontal lobotomy* became a widespread method of
dealing with aggressively disturbed human beings, including
children and convicts. In recent years other parts of the brain
have been treated in a similar way in order to subdue violent
symptoms, the removal of the *amygdala* and lesions to the
hypothalamus being two such methods.[18]

Critics of psychosurgery point out that although some
patients lose their aggressive tendencies while remaining
psychologically intact, others experience a general suppres-
sion of activity and a deadening of emotion – classic symp-

toms of brain damage. It has been described as a 'therapeutic weapon', used to contain otherwise intolerable patients, and sometimes misused as a 'cure' for arbitrary 'symptoms' such as 'latent homosexuality'. Nevertheless the search for more accurate psychosurgical methods continues. A technique currently being developed in England is known as *stereotactic brain surgery*,[19] which aims at a more precise removal of tissue using heat rather than the surgeon's knife. The same research team will investigate the use of radioactive implants to destroy selected areas of brain tissue.

Many fascinating discoveries about the functioning of the brain have been made by way of 'spin offs' from brain surgery, such as the pioneering work of Wilder Penfield, which we have already encountered. Another bizarre example has been so-called 'split brain' research, which has been made possible because of an experimental treatment of epilepsy during the 1950s which required the physical separation of the two apparently symmetrical halves of the brain.[2, 18] It was found that patients who had undergone this operation showed no outward signs of abnormality, but on closer inspection it was discovered that some of their mental functions had become 'separated'. For instance, a blindfold patient holding a familiar object in the *left* hand would be unable to utter its name, because speech is activated by the *left hemisphere*, which is connected to the *right* side of the body. Exhaustive studies of the aptitudes of these split brain patients concluded that many higher brain functions are located in either one hemisphere or the other, contradicting the brain's apparent physical symmetry. The left hemisphere is responsible for rational, abstract skills such as speech, reading, writing, logic and so on, while the right hemisphere processes the awareness of space, shape, environment and intuition. In response to this discovery it has been argued that western civilization places insufficient emphasis on the functions associated with the right hemisphere, and that 'non-verbal' skills should be given greater encouragement in education.

There are many former mental patients who are grateful

for the use of drugs, ECT and psychosurgery in relieving them of their misery and enabling them to resume normal, healthy lives. But at the same time there are many others whose symptoms were not alleviated by these methods, who may, indeed, have been adversely affected by them. It has been argued that the problem arises out of a total misconception of the root causes of mental illness, and that clinical techniques simply tamper with the outward manifestations of disturbance without in any way changing the social circumstances which cause it.

This contradiction between 'organic' and 'social' theories of mental illness show no signs of being reconciled, unless, as some would argue, it is an unreal contradiction in the first place. There is a bewildering range of non-clinical psychiatries which emphasize the origin of disturbance in 'human' factors such as economic stress, family tension, sexual frustration, childhood trauma, or the superficial, incomplete relationships with others which arise out of the giddy round of modern urban life. Most of them recommend therapies which encourage the patient to express his or her problems outwardly, either bilaterally, as with psychoanalysis, or among a larger number of people, as with group therapy, encounter, psychodrama and many others. This proceeds from the assumption that mental unease accumulates over time as a result of people being unable to express themselves, their experience or their problems due to the absence of the kind of sympathetic relationship which validates such expression. These social therapies enable the patient to learn the necessary techniques both of forming relationships and of externalizing their inner experience to develop self-understanding.

In the late 1960s there was a widespread, chaotic but none the less influential movement called *antipsychiatry*,[20, 21] which was opposed to the use of the concept of 'insanity' as a 'condition' from which people are alleged to 'suffer'. By taking a societal perspective, antipsychiatrists (the movement was associated with such people as R. D. Laing, David Cooper and Thomas Szasz) came to regard 'lunatic' as a *social role* not in principle different from, say 'mother', 'politician', 'artist' or 'psychiatrist'. A sick society needs the insane in order to shore up its repulsive ideas of normality, and mental patients are just those people unfortunate enough to fall into the category. Their strongest criticism was reserved for the mental health professions themselves; practitioners were seen as imposing quite arbitrary ethical and behavioral standards on to helpless people in an almost religious assertion of the norm. In the aftermath of this movement there have arisen a number of 'deprofessionalized therapeutic communities' which aim to help disturbed people not by focusing on, and trying to eradicate, their 'symptoms', but by establishing tolerant social relationships in which the pressure to conform is minimized.

Clinical behavioral therapy

We have already encountered the use of behavior modification techniques in prison experiments. The manipulation of aversion and reinforcement works best in 'closed' environments such as prisons, boarding schools and hospitals, because authorities in such institutions are able to control conditions to a far greater extent than authorities in the world at large. For example, a residential school for maladjusted children in Devon, England, operates a *token economy*, in which pupils are rewarded tangibly for clever, sociable or considerate behavior. In the course of time this 'prop' is gradually phased out so that pupils eventually come to use desirable behavior automatically, as they will have to in the 'outside world'.

This variant of behavior therapy is known as *operant conditioning*. It rests on the assumption that 'the consequences which follow a particular piece of behavior influence the future occurrence of that behavior'.[22] If a child in a class cries excessively and is 'rewarded' after each crying bout with kindly attention from the teacher then the crying is likely to persist. Operant conditioning requires that the teacher ignores attention-seeking tantrums but responds favorably to acceptable behavior. The technique has also been used to deal with various sorts of behavior abnormalities such as stuttering.

Another form of behavior modification used medically is *desensitization*, which is designed to help neurotic patients overcome anxieties and phobias. The principle is quite simple : through discussion with the patient the therapist tries to construct an *anxiety hierarchy*, listing undesired phenomena and circumstances in order of increasing anxiety. Over a period of time the patient is slowly led through this hierarchy and rewarded with encouragement every time a potentially anxious response is avoided or overcome. A patient with an abnormal fear of crowds, for example, will be introduced gradually into increasingly crowded situations until the old behavior pattern of phobic anxiety is replaced with a new behavior pattern of calm acceptance. There are obviously many more ramifications and subtleties to this process; it is the most widely used and the most successful clinical application of behaviorism.

The use of *aversion therapy* as a form of treatment has always been controversial, particularly because it is often regarded as reinforcing a social ethos which is intolerant of harmless behavioral variations. The usual method of applying this technique is to subject the patient to pain or discomfort while he or she is indulging in the trait which is to be eradicated. The literature of aversion therapy abounds with descriptions of, for example, transvestites 'treated' with electric shocks until they lose the desire to cross-dress, or homosexuals treated similarly until they lose their desire for members of the same sex. Such methods have also been used in other areas, such as the treatment of alcoholics and overeaters. Proponents of this technique argue that people's lives are often ruined by such abnormal inclinations, and that they have a right to be cured of them if that is what they desire. Opponents argue that people are oppressed into 'killing' a valid part of their selves through social intolerance, of which aversion therapy is only a refined example, and that a better form of therapy is to help people accept their own personalities fully without feeling ashamed or persecuted.

The crux of this issue is whether people are 'pressured' into undergoing aversive treatment, or whether they choose to do so having considered all the alternatives and ramifications. In an imperfect world, of course, not all the alternatives are available.

Controlling one's own mind

All the methods of mind control discussed so far have involved the intervention of a 'controller' into the domain of a 'controllee', with all the attendant ethical problems about interference with the freedom of others. But there is a developing field of mind-control which avoids these objections and which, in the long run, promises many more potential benefits to the human race than the techniques of manipulation. This is the field of mental self-control.

One of the traditional mainstays of western physiology is the division of bodily functions into those which are voluntary and those which are automatic. Voluntary functions include all the normal muscular movements which people can perform at will; automatic functions are those controlled by the *autonomic nervous system*, and which take place without any need of premeditation. In fact, it has been generally assumed that this latter category are actually immune to mental intentions, that there is no way in which people can control such functions even if they wanted to. Examples include heart beat, blood circulation, the 'knee-jerk reflex' and many more.

In recent years, however, it has become clear that many of the so-called autonomic functions, maybe, eventually, all of them, can in fact be controlled by an individual using such techniques. The root of this trend has been a growing awareness in the west of the capabilities of various eastern techniques of meditation and self-discipline, of which *yoga* is the best known.

Yoga

There has been a growing awareness in the west of the capabilities of various eastern techniques of meditation and self-discipline of which Yoga is the best known.

Yoga is a doctrine and a practice that originated in India. Its main purpose is to gain for the individual complete independence; not only an independence from external influences of nature and man but also an independence from the internal emotions towards an ideal state of self-sufficiency.

The Yogin goes from the control of his senses to studying the internal activity of his own mind. The subject lies in an environment with no sound, light and a fixed temperature. Once the noise of the outside world dies down, the Yogin suddenly becomes aware of the internal noise of his body. These abilities can only be acquired as a result of rigorous training in self-awareness and control, but they prove that people can, under certain circumstances, exercise considerable influence over their autonomic functions.

In addition, the Yogin spends a great deal of effort in exercising his memory function. He is fully aware of the close link between the memory function and the mind itself.

An experienced Yogi can control his heart rate, avoid the normally automatic flinching response to pain, remain apparently unmoved by hunger, and many other extraordinary things besides.

For instance, the Yogin achieves remarkable control over the gastro-intestinal tract. An often-quoted example is that of swallowing a long bandage and propelling it throughout the entire twenty-eight feet of the gastro-intestinal canal and having it emerge at the rectum. The Yogin develops this technique in order to clear out his gastro-intestinal tract. A Yogin can also revoke the processes of nature. One of the simpler feats of the Hatha Yogin is the ability to draw fluid to flush the bladder. Another feat is the ability to change his respiratory rate. Yogins have been known to demonstrate this ability by performing the feat of being buried alive. In some cases burials have lasted as long as thirty days. When the Yogin arises from his long deep sleep, although having lost considerable weight and become dehydrated, he is nevertheless in good health, a feat almost comparable to hibernating animals.

Classes in Yoga and meditation are now commonplace in the west, but a characteristic fascination with gadgetry and technology has led some scientists to look for short cuts around the immensely demanding discipline which yoga requires. One of the obstacles to individual control over the autonomic nervous system is that people are not generally aware of its functioning. You do not normally notice your blood circulation, for example, so you would not know if it was responding to your efforts to control it. *Biofeedback* is a system which monitors the activities of automatic body functions and relays the information back to the subject, so that he or she is aware of how their body is reacting. For example, skin resistance is often an indicator of relaxation or arousal (when aroused, minute increases in the rate of perspiration cause the subject's skin resistance to decline), so a typical biofeedback device would measure skin resistance and convey the information either in the form of a dial, like a speedometer, or in the form of a tone of varying pitch heard through earphones. The subject can thereby observe the evidence of relaxation directly and, so the theory goes, develop more efficient relaxation techniques as a result of such assisted self-awareness.

There is overwhelming evidence that biofeedback can indeed be used to exert control over body functions which were previously considered to be uncontrollable. In an experiment carried out by psychologists at Oxford[23] subjects were asked to try to make one ear-lobe warmer than the other using a biofeedback device which displayed relative ear-lobe temperatures on a meter. This was designed to test subjects' control over a tiny area of their blood flow : the faster the blood flow the warmer the skin becomes. After a few false starts during which people contrived an effect *opposite* to what was required, they began to learn how to succeed in this task to a significant extent. Although the experimental task is in itself trivial it indicates the principle that this sort of self-control is possible providing the subject has the necessary self-awareness. The technique is at such an early stage in its development that it is difficult to speculate what future uses it might prove to have, but the field of self-medication is an obvious possibility. If people were able to detect organic malfunctions and to cure them by the simple application of brain power then a lot of doctors' time would be liberated. Certainly headache control looks like a real possibility.

One area in which biofeedback has been used is the development of voluntary control over one's *alpha-wave*. The alpha-wave is one of a number of tiny electrical signals detectable in the brain using an *electroencephalogram* (EEG), having a frequency of 8 to 13 cycles per second and a voltage level of 5 to 150 microvolts. It was first discovered in 1924 by Hans Berger while conducting experiments on his son Klaus, but controversy still surrounds its functions and implications. In some people the alpha-wave is very strong, although it is undetectable in others. It is generally at its most powerful when the subject's eyes are closed. It has been argued that alpha-wave strength is a positive indicator of 'wakeful relaxation', and that the use of biofeedback to control one's alpha rhythm is a useful aid to inner tranquillity. However, in an experiment conducted at the University of Otago, New Zealand,[24] it was discovered that there was a wide spread of subjective states with which people associated heightened alpha activity, ranging from contentment at one extreme to tension and anxiety at the other, so the technique does not, at the moment, seem to be universally useful.

That biofeedback can measure and record brain waves, however, is not in doubt. The most dramatic example, if frivolous, was the concert given at New York's Automation House in 1972 by composer David Rosenbloom. The music was produced entirely by regulating his mental state to generate varying brain waves, linking these to a Moog synthesizer.

Although psychologists are becoming increasingly aware

of the bizarre attributes of the human mind under abnormal conditions a satisfactory scheme of explanation is slow to emerge. This applies to the heightened self-awareness associated with yoga and biofeedback, and even more so to the heightened suggestibility associated with hypnosis. We have already seen how hypnosis has been used in police investigation to aid the memory of witnesses, although this practice is extremely rare. Before the discovery of chemical anaesthetics hypnosis was occasionally used on patients under going surgery, and there are reports of people having limbs amputated in this way without apparent discomfort. And, of course, everyone knows about the party tricks of stage hypnotists who can make their subjects behave like zombies in unquestioning obedience to their instructions.

Yet in spite of being long established as an effective technique in certain circumstances, hypnosis is not well understood, nor is it widely practised. A few therapists have successfully used hypnotic suggestion to deal with behavioral disorders, such as overeating, kleptomania, fear of social situations and so on, but the most that can be said is that sometimes such techniques work and sometimes they do not, and the majority of practicing psychiatrists prefer other methods. Suggestion in this sense is clearly comparable with the effects of subliminal advertising: the therapist gains access to the patient's unconscious mind by suspending the normal series of checks and filters, and implants the seed of an altered behavior pattern.

Another aspect of hypnosis is autosuggestion, which has been advocated as an aid to self-discipline, perhaps in order to cut down smoking or to gain in confidence and assertion. The subject achieves a highly suggestible state by concentrating on a single object, and then repeats significant words or phrases over and over, such as 'I *can* succeed' or 'I *will* stop smoking'. It is in the nature of such techniques that one hears more about the successes than the failures.

The psychological environment

The insights of the mind sciences are supplementary to well established ideas and practices. In much the same way as the human race grew plants before the development of botany as a science, or drove vehicles before Newton's laws of motion were ever dreamt of, so the tools of psychology are used as a more scientific refinement of functions which began in prehistory. Architecture and town-planning, for example, are as old as civilization itself, and there is an immense diversity of ways in which habitations have been designed to fulfil various human needs. Apart from physical considerations, such as the availability of materials and structural safety, buildings have been traditionally designed according to criteria which are partly aesthetic and partly the result of long experience. People grew to understand which types of houses were most pleasant to live in, and these, through constant imitation, came to constitute the staple 'vernacular' architecture of a locality.

But in the modern age the profusion of new building techniques and materials, and a consistent urge to innovate, has produced an architecture devoid of traditional ratification. Unless we were prepared to build tower blocks and new towns on a trial and error basis in the hope of finding design criteria which suited the inhabitants, we need a scientific method of gauging the desirability of buildings in advance.[25] Through the study of people's psychological needs and responses it is possible to determine necessary facilities, the amount and juxtaposition of open space, the counter-considerations of work, transport, leisure, education, privacy, formal and informal contact and so on. There is continuing research into such details as the ideal shapes of rooms, the size of windows, the slope of ceilings and roofs and the color of exteriors and interiors, calculated to give the inhabitants a feeling of satisfaction with their environment. In many cases this process constitutes the rationalization of what is already felt intuitively, but at the same time establishing a scientifically tenable basis for such judgements contributes considerably to their usefulness.

The archetypal example of this is the classic social-psychological concept of *defensible space*,[26] which distinguishes several graduations of territory ranging from 'public' through 'semi-public', 'semi-private' to 'private'. A strong connection was discovered between a high crime rate in certain modern developments and the absence of a feeling of territorial 'belonging'. Buildings had been designed so that the non-private areas (hallways, lifts, corridors, green spaces) were neither visible to residents, nor did they 'feel' as if they were the residents' responsibility. As a result there was no social or psychological control over people's behavior within them. The supporters of this concept recommend that urban housing is in future designed not so much to gratify the architects' preoccupation with abstractions of form and line, but to reflect the need of residents to use, monitor and control their public environment, and keep it safe and healthy. They are confident that such considerations will not only reduce crime, but will lead to all-round improvements in residents' social and emotional lives.

A hopeful future?

Although the contemporary urban landscape is sometimes branded as a form of crystallized madness, we must necessarily find a viable way of using it. Most of us suffer from neuroses of modernity to a greater or lesser degree, and while philosophers can make general judgements about the stress and artificiality of our lives, philosophy alone is no relief. Having first encountered psychology in one of its more sinister guises we should, perhaps, conclude on a more positive and more hopeful note. Persuasion, manipulation and control can, like all scientific discoveries, be used for good or evil, but they are necessary and inevitable. Perhaps in architectural psychology we have a model for ideal application: the astute use of insight, knowledge and technique to persuade people to enjoy their world and their lives.

One thing is certain, psychology and psychiatry are changing quickly and building on the work we have described, and the future holds the certainty of greater hope for the sick and a greater threat to the community.

REFERENCES

1. B. F. Skinner, *Beyond Freedom and Dignity* (Jonathan Cape, 1972).
2. Steven Rose, *The Conscious Brain* (Weidenfeld & Nicolson, 1973).
3. *People's News Service* (May 4th, 1974).
4. J. D. McConnell, *A Psychologist Looks at Crime and Punishment* (University of Michigan, 1974).
5. Dr Bernard Rubin in *New York Review of Books* (March 7th, 1974).
6. John Camp in *World Medicine* (September 25th, 1974).
7. Stan Cohen in *Humpty Dumpty 5* (1974).
8. Keith Oatley, *Brain Mechanisms and Mind* (Thames & Hudson, 1972).
9. *Daily Mail* (November 22nd, 1976).
10. *Forensic Science and the Police* (British Association for the Advancement of Science, 1974).
11. Dent and Gray in *New Behaviour* (September 4th, 1975).
12. Loftus and Palmer, *Journal of Verbal Learning and Verbal Behaviour* (1974).
13. Janis and Feshbach, *Journal of Abnormal and Social Psychology* (1953).
14. John Rowan in *New Behaviour* (July 3rd, 1975).
15. N. F. Dixon, *Subliminal Perception* (McGraw-Hill, 1971).
16. Feshbach and Singer, *Television Aggression* (Jossey & Bass, San Francisco, 1971).
17. Colin Blakemore in *The Listener* (November 18th, 1976).
18. Colin Blakemore in *The Listener* (December 16th, 1976).
19. *New Scientist* (February 26th, 1976).
20. David Cooper, *Psychiatry and Antipsychiatry* (Paladin, 1972).
21. *The Radical Therapist* (Penguin, 1974).
22. H. R. Beech, *Changing Man's Behaviour* (Pelican, 1969).
23. Derek Johnston in *New Behaviour* (July 24th, 1975).
24. David F. Marks in *New Behaviour* (September 11th, 1975).
25. David Canter, *Psychology for Architects* (Applied Science Pubs, 1974).
26. Oscar Newman, *Defensible Space* (Architectural Press, 1973).

THE SCIENCE OF THE PARANORMAL

Roy Stemman

A freelance journalist and author, he has been interested in the paranormal for twenty years. He was assistant editor of *Psychic News* for eight years during which time he witnessed many types of paranormal phenomena. After leaving *Psychic News*, he worked as an industrial journalist in publishing, aircraft, security, oil and printing fields. Books written by this author include *Medium Rare, the psychic life of Ena Twigg* (1971), *100 Years of Spiritualism* (1972), *Spirits and Spirit Worlds* (1975), *Visitors from Outer Space* (1977), and *Atlantis and the Lost Lands* (1977).

The communication breakthrough

from wireless to telepathy

Sir Oliver Lodge, the distinguished Victorian physicist, gave the first public demonstration of wireless telegraphy – without realizing the implications of his discovery. At a meeting of the British Association in Oxford, on August 14th, 1894, he transmitted electromagnetic waves from one room to another. His equipment was a standard telegraphy instrument in which a spot of light was deflected by the received signal. The transmitter had a key which could be held down for long or short periods. Lodge transmitted a few letters of the alphabet by tapping them out in Morse code on the equipment. The purpose of his impressive and historic demonstration was to show that the way in which electromagnetic waves were transmitted with his invention might be analogous to the way in which the eye converts incident light into an impulse to the brain. He did not foresee its potential for radio communication.[6]

After this successful demonstration Lodge went off to Europe to combine a holiday with another of his many pursuits: psychic research. He joined other investigators in France for experiments with the Italian medium Eusapia Palladino, in whose presence paranormal phenomena were said to occur. Meanwhile, another Italian – 20-year-old Guglielmo Marconi – was busy working on his own development of wireless telegraphy. Just a year after Lodge's demonstration the self-taught Marconi was transmitting long-wave signals over a distance of a mile, and by 1899 he had established a company in Britain and was at sea with the British fleet producing reliable radio communication between ships 20 miles apart. Lodge said some years later that he had not pursued his discovery into telegraphic application because he was 'unaware that there would be any demand for this kind of telegraphy'. But wireless communication was a world sensation and it now plays, in its many guises, an important role in our world. And it must seem to many people to be the ultimate method of communication ... until they learn of the latest, astonishing experiments of the world's parapsychologists, the scientists who probe the paranormal. They can now send messages across vast distances without *any* equipment. Human brains alone act as transmitter and receiver. Mental radio has arrived.

Over 50 years after Lodge's demonstration, Dr Stepan Figar of Prague, Czechoslovakia, made a surprising, accidental discovery. Looking at readings from a device which measures blood volume – a plethysmograph – he realized that intense thought about a person caused a measurable increase in that individual's blood volume, though the subject had no idea that this was happening. It was a British-born electrochemist and computer professor, Douglas Dean, now working in America, who saw the potential of this discovery in 1960. His subsequent research led to an even more remarkable discovery. If a telepathic sender concentrates on the name of any person with whom the receiver has an emotional tie, that person, lying in a relaxed state, may 'pick up' the thought and respond with a change of blood volume. About a quarter of the population are said to show this unconscious telepathic response.[8]

Dean, with two engineers of the Newark College of Engineering in New Jersey, designed a system around the plethysmograph that could one day be used to provide astronauts with rapid telepathic communication. The Dean system, which he presented to the Canaveral Council of Technological Societies at the First Space Conference in 1964, involves the sender concentrating on a name which has emotional significance for the receiver. A registered response stands for a dot in Morse code. If no signal is sent for a specified time, this is registered as a dash. Using this method Dean has successfully communicated over short distances, room to room, and enormous distances – New York to Florida, 1200 miles.[8]

Researchers in other countries have been working along similar lines, with equally startling results. The Russians in particular have had outstanding success with telepathy. Using two talented psychics Karl Nikolaiev, an actor, and Yuri Kamensky, a biophysicist, Russian parapsychologists have been able to send a Morse message across the 400 miles which separate Moscow and Leningrad. This achievement stemmed from their discovery that telepathic reception could be monitored with an electroencephalograph (EEG). They found that as soon as Kamensky began transmitting his thoughts the brain waves of Nikolaiev – who was wired to a large array of monitoring equipment in a soundproof room in Leningrad – changed drastically. What is more, the pattern of these waves corresponded to the type of image being transmitted.

Dr Lutsia Pavlova, an electrophysiologist at the University of Leningrad's Physiology of Labour Laboratory, observed : 'We detected this unusual activation of the brain within one to five seconds after the beginning of telepathic transmission. We always detected it a few seconds before Nikolaiev was consciously aware of receiving a telepathic message. At first, there is a general nonspecific activation in the front and mid (motor-logical) sections of the brain. If Nikolaiev is going to get the telepathic message consciously, the brain activation quickly becomes specific and switches to the rear, afferent regions of the brain.'

Thus, when Kamensky sent an image of a cigarette box, Nikolaiev's brain localized in the back occipital region, which is normally involved with sight. Nikolaiev's telepathic impression was : 'Something that appears like cigarettes. It is the lid, inside it is empty. The surface is not cold ... it's cardboard.' When Kamensky tried to convey sound, like buzzes and whistles, it was the temporal section of Nikolaiev's brain, which is normally associated with sound, which became the focus of activity.

Using this knowledge the Russians devised a system in which mental images were not used as targets. Instead, in March 1967, Kamensky had to imagine he was fighting

Nikolaiev : punching his face, kicking his legs and wrestling with him on the ground. A 45-second bout of mental pugilism represented a Morse dash. A 15-second round was a dot. Using this method the Moscow telepath's brain was able to transmit the word MIG, meaning instant, to the brain of his colleague in Leningrad. And so it was ... *instant* communication using only a human transmitter and receiver. Though it was a brilliant success, it would clearly take a long time to convey a comparatively simple message. But Bulgarian scientists who had succeeded in sending a telepathic Morse message a year earlier had developed a simpler and faster technique.

Dr Georgi Lozanov, head of Bulgaria's Institute of Suggestology and Parapsychology at Sofia, explained the method at the 1966 Parapsychology Conference in Moscow. It involved none of the expensive and cumbersome monitoring systems used in other experiments. Lozanov's first subject, a young Bulgarian man, sat in front of two telegraph keys, one by each hand. He was the receiver and he had to press either key when he received a telepathic command. The sender repeated each command ten times, to the clicks of a metronome, and the receiver had to register six of these for the symbol to be considered received. With this system, the left key might represent a dash and the right a dot in the Morse code. Dr Lozanov told the conference that of 1766 individual commands sent telepathically, his subject received 1215 – about 70% – correctly. 'Chances for this to have happened by coincidence are less than one in a million,' he said, adding : 'With this telegraph method, we sent not only individual words but also phrases and entire sentences. Sender and receiver were separated by several rooms.'[8]

These astonishing experiments will surprise many people who imagine telepathy and extra-sensory perception (ESP) research has not progressed since the 1920s and Dr J. B. Rhine's famous card-guessing experiments at America's Duke University, North Carolina. Theoretically, it should have been comparatively easy for the early psychic research pioneers to establish whether it was possible for one mind to

send an image to another. In reality, the task was very difficult. As each set of results was published showing positive
results above chance, so the skeptics demanded tighter controls. The non-believers charged that the subjects and/or
the experimenters may have resorted to fraud to achieve
the results, or that their calculations were incorrect. While
the parapsychologists endeavored to devise more convincing
tests, they were also up against other problems, too. ESP,
they discovered, was far more complicated than many had
anticipated. While some individuals scored above chance
consistently, others scored below. Zener cards were usually
used in these tests, depicting five symbols : circle, square,
cross, triangle, and wavy lines. A pack of 25 cards contained
five of each. If a subject relied only on guesswork, he could
expect to get five out of 25 right, on average. So the people
who consistently scored *less* than that seemed to have extra-
sensory power which operated negatively.

Examination of hundreds of results revealed another hidden factor. Some subjects who appeared to be doing no better
than chance were found to be one jump ahead of the researchers. They were predicting the next card before it was
revealed. Others produced significant results when they 'read'
a pack of cards which was left undisturbed, face down on a
table. This was not telepathy but clairvoyance. Further tests
have verified this observation, though not all experimenters
achieve the same results. Meanwhile, Dr Rhine was investigating another paranormal phenomenon, now known as
psychokinesis (PK) : mind over matter. He and fellow researchers found some individuals were capable of influencing
the fall of a dice with odds well above chance. Criticism was
still levelled, particularly about the randomness of some experiments. But Dr Helmut Schmidt, who was then a senior
research scientist at the Boeing Research Laboratories in
Seattle, overcame these objections with the invention of a
machine which used a radioactive source, strontium-90. The
radioactive decay of atomic nuclei is entirely random and
cannot be predicted ... unless, of course, you have ESP
talents. Very soon Schmidt was getting positive psychic

results from subjects, who were asked to predict which of various lights would be activated by the machine. The verdict after a set of 63,066 trials with three subjects was that they had scored at a rate of 4·4% higher than chance. The odds were 500,000,000 to one!

British researcher John L. Randall remarks: 'In so far as it is humanly possible to prove anything in this uncertain world, the Schmidt experiments provide us with the final proof of both ESP and PK.'[11] Taking the ESP field as a whole, Dr Hans Eysenck, who runs the parapsychology laboratory at Edinburgh University, has observed: 'Unless there is a gigantic conspiracy involving some thirty University departments all over the world, and several hundred highly respected scientists in various fields, many of them originally hostile to the claims of the psychical researchers, the only conclusion the unbiased observer can come to must be that there does exist a small number of people who obtain knowledge existing either in other people's minds, or in the outer world, by means yet unknown to science.'[3]

A change of emphasis has crept into modern ESP research. The parapsychologists no longer concentrate on amassing more and more statistics to prove the existence of ESP but instead are working on methods to develop practical uses of the phenomena and to understand the *modus operandi* of psi – the all-embracing term used to describe psychic manifestations. It seemed at one time that the only interest in psi was in the West, but while Rhine was grappling with telepathy in America, a Russian physiologist, Dr Leonid Vasiliev, was investigating 'mental radio'. His approach was to hypnotize subjects and give them mental commands. They responded. He was even able to put people into a hypnotic state and wake them again simply by mental suggestion. His work was kept secret until the 1960s when there seems to have been a sudden explosion of psychic interest behind the so-called Iron Curtain.

The reason for official Soviet interest in the paranormal was said to have stemmed from French newspaper reports in 1959 that the Americans had succeeded in telepathic communication between the submerged atomic submarine *Nautilus* and the shore. This is something which is impossible with conventional communications unless the sub comes to the surface. The American Navy denied the reports, but Russia was well aware that if a form of sub-to-shore communication *had* been established, the United States would have a great military advantage in any conflict. If the USA did *not* conduct a successful submarine telepathy test, Russia apparently did, according to leading Soviet parapsychologist and biologist Edward Naumov. He told Western observers at a parapsychological conference in Moscow that researchers had placed baby rabbits in a submarine, keeping their mother in a laboratory with electrodes implanted in her brain. When the submarine had dropped to a considerable depth the babies were killed one at a time. Naumov said : 'At each synchronized instant of death, her brain reacted. There was communication.'[8]

If this report is accurate then water seems not to be a barrier to telepathy. Nor, it would appear, is space. When American astronaut Edgar Mitchell went to the Moon on Apollo 14 in 1971 he smuggled aboard a set of Zener cards. While on the Moon he sent back mental images of the cards which were picked up by an American psychic. The result is said to have been above chance, though one test alone cannot be regarded as proof of telepathy across nearly 240,000 miles of space. Since retiring from the US Navy, Mitchell has devoted himself to psychic research.

Spontaneous telepathy has been reported for centuries. In the face of the latest evidence, however, it would seem that laboratory experiments have now proved its existence beyond doubt, except to a handful of hardened skeptics. The question that now has to be answered is : How does it occur?

Many researchers believe that other recent parapsychological discoveries are already throwing new light on the very nature of life, and that in time a new concept of human and animal existence will evolve to encompass these exciting psychic powers.

Photographs of the aura

Russian machine films invisible 'body'

We each possess an unseen force which radiates around us. That is the astonishing discovery of a Russian electrician. Throughout our life it teems with activity and bursts with color: an incredible, invisible world which appears to reflect our emotions and actions.

The concept of a human aura is not new. Ancient paintings from many countries depict holy figures surrounded by a luminescent halo. Seers and psychics over many centuries have claimed to be able to see the aura and to learn about a person's emotions or character from its colors. But attempts to analyze the invisible force, if it really existed, or to enable non-psychics to see it, were largely unsuccessful.

Then, in 1939, Semyon Kirlian went to a local research institute at Krasnodar in the south of Russia to collect a piece of equipment he had been asked to repair. During his visit he saw a demonstration of a high-frequency electrotherapy instrument and noticed a tiny flash of light between the electrodes and the patient's skin. Soon he was experimenting in an attempt to photograph this phenomenon. His research produced a severely burnt hand, a photograph of luminescence surrounding his fingers and, eventually, 14 patents to cover the developments and refinements which he and his wife, Valentina, a teacher and journalist, had invented to perfect their entirely new method of photography.

The results are achieved by a specially constructed high-frequency spark generator or oscillator, generating 75,000 to 200,000 electrical oscillations per second. It can be connected to various instruments, including an electron microscope. Kirlian's method is to insert an object, such as a hand or a leaf, between clamps together with photographic paper. No camera is used. When the generator is switched on it causes the object to radiate a luminescence. At first the Kirlians produced only black and white still photographs, but later they developed an optical instrument which enabled them to see the effect in motion and to film it. This is how American writers Sheila Ostrander and Lynn Schroeder describe the moment when Kirlian first looked at his hand through the new instrument :

'The hand itself looked like the Milky Way in a starry sky. Against a background of blue and gold, something was taking place in the hand that looked like a fireworks display. Multi-colored flares lit up, then sparks, twinkles, flashes. Some lights glowed steadily like Roman candles, others flashed out then dimmed. Still others sparkled at intervals. In parts of his hand there was little dim clouds. Certain glittering flares meandered along sparkling labyrinths like spaceships travelling to other galaxies.'[8]

Years of experiments yielded fascinating results. Living things, for example, had totally different structural details to inanimate objects. Among the many eminent scientists who visited the Kirlians' laboratory to see the discovery in action was a botanist from a major scientific research institute. He brought with him two identical leaves to be photographed. The Kirlians knew from past experience that every species of leaf had its own energy pattern, but they were baffled by their results with the two specimens. The photographs were not similar. Believing their instrument to be faulty the Kirlians made adjustments and took more pictures of the identical leaves, but always with the same result. When they told their visitor what had happened he was delighted. Though they looked the same and came from the same species, he revealed, one of the leaves was from a plant that was contaminated with a serious disease. There were no physical tests that would show it would soon die, but Kirlian photography had detected the condition. Subsequent tests with human subjects has also enabled researchers to detect the effects of disease before conventional methods of diagnosis had succeeded.

Even more fascinating, one of the original leaf pictures showed the normal energy pattern but with a line down the middle of the right side of the leaf. In fact it was a picture of a leaf with approximately one third cut away (beyond the line). Yet the energy pattern remained intact even though part of the physical leaf was removed! An interesting side-light on the phantom pains many amputees remark on.

Kirlian research in the West has been difficult because the apparatus was not available. Former actress Thelma Moss, who became assistant professor of medical psychology at the Neuropsychiatric Institute of the University of California in Los Angeles visited Russia to see psychic developments for herself, but she was not allowed inside the Kirlian labora-tories. However the information and drawings she obtained enabled her and colleague Kendall Johnson to build a workable Kirlian-type apparatus, using a lower frequency range of 100 to 4000 cycles per second. She has used the machine to study different states of consciousness and the effects of alcohol and drugs, and has also photographed the hands of healers and patients during treatment. The results indicate that, at least with the people she studied, the patients draw power out of the healer. The first of her Kirlian-type photograps was published in 1972.

Another American scientist who has experimented with the phenomenon is Douglas Dean, whose Morse telepathy tests have already been discussed. He managed to obtain a Kirlian machine from a Czech source in 1972 and, like Dr Moss, has used it to examine the 'aura' of healers and patients. Before obtaining the apparatus he had produced Kirlian-like pictures with an ordinary $39.95, 25,000-volt machine that hairdressers use to make hair stand on end for brush-cuts.[4] His research agrees with that of the Russians: 'Healing generates increased energy emanations.'

It is too soon to say, as some enthusiasts have predicted, that the Kirlians are photographing the soul – the in-destructible spirit which many believe survives death. John Taylor, Professor of Mathematics at King's College, London, is skeptical too of the claim by some Russian scientists that Kirlian photography has revealed a 'bio-plasmic energy, a definite system of ionized particles which both surrounds and interpenetrates the physical body.' He argues that there is no evidence for this, outside Kirlian photography, but he admits that it is 'undoubtedly ... a new method of obtaining information about the physical and emotional state of a living organism.'[14]

Is the Kirlian effect, then, just a fascinating manifestation that occurs, as Professor Taylor puts it, by dragging electrons out of the skin with a high voltage? Or is it a vital, ever-present force field on which life depends? We will speculate on the possibilities in a later section, but it should not be too long before the intense research that is now going on in laboratories around the world comes up with the answer.

Seeing is believing

Reading with fingers and finding missing people with pendulums

The patient who consulted Dr Iosif M. Goldberg in the Spring of 1962 had a most unusual 'complaint'. Rosa Kuleshova told the doctor, in the Ural mountain city of Nizhniy Tagil, that she could 'see' with her fingers. He was, quite naturally, a skeptic until Rosa demonstrated her powers. She was securely blindfolded, then Dr Goldberg slipped colored papers and newspapers under her hands. She was able to differentiate between the colors and read the newspaper by running her finger over them. After conducting tests and having satisfied himself of his patient's genuineness, the once-skeptical doctor arranged for her to go to Moscow where she was exhaustively examined by the Soviet Academy of Science laboratories and pronounced genuine.

Any conjuror could duplicate Rosa's blindfold feat, so the scientists had to ensure that the young girl with strange powers was not tricking them. As well as blindfolding her they placed a large sheet of cardboard around her neck to obscure her hands from any glimpse that may have been possible through the blindfold or down its edges, and they covered the objects to be read with glass to prevent any tactile clues. Rosa still succeeded, and with the Academy's seal of approval she also became a celebrity. She was detected in fraud later, though not under test conditions, but she went on to give conclusive proof of her talents by allowing examiners to press on her eyeballs while she read small print with her elbow. As Lyall Watson comments : 'Nobody can cheat under this pressure; it is even difficult to see clearly for minutes after it is released.'[16]

One of the testimonies to Rosa's authenticity comes from Dr Gregory Razran, head of the psychology department of Queens College in New York. Having examined her in Russia and spoken to Soviet researchers, Dr Razran told *Life* magazine : 'It is, after all, the kind of thing one automatically disbelieves. But there is no longer any doubt in my mind that this work is valid.'

A Bulgarian parapsychologist has found a remarkable and obvious application for this paranormal power. He is teaching blind people to see. In 1964 Dr Georgi Lozanov started tests with sixty children who were born blind or became blind in infancy. To rule out fraud he blindfolded them all. He discovered in more than 400 tests that three of the children already possessed what is known as 'skin sight'. But more important, he was able to *train* the other 57 children to do the same. 'Little by little,' he reported, 'these blind children were trained to know colors, geometrical figures, and even to read.'[8]

Why should we find the prospect of 'eyeless vision' startling? After all, many people take for granted an even more astonishing gift : the ability to detect objects hidden under ground. Almost every water authority in the world has called upon the services of a water diviner, or dowser, usually with successful results. Some dowsers are on water companies' payrolls, being used regularly to find water or trace the paths of subterranean pipes. Dowsers have used their sensory powers to detect almost every conceivable object underground. The University of Toronto, for example, has a unit specializing in archaeological research with dowsing.

Some dowsers use twigs, others use pendulums, to pick up signals. But it is the mind which performs the puzzling feat – the twig and pendulum merely react when the mind finds what it is looking for. That would seem to be the only explanation for a particularly astonishing ability which some people possess : map dowsing. They can find buried objects or missing people simply by holding a pendulum over a map. Belgian police consulted Irish dowser Thomas Trench after one of their colleagues had been killed in the Brussels riots of February 1966, and his body taken away by his murderers. With a photograph of the dead man in front of him Trench held a pendulum over a small-scale map of Belgium. It indicated a spot near Blankenberghe. Then, with a large-scale map of this area, Trench was able to pin-point, from 500 miles away, the place where the man's body was eventually found – to within about 50 yards.[5]

A Welsh water-diviner, Bill Lewis, discovered that a megalithic stone possessed strong powers. The effect the stone had on Lewis was put to the test by Prof. John Taylor, King's College mathematician, and Dr Eduardo Balanovski, a physicist from Imperial College, London. Using a gauss-meter to measure the stone's electromagnetic properties they found that it was far greater than they could account for. Whatever the cause, this finding – if confirmed by more research with other standing stones – might give us a clue not only about the purpose of the strange megalithic structures which cover Europe, but also about the forces which influence the dowser's mind. It may prove to be the same force that is at work in telepathy.

The next important question is to ask whether the same, or associated powers, can enable man to see not only beyond the normal range of his senses but also beyond time and into the future. However fanciful this notion may seem, experiments in the world's parapsychology labs appear to confirm that the vivid and dramatic premonitions which are frequently reported in the Press may well be based on fact. The early card-guessing ESP tests provided evidence for precognition, and in 1968 two French scientists conducted experiments with mice to see if they could provide even more impressive evidence. A mouse was put into a cage, the floor of which was divided in two by a low barrier. The cage was linked to a binary random number generator which every so often would electrify one half of the cage floor, giving the mouse a mild but unpleasant electric shock. The scientists wanted to see if mice were able to avoid the shocks by knowing psychically which half of the floor would be electrified and leaping the barrier to the other side in advance. They found that the mice had, apparently, displayed precognitive powers. The odds were 1000 to one against the results being due to chance.

Fearing the reaction of the scientific community when they published their paper, the scientists used pseudonyms: Duval and Montredon. It has since become known that the senior author of the report is Prof. Remy Chauvin, a distinguished Sorbonne zoologist who is famous for his work on animal behavior.[11]

The evidence then, at this early stage of investigation, indicates that if man chooses to rely solely on his eyes for vision, he is a very near-sighted organism. If he knew how, man might be able to 'extend' his vision to see almost anything.

The healing touch

A new approach to treating the sick

No one who has studied spiritual healing would deny that patients often derive benefit from healers. The issue is not '*Do* people sometimes get better?' but '*Why* do they get better?' Skeptics argue that self-suggestion explains most cases. If a patient imagines he will get better, he will. The mind is a powerful healer. This theory suggests that the healer possesses no special powers; he merely gives the patient the will to improve. Healers, on the other hand, often disagree about who or what enables them to cure the sick. Some, like famous Spiritualist healer Harry Edwards, attribute their successes to spirit doctors in the next world. Others give thanks to God or Jesus. Many speak of a 'universal' force which they direct to their patients. On one thing they nearly all agree : they are only channels. The healing does not come *from* them but *through* them.

Who is right? Recent research enables us to convert some of the speculation into knowledge. Though suggestion may play a part in many cases some extraordinary laboratory experiments have shown that positive changes do occur in organic substances held close to a healer's hand.

Sister Justa Smith, a biochemist and chairman of the Natural Sciences concentration at Rosary Hill College, Buffalo, N.Y., tested the powers of a Hungarian healer, Oskar Estebany, now living in Canada. He was asked to hold a small flask containing a solution of the enzyme trypsin in hydrochloric acid for seventy-five minutes each day. (Enzymes catalyze the metabolic reactions of each cell in our bodies.) Other flasks of the same solution were kept as controls for subsequent comparison. When she analysed the treated solutions Sister Justa was startled. 'Estebany's hands had increased the activity of the enzyme to a degree comparable to that obtained in a magnetic field of 13,000 gauss.'[4] (The earth's magnetic field is ·5 gauss.) The experiment showed that a healer can have a dramatic effect on an organic substance which is critical to the body's healing processes.

Dr Bernard Grad, assistant professor in the Department of Psychiatry at McGill University, Montreal, Canada, has also used Estebany in experiments. First he induced enlarged thyroid glands in mice by removing the iodine from their diets. Estebany's healing effectively counteracted this in the mice he treated, while the thyroids of the others grew larger. Grad then experimented further by removing small pieces of skin from the mice. Those who received healing from the Hungarian were completely healed in 14 days, whereas the others were only partially healed.[11]

How, then, does the healing process work? The USSR offers a clue. As I have already mentioned, Kirlian photography has beeen used to study American healers, but there are healers in Russia, too, and the Soviets have conducted their own experiments with them. Of special interest is Alexei Krivorotov, a retired colonel, who practises in the same office as his son – a medical doctor who refers patients to him!

Kirlian photographs were taken of Krivorotov's hands at the very moment a patient reported feeling a sensation of intense heat from them. The general brightness of his hands in earlier pictures was found to have faded and instead, in one small area, a narrow channel of intense brilliance had developed as though energy were being focused like a laser beam.[8]

Kirlian photography has also shed new light on the ancient and mysterious Chinese healing method of acupuncture in which thin needles are inserted into a patient's body to restore his health or to anesthetize specific areas. Acupuncturists believe that there are twelve channels in the body through which energy flows and, during the 5000 years that acupuncture has been practiced, over seven hundred spots on the surface of the body have been plotted where these channels come close enough to the surface to be manipulated.[16] The greatest difficulty has been finding these points exactly, since there are no physiological clues. But two Russians have invented an electronic device based on Kirlian photography, called a 'tobiscope', which locates the acupuncture points to within a tenth of a millimetre. The Soviet government proudly displayed the invention at Expo '67 in Montreal, and its use will doubtless enhance the cause of acupuncturists and their patients.

No examination of healing, however brief, would be complete without mention of the startling psychic surgeons, in Brazil and the Philippines in particular, who can apparently open up the body, remove diseased tissue with their bare hands, then close the wound without leaving a scar. There is evidence of blatant fraud in many cases, and sadly patients often derive no benefit from this rather gruesome treatment. But at its best it may be one of the most startling proofs of a dynamic psychic force with tremendous potential. Perhaps the best-known practitioner was Jose Arigo, a Brazilian, who died in 1971. In the last 15 years of his life he is said to have treated more than two million people.

Dr Andrija Puharich, a New York neurologist, took a team of investigators to Brazil, including six doctors and eight other scientists, to watch Arigo perform operations, diagnose illness and prescribe treatment. Puharich himself had a small benign tumor removed from his arm by the Brazilian medium who used an unsterilized knife borrowed from a spectator. At the end of his study Puharich announced : 'He does it. I can't tell you how. His one-man output per week is equivalent to that of a fairly large hospital, and I suspect that its batting average is just as good. At the moment we are preparing our material in the hope that some medical journal will accept our evidence.'[17]

That was in 1968. It still awaits publication.

Plants never lie

The rubber plant which identified a 'murderer'

If you plan to do something you should not, then put your plants out of sight or cover them. Otherwise, they might tell on you! So says Cleve Backster, who caused a sensation in 1968 when he announced that he had detected primary perception in plants.

Backster, a top American lie-detector expert, runs a school on lie-detection for policemen and security agents near Times Square, New York. One day in 1966, on an impulse, he attached a polygraph to a tropical plant, a dracaena, on his desk. He wanted to see if the plant would react measurably to being watered, and if so how soon. The plant's behavior was not what Backster had expected. The saw-tooth plot on the graph paper was similar to that produced by a human being experiencing a brief emotional stimulus. Intrigued, Backster plunged into further experiments, discovering on the way that just the thought of putting a flame to the plant's leaves caused an immediate change in the tracing pattern. The plant, it seemed, was telepathic.[15]

More experiments showed that plants reacted not only to threats to their own well-being but to those of other living organisms. Using automatic randomizing equipment to drop live shrimp into boiling water he found that a potted plant in another room showed significant electrical changes at the precise moment the shrimp fell. When dead shrimp were dropped there were no reactions from the plant.[17]

Backster has been criticized by some researchers who claim that he has misinterpreted the data. The humidity caused by boiling water in the shrimp test for example, could be the reason for the changes in the polygraph readings without needing to resort to telepathy for an explanation. The failure of some other researchers to replicate Backster's experiments and achieve significant results has also reinforced the scepticism of many investigators. But some claim to have produced positive results confirming Backster's findings. Among them is Lyall Watson, the biologist and author of the best-selling *Supernature*. He has been able to repeat one of Backster's most publicized tests :

'On a number of occasions, in different laboratories and with different equipment, I have played a botanical version of the old parlor game called "Murder". Six persons are chosen at random and told the rules of the game. They draw lots and the one who receives the marked card becomes the culprit, but keeps his identity secret. Two potted plants of any species, although they must be of the same species, are set up in a room and each of the six subjects is allowed ten minutes alone with them. During this period the culprit attacks one of the plants in any way he likes. So at the end of the test hour, the foul deed has been done and one of the plants lies mortally wounded, perhaps torn from its pot and trampled into the floor. But there is a witness. The surviving plant is attached to an electroencephalograph or a polygraph and each of the six subjects is brought briefly to stand near the witness. To five of these, the plant shows no response despite the fact that some of them may have spent their periods in the room after the deed has been done; but when confronted with the guilty party, the plant will almost always produce a measurably different response on the recording tape.'[17]

Not only do plants apparently have emotions, but they also respond to love and care – as many 'green-fingered' gardeners have believed for a long time. Laboratory experiments have shown that plants treated by a healer will grow at a greater rate than non-treated plants. Plants may even be able to exert an influence over particles of matter. For example, tests have been carried out to see if they can influence a random number generator in order to obtain light for their photo-synthetic process.[11] The results were statistically significant, though many more such tests need to be conducted by other researchers before any firm conclusions can be drawn.

Despite the criticisms of his work Backster has taken his work one step further and conducted similar experiments with other living organisms. On December 3rd, 1972, the polygraph expert put electrodes into a sample of fresh human semen. The donor, sitting forty feet away, then crushed a phial of amyl nitrite and sniffed the contents. Two seconds later, as the chemical damaged the mucous membrane cells in the man's nose, his isolated sperm showed a reaction on the recording instruments. Further tests showed that this did not happen with sperm from non-related humans.[17]

It may be that before too long man will have to revise his thinking about the depth and scale of his relationship with other living organisms.

Mind over matter

Making metal bend and objects disappear

The name of Uri Geller is likely to evoke one of two strong reactions: wonder or derision. To some he is a psychic saviour with the power to bend metal without touching it, read minds and heal the sick. To others he is no more than a dynamic and gifted conjurer who has cloaked his sleight-of-hand tricks with an aura of mystery. In an attempt to decide which view is true, scientists have subjected the willing young Israeli to numerous laboratory tests which, for the most part, have confirmed Geller's claim that he possesses extraordinary paranormal powers. The sceptics, on the other hand, are scornful and conjurers insist that scientific qualifications are no match for a skilful magician.

Now judge for yourself. After witnessing a Geller performance in November 1973, John Taylor, Professor of Mathematics at King's College, London, decided he had encountered a hitherto unknown force at work (or a known force that had more qualities than realized) and so he set about exploring it with his colleagues. Since Geller was in great demand Prof. Taylor had to find others with similar metal-bending powers, and surprisingly there was no shortage of offers. Geller, it seemed, had triggered a spate of spoon-bending phenomena, particularly among the young. Within two years Taylor had produced a book, *Superminds*,[14] giving his reasons for believing the 'Geller effect' is genuine. He confirmed that he had seen metal bend without it being touched. A 16-year-old boy, for example, bent a strip of straight aluminium into an S-shape inside a sealed Perspex tube. Some subjects caused a geiger counter to register non-existent high radiation and made pieces of metal rotate inside transparent cylinders.

But far more astonishing phenomena are said to be produced by Geller. During a three-hour visit to King's College on June 20th, 1974, Geller was asked to try to bend a strip of metal fixed to a letter scale (to measure the pressure he used). The metal bent and so did the needle on the balance – through 70°. In the next test a pressure-sensitive device in a cylinder stopped working at the moment the metal began to bend. The 'Geller effect' was apparently interfering with the monitoring apparatus as well as the metal. Prof. Taylor immediately took it away from Geller in time to see a diaphragm on the device disintegrate. 'After another three minutes the strip in which the cylinder was embedded had bent a further 30°. The Geller effect had been validated, but at the cost of $360 worth of equipment!'[14]

The professor and the psychic then went into another room for more tests. During one experiment, Taylor reports, he noticed 'that a strip of brass on the other side of the laboratory had also become bent. I had placed that strip there a few minutes before, making sure at that time that it was quite straight. I pointed out to Geller what had happened only to hear a metallic crash from the far end of the laboratory, twenty feet away. There, on the floor by the far door, was the bent piece of brass. Again I turned back, whereupon there was another crash. A small piece of copper which had earlier been lying near the bent brass strip on the table had followed its companion to the far door. Before I knew what had happened I was struck on the back of the legs by a Perspex tube in which had been sealed an iron rod. The tube had also been lying on the table. It was now lying at my feet with the rod bent as much as the container would allow.'

London physicist Prof. John Hasted of Birkbeck College, London, has also carried out metal-bending experiments with nine children. In addition to the distortion of untouched metal the professor also reports nearly one hundred cases of objects disappearing and reappearing supernormally 'from one place to another'. There have been 'several transferences of electron microscope foils out of and into plastic capsules ... observed under good conditions.'[10]

Magician James Randi accuses Geller of being a charlatan and has written a book[12] exposing how he believes the tricks are done. But it is hard to see how trickery could explain many of the laboratory results in which the subjects (many too young to be skilled conjurers) were not even allowed to touch the test objects which bent. Magician Artur Zorka, on the other hand, testifies that he and a colleague saw Geller make the nylon reinforced handle of a fork explode and saw him accurately guess several concealed drawings they had made. There is no way, to their expert knowledge, 'that any method of trickery could have been used to produce these effects under the conditions to which Uri Geller was subjected'.[14]

There is nothing new about mind-over-matter powers. The famous 19th-century psychic Daniel Dunglas Home allowed Sir William Crookes to conduct convincing scientific tests of his paranormal powers over one hundred years ago.[7] There are also many remarkable individuals around the world with similar talents to Geller, such as Nelya Mikhailova, a Russian woman who can make objects move by passing her hands over them and clocks stop by looking at them.[8]

So even if Geller is ever proved to be no more than a cunning trickster, the paranormal powers he claims to possess will still continue to come under the scrutiny of top scientists. Meanwhile Geller, in collaboration with Moon-walker Edgar Mitchell, is trying to bring back from the Moon by psychic means the camera case which the astronaut left there on his Apollo 14 mission. That, if he succeeds, will be the best 'sleight-of-hand' trick ever performed.

Psychic machines

Has inventor harnessed psychic energy?

Having satisfied himself that metal-bending powers exist, and having suggested a tentative electromagnetic hypothesis to explain the Geller effect, where will Prof. John Taylor's work lead him? The problem, he wrote in 1975, is 'to construct a machine which will achieve the same effects of metal-bending. That is still being pursued; hopefully with success within the next year or so.' Prof. Taylor's optimism probably stems from the fact that such a machine already exists. The space program uses a device called 'the electromagnetic hammer' which distorts metal without contact. 'This achieves its effects by setting up stresses in metals by very strong magnetic material, stresses which exceed the yield strength of the material so that it flows in a plastic fashion yet is cold.'[14]

But if reports of recent Czechoslovakian research are confirmed then even a metal-bending machine will be totally overshadowed by Robert Pavlita's amazing psychotronic generators. The inventor, the design director of a large Czech textile plant, claims that his strange-looking devices draw psychic energy from humans, accumulates it and then uses it to perform paranormal effects. Here, it seems, we are leaving science fact behind and stepping into a science fiction world. It has to be admitted, too, that the Czechs, although they have demonstrated the generators to a few Western observers and have published scientific papers on the experiments they have conducted, are keeping the secret of the invention to themselves. Unlike the Kirlian apparatus which is now available in the West, the psychotronic generators have never been examined or tested by Western scientists. Some believe Pavlita is deluding himself and others, but observers who see a generator in action soon lose their skepticism. So I make no apology for including a brief account of the so-far unsubstantiated Czech claims, for if Pavlita's work is genuine it could have an impact beyond our wildest dreams.

The generators come in different shapes and sizes depending on their purpose. They are 'charged' by the human subject staring at them. The mind of the person can then control their function. Pavlita worked on his invention for thirty years before taking it to Hradec Králové University, east of Prague, for tests. An electric motor which turned a spike on which was balanced a copper strip was sealed in a metal box, together with a small charged psychotronic generator which was not linked to anything. Six feet away Pavlita stared at the box while the scientists watched the electronically monitored movements of the spinning device. Suddenly the copper strip stopped revolving. Then it began to spin in the opposite direction. The scientists tested the generator for two years and ruled out fraud. When it learned of the experimental reports the Central Committee of the Czech Communist Party approved research, and the Pavlita generators also received the backing of the Czech Academy of Science.

A film about the invention was shown at the 1968 Parapsychology Conference in Moscow, and Dr Zdenek Rejdak, the Czech delegation leader, explained : 'Everybody has psychic abilities ... (but) the psychic force lies dormant or is blocked, making telepathy or PK a rarity.... If we assume human or other living things give off a certain energy, then we might be able to accumulate it. If so, we can have work carried out by the energy. ESP needn't be a rarity then. It could work all the time under any conditions.'[8]

Dr Genady Sergeyev, a Leningrad neurophysicist and one of the scientists involved in the remarkable Nikolaiev–Kamensky telepathy tests already described, has said: 'The Pavlita work shows it is possible to transfer energy from living bodies to nonliving matter. The most important influence of this energy is on water.' Experiments, say the Czechs, have shown that a Pavlita generator can clear polluted water in twelve hours, apparently by changing the water's molecular structure.

Only time will tell whether the Pavlita work will live up to the Czech scientists' expectations or fade into oblivion.

Meanwhile, some parapsychologists assure us that a common 'psychic' machine is available to all of us for experiments: the tape recorder. Intense research has been carried on since Friedrich Jürgenson, a Russian-born writer and film-producer, discovered the voice of his dead mother on a recording of bird song which he had made in a forest in 1959. Jürgenson, who lives in Sweden, experimented and picked up more voices. Others began to report the same phenomenon and by 1968 Dr Konstantin Raudive, a Latvian-born psychologist, claimed to have recorded 70,000 voice effects.[13] As a result of his work the phenomenon is dubbed 'Raudive voices' by researchers. All you need do to get results is to run a blank tape through a tape recorder, then play it back and listen for voices. The parapsychologists have developed more sophisticated apparatus, however, in their attempts to unravel the voice mystery.

'In one 10-minute recording I got 200 voices,' Raudive once said. 'With patience there is no reason at all why anyone cannot tape the voice phenomena. But the experimenter must develop his hearing by constant listening to tapes. What at first seems like atmospheric buzzing is often many voices. They have to be analysed and amplified.' Raudive often makes sense of these voices by assuming the 'communicators' use many different languages, so a sentence may be multilingual. The subject is further complicated by the fact that not every listener interprets the voices in the same way. Some scientists dismiss the phenomenon as being no more than radio waves picked up from a wide variety of transmitters, while others insist that when every possible means has been used to screen the equipment from extraneous radio sources, the voices still occur. Besides, say the experimenters, it is sometimes possible to hold a conversation with the voices.

Despite the criticisms and the obvious difficulties and frustrations inherent in such investigations, researchers hope that their work may eventually open up a pure channel of communication between this world and the next.

The indestructible element

Has science detected the soul?

If the gene that governs eye formation in the fruit fly *Drosophila* is altered by mutation, a fly without eyes is produced. If similar mutants are bred they produce a strain of eyeless flies. But after a while a rearrangement of the gene complex takes place and other genes deputize for the damaged gene with the result that, once more, perfect *sighted* fruit flies are born.[16] This and other biological phenomena have caused scientists to ponder the possibility of a so-far undetected organizer in living organisms : an invisible matrix which controls physical shape and development. Prof. Hans Driesch, the eminent German biologist, said in an address to the Society for Psychical Research in 1926 : 'The forces of matter are at work in an organism; there is no doubt about this. But something else is at work in it also, directing the material forces....' He described that something else as 'a unifying, non-material, mind-like something' and borrowed Aristotle's word *entelechy* to describe it.'

Parapsychologists are now theorizing along much the same lines in the light of current psychic knowledge. Brazilian researcher Hernani G. Andrade, for example, suggests that we possess biological organizing models, BOMs (or that they possess us). The BOM is composed of fourth dimensional atoms and it conserves the biological experience of its own species within some form of space–time structure. When it interacts with a seed or an egg, for example, the BOM regulates its growth and development. But to do this the 4D BOM needs an intermediary through which it can influence the earth's 3D world. That agent, he suggests, could be a bio-magnetic field, using the human brain as a two-way communicator between the BOM and the body.[9]

British biologist John L. Randall points out that 'according to physics matter ought not to organize itself into increasingly complex structures; yet in the course of the evolution of life upon this planet it has quite definitely done so'.[11] He goes on to suggest that there are two distinct entities in the universe : matter and mind. Matter has a tendency towards increasing disorder but when mind and matter interact then order comes out of chaos.

Lyall Watson draws attention to another remarkable biological phenomenon. If the fertilized egg of a salamander is allowed to develop until it begins to look like a young amphibian, and the embryo is then put into a saline solution, it will be reduced to a mound of separate cells in five minutes. If some of these cells are put back into normal acidity, and if all the embryonic parts are adequately represented, 'the cells succeed in getting back into proper shape and go on to fulfil their collective destiny and become a salamander'.[17] Again, an organizer seems to be present. The late Dr Harold Burr, Professor of Neuroanatomy at Yale University, discovered in 1935 that all living matter is surrounded and controlled by electrodynamic fields. These, he believed, were the invisible organizers. He used the analogy of iron filings scattered on a card. When held over a magnet they arrange themselves in the pattern of the 'lines of force' of the magnet's field. Throw away the filings and scatter fresh ones on the card and they, too, will form the same pattern as the old. The same phenomenon occurs in life, he maintained.

If such a force field surrounds living things, the most pertinent question is whether it is dependent on the physical form for its existence, or whether living things owe their existence to its presence. And if the latter is the case, does it continue to exist after the death of the physical bodies which it animates? In other words, is this 'electrodynamic force' or 'bio-magnetic field' the spirit or soul?

Among the many mediums who have provided outstanding evidence for life after death is Geraldine Cummins, an Irish woman who produced automatic writing scripts. In the 1950s she received the impressive Cummins–Willett communications which were said to come from the 'dead' Mrs Charles Coombe Tennant. The medium was given only the name of the dead woman's son when, in 1957, she was asked to try to get a 'message' from the mother. The message developed into six astonishingly detailed scripts containing information far beyond what Miss Cummins could have gleaned by normal means. Prof. C. D. Broad, philosopher and Fellow of Trinity College, Cambridge, says in his foreword to the book containing the Cummins–Willett case: 'I believe that these automatic scripts are a very important addition to the vast mass of such material which *prima facie* suggests rather strongly that certain human beings have survived the death of their physical bodies and have been able to communicate with certain others who are still in the flesh.'[1]

Interestingly, it was Geraldine Cummins who also received this script in the 1930s: 'Mind does not work directly on the brain. There is an etheric body which is the link between mind and the cells of the brain.... Far more minute corpuscular particles than scientists are yet aware of travel along threads from the etheric body, or double, to certain regions of the body and to the brain.... The invisible body – called by me the double or unifying mechanism – is the only channel through which mind and life may communicate with the physical shape.... It should be possible to devise in time an instrument whereby this body can be perceived.'[2]

Has that prophecy been realized in the Kirlian discovery? If so, we are one step nearer discerning the soul.

Into the future

Healing machines and manipulated thoughts

What impact will today's parapsychological discoveries have on mankind by the turn of the century? I claim no psychic insight into the future but many of the findings reported in this chapter would have tremendous implications and applications; not all of them good.

Let us look first at telepathy. However exciting the Morse code thought transference experiments may be they are still a very primitive form of mental radio. But acording to an American scientist writing in *Analog*, the Czech telepathic researchers have applied information theory to telepathy with very successful results. They assumed telepathy was a communication channel with a noise level so high that nearly all the messages were drowned, so they applied the information theory solution of calculating, among other things, how many repetitions of a single piece of information are necessary for proper reception. Two people were then asked to send two-symbol coded messages back and forth with a computer calculating the formulae for information theory. According to the US scientist, the Czechs 'demonstrated something like 98% reliability of pure telepathy communication. In other words, something better than the reliability of field communication by field telephone or radio transmitters.'[8]

Better still, the mysterious Pavlita generators are credited with 100% telepathic success. Czech scientists report that a psychotronic generator with a rotating pointer on top can be placed in a circle of ESP (Zener) cards. In another room a person shuffles an identical pack then places them one at a time face up on a table, and concentrates on each pattern. As he does so the generator's pointer swings from card to card indicating its 'choice'. The scientists report : 'It is always 100% correct. The generator never makes a mistake.'

If the success rates really are so high then telepathy could be a valuable means of transmitting secret information. The Russians, not surprisingly, have explored that possibility with astonishing results. Edward Naumov tells of an experiment carried out during Nikolaiev–Kamensky thought transmissions between Moscow and Leningrad : 'Unknown to both of them, there was an interceptor involved in Moscow, in a different building to Nikolaiev.... This man, Victor Milodan – I was the only person who knew about him – successfully cut into the telepathic communication.' Nikolaiev correctly perceived three of the five images sent by Kamensky, and Milodan got two of them. What is more, Milodan was able to tell psychically at what time the telepathic transmissions were taking place between the other two men.[8] So telepathy can be bugged !

In the field of Kirlian photography one of the exciting possibilities is that the equipment could be used to give early diagnosis of illness. Perhaps one day we will each own a miniature Kirlian device which will give us a daily check-up. Should we find something wrong, another machine may be able to carry out simple adjustments to our electromagnetic force fields in order to put matters right.

'Skin sight' research holds hope for the blind. They may one day be trained to experience some form of vision, however limited. And the gifts of clairvoyance, if refined to give high success rates, could be invaluable in locating kidnappers, for example, or missing people.

The metal-bending research may lead to new techniques for industries handling metal, and it may even prove possible to change other substances, too, with far quicker and simpler methods than those now in use. If Prof. Hasted's reports of objects being materialized and de-materialized into and out of plastic capsules are verified and understood then it might be possible to pass matter through matter. That could have mind-boggling uses, the most obvious of which would be to transport large loads, or even people, across vast distances in an instant. And if Pavlita's psychotronic generators were mass produced the whole world would be in a spin.

The dangers are obvious. If the means by which thought is transferred is mastered, it could prove possible to radiate thoughts to large sections of the population, manipulating their thoughts and actions without their knowledge. Mind over matter powers could have disastrous effects on electronic equipment. USA scientists have already reported that both Uri Geller and Ingo Swann, a powerful psychic, have been able to influence the activity of computers by their thoughts.

The greatest achievement of parapsychology, however, might be the scientific confirmation that man has a spirit or soul : a thinking, seeing, loving entity which is indestructible. That would give mankind a new perspective. Life would have new meaning. It would surely have a greater impact than almost any other conceivable discovery. Whether current research will provide that confirmation remains to be seen, but it is certain that scientists will continue to search for proof of immortality, making further exciting discoveries about the nature of man on the way. Perhaps they will even produce an instrument that will enable man to look into the future.

Paranormal conclusions

Despite the positive results which have been achieved in the world's parapsychology laboratories, scientists in general still tend to treat the paranormal with skepticism. Although American, Russian and Bulgarian researchers have repeatedly demonstrated telepathy over many miles, the accusation can still be heard that psychic experiments cannot be replicated. The work of Dr Milan Ryzl, a Czechoslovakian who defected to the West and now works in America, may change the climate of opinion, for he has trained people to be psychic. His method, using hypnosis, resulted in 50 out of 500 students cultivating a positive, measurable psychic ability. One, Pavel Stepanek, is outstanding and is now probably the most exhaustively tested subject in the history of parapsychology. In addition to scoring significantly above chance in normal extra-sensory tests using Zener cards, the Czech student was also able to distinguish whether cards in sealed, opaque envelopes were white- or black-side up. Out of 2000 guesses of randomly

presented cards he gave the correct answers 1114 times –
odds of a billion to one. The experiments conducted with this
willing subject have supplied many clues to the *modus
operandi* of telepathy, and if Dr Ryzl's training techniques
succeed in developing similarly strong powers in others then
the ESP laboratories should have no shortage of subjects for
further investigation into the fascinating fields of mental
radio and associated psi faculties. And the repeatable experiments, demanded by the skeptics, may be readily available to those wishing to probe deeper into man's mysterious
but significant psychic powers.[8]

But what of the implications which some investigators believe lie beyond paranormal powers : the belief that telepathy, clairvoyance, precognition and psychokinesis are
evidence that man has a spiritual body which is independent
of the physical self, and which survives death? Attempts to
prove this have usually centered largely on alleged contact
with the dead. Sometimes the research has produced
astonishing results, but more often it is inconclusive. But it
could be that Nature provides us with a repeatable experiment in life that furnishes us with all the evidence we need,
without recourse to mediums or the dead. To establish that
we survive death it is not necessary to prove that someone
who once lived is now continuing to exist, though dead. The
same proof is offered if it can be established that someone
who once died, now lives. In other words, reincarnation.
Ian Stevenson, a psychiatrist at the University of Virginia,
has investigated many cases of individuals who claim to have
memories of previous lives on earth. The best are contained
in his book *Twenty Cases Suggestive of Reincarnation*.[18]

In most of these, a very young child begins describing
people and events in a previous life and on subsequent investigation the facts are proved to be correct. In some cases
the child even recognizes the relatives and friends he knew
in that earlier life. Frequently these children claim to have
died violent deaths. Ravi Shankar, for example, born in 1951,
gave details of his murder. He had been beheaded at the age
of six by a relative and an accomplice. Perhaps significantly,

Ravi had a scar on his neck when he was born which resembled a long knife wound. H. A. Wijeratne, born in Ceylon in 1947, could recall the life of his paternal uncle who had been hanged for his wife's murder. A Tlingit Indian of Alaska, Jimmy Svenson, claimed at the age of two that he was his maternal uncle, Jimmy Cisko, who was probably murdered. The young boy had abdominal marks resembling gunshot wounds.[19]

It has been argued that some of these cases could be explained in terms of genetic memory – the as yet unproven ability of genes to pass memory down through successive generations of a family. The problem is that in many cases of apparent reincarnation there are no blood ties to make this possible. Even more puzzling is the case of Jasbir Lal Jat who, at the age of three-and-a-half, claimed to be Sobha Ram who had fallen from a cart and died from head injuries. Investigation revealed that such a man had lived and died in the manner Jasbir described, but he had not been killed until Jasbir was three-and-a-half, the age when the boy first began recalling his 'earlier' life. This, instead of being reincarnation, appears to be spirit possession. Having ruled out chance, fraud or normal memory as explanations in such strange cases, we appear to be left with only three alternatives, all of them supernormal : (a) the subjects have extrasensory access to information beyond their normal perception which they then dress up in a personal form; (b) spirits of the dead can possess the bodies of young children, at birth or later; (c) individuals *do* live more than one life on earth. Two of the three alternatives support a belief in life after death while the third indicates a strong ESP faculty in some people.

If we accept Dr Stevenson's cases (and those of other researchers) as evidence for reincarnation, the question that then arises is : Are these rare incidents or do we all reincarnate but without usually remembering our previous lives? The answer could well lie dormant within each of us, waiting for a sympathetic hypnotist like Arnall Bloxham to penetrate the depths of our subconscious and uncover our earlier incarnations. He has put many subjects into a trance and regressed them back to the moment of their birth ... and then back further in time. Suddenly they appear to become someone else. In Jeffrey Iverson's book about Bloxham's work, *More Lives Than One?*[20] several case histories are given and examined in detail. A young married woman living an uneventful life in Wales this time round, has, under hypnosis, described in fascinating detail her lives as the wife of a tutor in York (then Eboracum) nearly 1700 years ago; as a Jew, again in York, in the 12th century; as the servant of a 15th-century French merchant prince, Jacques Coeur; as a Spanish handmaiden who came to England with Catherine of Aragon in the 16th century; as a sewing girl in the reign of Queen Anne a century later; and as a nun in an American convent at the turn of the present century.

It is argued by some that the Welshwoman, and other subjects of these experiments, is merely dramatizing knowledge gained normally but forgotten by the conscious mind. But some experts have been impressed by the accuracy of many of her statements which, they say, could not be known to anyone who had not spent many years studying those specific areas of history. Bloxham's experiments indicate that we all live many lives; and if future research along similar lines is able to provide indisputable evidence of rebirth it will have a tremendous impact, not only on world religions, but in changing the attitudes to life of most men and women.

References and bibliography

1. Geraldine Cummins, *Swan on a Black Sea* (Routledge & Kegan Paul, 1965).
2. Geraldine Cummins, *Beyond Human Personality* (Ivor Nicholson & Watson, 1935).
3. H. J. Eysenck *Sense and Nonsense in Psychology* (Penguin Books, 1957).
4. Sally Hammond, *We Are All Healers* (Harper & Row, New York, 1973).
5. F. Hitching, *Earth Magic* (Macmillan Pub. Co., New York).

6. W. P. Jolly, *Sir Oliver Lodge* (Constable & Company, 1974).
7. R. G. Medhurst, *Crookes and the Spirit World* (Taplinger Publishing Company, New York, 1972).
8. S. Ostrander and L. Schroeder, *Psychic Discoveries: Behind the Iron Curtain* (Prentiss Hall, New Jersey, 1970).
9. G. L. Playfair, *The Flying Cow* (J. V. Dent, Canada, 1975).
10. *Psychic News* (December 11th, 1976).
11. J. L. Randall, *Parapsychology and the Nature of Life* (Souvenir Press, 1975).
12. J. Randi, *The Magic of Uri Geller* (Ballantine Books, New York, 1975).
13. R. Stemman and C. Wilson, *Spirits and Spirit Worlds, Mysterious Powers* (jointly) (Doubleday & Co. Inc., New York, 1976).
14. J. Taylor, *Superminds* (Picador, 1976).
15. P. Tompkins and C. Bird, *The Secret Life of Plants* (Harper & Row, New York, 1973).
16. L. Watson, *Supernature* (Doubleday & Co. Inc., New York, 1974).
17. L. Watson, *The Romeo Error* (Doubleday & Co. Inc., New York, 1975).
18. I. Stevenson, *Twenty Cases Suggestive of Reincarnation* (Proceedings of the American Society for Psychical Research 26: 1, 1966).
19. C. Wilson, *The Occult* (Random House, New York).
20. J. Iverson, *More Lives Than One* (Methuen, Ontario, Canada, 1976).

DEFENCE AND WEAPON RESEARCH AND DEVELOPMENT

Denis Archer

Denis Archer served as a radar officer in the British Army during the Second World War and has since divided his time between manufacturing industry and journalism. He was joint editor of *Jane's Weapon Systems* from its inception in 1969 until 1975 when he became editor of *Jane's Infantry Weapons*; and he has been editor of the monthly journal *Defence* since 1972.

When, in 1917, President Wilson committed the United States to participation in the First World War, a representative of the American Chemical Society called on the Secretary of War to offer the services of the nation's chemists to the government. He was thanked and asked to come back the next day, when he was told that the offer was unnecessary since the War Department already had a chemist.[1]

Today, like most other countries, the United States no longer has a War Department; but its successor, the Department of Defense, directly and indirectly employs more scientists and technologists than any other government department in the world, with the possible – though unconfirmable – exception of its opposite number in Moscow. The nature and significance of this technological explosion are the subjects of this study.

INTRODUCTORY SURVEY

By way of introduction to a vast and complex subject, the scale and rate of change of which almost passes human comprehension, it may be helpful to list a few of the more important areas of current and recent military research and development in the USA and elsewhere.

Strategic missiles

In the armouries of both Russia and the USA there are intercontinental ballistic missiles (ICBM) which, launched from heavily-protected fixed sites, can carry nuclear warheads to targets 8000 miles or more away at speeds in the order of four miles a second. The largest (Russian) warheads are credited with an explosive power of 50 megatons (notionally equivalent to 50 million tons of TNT) or more but most missiles carry much smaller warheads. The Russians also have land-mobile ICBM with significantly shorter ranges and both countries have submarine-launched ballistic missiles

(SLBM) which can be launched from submerged nuclear submarines; the latest Russian missile of this type is reported to have a range of more than 4000 miles. China, France and the UK also have strategic missiles in service but on a scale that is dwarfed by the American and Russian deployments.

Of the ICBM currently deployed, that which is credited with the greatest accuracy is the American Minuteman III which has three MIRV warheads (see below) each of 170 kilotons (170 thousand tons of TNT) equivalent explosive power. Fired at a target some 7000 miles away it has a 'circular error probable' (CEP) of about 400 metres; the CEP being the radius of a circle round the point of aim within which there is a 50 per cent chance that the warhead will arrive. An inaccuracy of this magnitude is unimportant if the target is a population center: a warhead of this size will destroy ordinary brick buildings within a radius of three or four kilometres from the point of burst and reinforced concrete structures within a radius of about 1500 metres: it would also ignite wood or fabrics in the direct line of sight at distances of seven or eight kilometres. The inaccuracy is, however, very important if the target is a protected military installation: 400 metres is about the maximum permissible error if such a warhead is to have any chance of destroying an enemy missile in its silo.

Submarine-launched missiles are at present mainly significantly less accurate than those launched from fixed sites. The Russian missile mentioned above is believed to have a very advanced navigation system and has been credited with an accuracy comparable with that of Minuteman III; but even if this is true – and it may not be – it probably applies to the missile when equipped with only a single warhead. Most currently deployed SLBM are strictly suitable only for attacking population centres or industrial areas.

MIRV (Multiple Independently-Targeted Re-entry Vehicle)

In recent years emphasis has been placed on equipping ICBM and SLBM with a multiplicity of small warheads, which can be dispersed over a wide area, in preference to a single large warhead which concentrates excessive explosive power in one place. Early work on this project involved multiple warheads (MRV) which were dispersed in a fixed pattern or 'footprint'; but more recent systems permit the individual pre-programmed targeting of each sub-warhead. The warheads are separated out while the missile is outside the earth's atmosphere and pre-programmed adjustments are made to their trajectories by small rocket motors so that they fall back to earth as a programmed shower. Decoy devices to confuse enemy defences can also be dispensed at the same time. Each sub-warhead is a complete re-entry vehicle so that for a given total payload or 'throw-weight' a MIRV system has substantially less explosive effect than a single warhead: but its practical effectiveness may be greater. The largest known current MIRV capability is that of the US Poseidon SLBM which can carry 14 sub-warheads. In practice, however, several of the places are reserved for decoys.

Poseidon is currently deployed in Lafayette-class submarines of the US Navy, each of which carries 16 missiles ready in its launch tubes. The warhead package fitted at present contains 10 MIRV warheads each of 50 kilotons equivalent explosive power. A single submarine therefore can attack 160 separate targets with warheads each of which has about twice the explosive power of the bomb dropped on Hiroshima.

It is important to realize, however, that the results of the Hiroshima and Nagasaki explosions were not typical of what might be expected from a nuclear weapon attack on a major European or North American city. The two Japanese towns – densely-populated with very few substantial buildings and a great many lightly-constructed and inflammable dwellings – were exceptionally vulnerable to atomic weapons which could be dropped with care over the unprotected targets. On the other hand, because so little was known about the effects of atomic weapons at the time, both bombs were detonated at well above what would now be regarded as the optimum height. To give some idea of the probable effects of the modern weapons on European or North American targets the table on p. 215 shows the numbers of Poseidon-type 50 kiloton warheads required to produce various levels of destruction in Los Angeles, Moscow and a selection of important cities in the United Kingdom.

Several points need to be borne in mind when interpreting a table of this sort. First, although it is based on the best available unclassified information, the effects of nuclear explosives on large built-up areas are, mercifully, not precisely known; and all data, classified or otherwise, result from limited series of experiments on sample targets : the extrapolation to a target of the size and complexity of, say, Los Angeles necessarily involves something not far removed from guesswork. Secondly, although the table suggests that, in some instances, very large numbers of warheads would be required to produce the destructive effect specified, significantly smaller numbers would certainly produce appalling casualties and extensive disruption of the means of dealing with them.

Thirdly, the 50 kiloton warhead has been chosen as the typical agent, not merely because it is an actual operational weapon, but also because it represents a reasonable compromise size. Very much larger warheads produce less destruction in proportion to their notional explosive power because they waste so much energy on the area in the immediate vicinity of the explosion in which total destruction would result even if a very much smaller weapon were used (this is the original meaning of the much-misused term 'overkill'). Multi-megaton 'doomsday' warheads were appropriate to the time when MRV and MIRV techniques had not been developed and missile accuracies were low. On the other hand the indefinite multiplication of smaller MIRV warheads in a single missile tends to be counter-productive because the space required for the guidance apparatus and re-entry shield of the sub-warhead excessively reduces the explosive payload that can be carried.

Finally, it is interesting to note that, although London and Moscow have populations of similar size, it would take nearly twice as many warheads to produce a given effect in the former as it would in the latter. The point is made because some people are fond of bandying about simple arithmetical calculations relating delivered megatons to millions of casualties. There are no such simple relationships – as the table on page 215 shows.

MARV (Manoeuvrable Re-entry Vehicle)

This term is applied to an American development which gives an ICBM or SLBM warhead the ability to manoeuvre, in the last stages of its journey, with the primary purpose of evading interception by enemy defences. Conceivably the apparatus employed for this purpose could also be used to enable the warhead to home more accurately on its target but this appears not to be the intention at present. One reason for this is that the incorporation of appropriate terminal guidance equipment for this purpose must reduce the payload of any given missile or sub-missile; another is that all known methods of providing such guidance involve the use of magnetic or electromagnetic (including optical and infrared) techniques all of which are to some extent vulnerable to jamming or decoy measures. Such measures might serve no useful purpose in the protection of large scattered targets; but it would be sensible for a defender to deploy them round key command or weapon sites.

It should be noted, however, that some very sophisticated terminal guidance systems have been and are being developed for other types of missile, some of which are mentioned later in this survey.

Anti-Ballistic Missile (ABM) systems

Enormous sums of money have been spent, in the USA and the USSR, in the quest for a satisfactory defence against the ICBM. It would be wrong to say that no satisfactory defence has been found because the possibility of intercepting incoming missiles has been convincingly demonstrated. Nevertheless after a 20-year development operation and a small initial deployment the main American programme was shut down in 1976, although research work continues in the USA

NUCLEAR WARHEADS REQUIRED TO PRODUCE VARIOUS LEVELS OF DAMAGE TO SELECTED TOWNS

Number of 50 KT warheads to produce damage level:

	Wide-spread fires throughout area	Brick built buildings destroyed	Unprotected population killed or severely injured	Large buildings destroyed (unless reinforced concrete)	Effectively all buildings destroyed (inc. reinforced)
New York–Metropolitan*	339	612	1089	1201	1773
Los Angeles–California	154	278	495	546	806
Tokyo–Japan	83	150	267	295	436
Detroit–Michigan	65	118	210	232	343
London–England	46	84	150	164	242
Washington D.C. & C†	35	64	114	126	185
	(5)	(8)	(15)	(18)	(27)
New York City–N.Y.	28	51	93	105	159
Paris–France	28	50	89	98	145
Moscow–U.S.S.R.	27	49	87	96	141
Berlin–E. & W. Germany	23	43	76	85	125
Kiev–U.S.S.R.	23	42	74	82	121
Leningrad–U.S.S.R.	19	35	62	68	100
San Francisco–California	13	24	43	48	71
Mexico City–Mexico	8	14	25	27	40

* The metropolitan census area of New York and N.E. New Jersey.

† Including adjoining urban areas outside the District of Columbia. D.C. figures in parentheses.

Notes:

1. Numbers specified relate to warheads detonated at optimum height with positional accuracy consistent with current levels of control and guidance techniques. No allowance is made for malfunction or enemy action.

2. Damage levels relate to the whole area of the town. Damage in the vicinity of each explosion will be greater than that specified for the total area: a single warhead will produce virtually total destruction over an area of about 1·5 square miles.

3. Numbers may be roughly equated with missiles and submarines by taking 10 warheads as equal to one Poseidon missile and 16 Poseidon missiles as equivalent to one American nuclear submarine. In practice, however, some allowance would have to be made for enemy action: moreover, where high concentrations of missiles are needed, there is a possibility that earlier explosions might induce malfunctions in later warheads.

4. All figures in this table have been arrived at by extrapolation from data in unclassified publications. They are thus reliable to no greater extent than are the disclosures made in such publications.

and the Russians still retain a system deployed round Moscow. The reasons for the abandonment of the American system were primarily political and financial; nevertheless it is probable that the approach hitherto adopted – that of using one nuclear missile to intercept another – is unlikely ever to provide sufficient protection to justify the expense that must be incurred. The technological advances made by both countries in their endeavors are unlikely to be totally wasted, however; they include the most elaborate radar detection and tracking systems ever developed; a fast-reaction missile said to be able to climb to a height of ten miles in about four seconds and longer-range interceptor missiles which can be launched before the hostile missiles are within range and 'loiter' in space while the defence systems identify and select targets for interception.

Tactical nuclear weapons

Early developments of nuclear missiles and bombs which were intended for use in the combat theatre, as distinct from strategic weapons designed to destroy cities, ports and industries, carried warheads that were so destructive that it was scarcely credible that they could be used in defensive operations; they were likely to do more damage to the defended territory than would the enemy. Although many such weapons are still deployed in Europe, recent emphasis has been on the development of smaller and much less destructive nuclear devices ('mini-nukes') which are still very much more potent than conventional munitions but are more manageable in their effects. Efforts in this direction have been made mainly in the USA and a typical yield is one-tenth of a kiloton (equivalent to 100 tons of TNT).

Tactical nuclear weapons exist in very large numbers. Including guided and unguided missiles, bombs, artillery shells and fixed demolition charges, there are currently about 7000 such devices in NATO Europe. Nuclear depth bombs have also been developed for use with the American ASROC (anti-submarine rocket) system and may well also exist in the Russian armoury of tactical nuclear weapons – which is probably not much smaller than that of NATO and almost certainly has a greater emphasis on large warheads.

Anti-tank weapons

One of the most energetically-pursued lines of development, particularly in European countries, has been in the direction of countering the threat posed by the modern fast, powerful and heavily-armored tank. Important advances have included the development of man-portable guided missile systems with warheads which can penetrate a foot of steel armor at a range of two miles or more and can scarcely miss the tank if the soldier can see it and does not lose his nerve; 'instant minefields' which can be laid in a few seconds in the path of advancing tanks by using airborne dispensers or land-mobile rocket launchers; laser guidance systems which enable a forward observer to call down remarkably accurate artillery fire on moving tanks from a range of ten miles or more; missile-dispensed heat-seeking sub-missiles which can be scattered over vehicle concentrations from even greater distances; low-light vision systems which make it possible to engage tanks in near-zero visibility; and thermal sensing systems which can be used to identify and engage them in total darkness.

Laser weapons

Laser technology has made great strides in recent years and lasers are now widely used for rangefinding and target identification. An important line of development which is certainly being followed in the USA and, it is believed, in the USSR, is aimed at using the ability of the laser to concentrate energy in a narrow beam to destroy enemy equipment. There are many practical problems to be overcome but the possibility of destroying aircraft in this way has been demonstrated and the anti-tank application is currently the subject of development work in the USA.

Anti-aircraft and anti-missile defences

Of particular interest in the European theatre of operations and at sea has long been the provision of satisfactory defences against low-flying aircraft and tactical missiles. The latter are, of course, very much less of a problem, from the point of view of the defender, than strategic ballistic missiles because they travel so very much more slowly; but they are still difficult to intercept because they are usually small and may travel close to the surface of the earth or sea. Current approaches involve alternatively directly intercepting the aircraft or missile with a well-aimed fast-reaction missile or filling the defended airspace with large numbers of projectiles from bursting rockets or from guns with very high rates of fire. An important series of related developments has produced several types of man-portable anti-aircraft weapons, some of which have been used with considerable success in recent wars.

Some of the guns used in close-in defence systems were originally developed for use in aircraft – in particular, multi-barrel guns which derive ultimately from the concept of the famous Gatling gun. One such gun is the American Vulcan six-barrel 20 mm weapon which in its airborne role fires at the rate of 6000 rounds per minute; even more impressive is the GAU–8/A 30-mm gun which is being installed in the USAF A–10 close-support aircraft. Over 20 feet long and weighing more than 4000 pounds this big seven-barrel weapon can fire its 30-mm projectiles at the rate of 70 per second with a muzzle velocity of 3500 feet per second and has demonstrated its ability to knock out main battle tanks. The largest gun ever mounted in a US attack aircraft, it may yet find other applications.

Reconnaissance and surveillance

Since the launching of artificial earth satellites became a practical possibility, extensive use has been made of such devices for reconnaissance purposes. The Americans and the Russians closely monitor each other's military activities in this way, using a variety of optical and electronic sensors, the information being either transmitted back to earth or recovered as a package at the end of the satellite's useful life. Little is known of Chinese satellite reconnaissance activities but it may be assumed that they are following a similar pattern. Satellites can also be used as early warning devices to detect the launching of, say, an ICBM attack; and an important related activity is that which could be aimed at neutralizing such satellites by electronic jamming or physical destruction. It is known that the Russians have carried out experiments along these lines and highly likely that similar action has been taken in the USA. Certainly the problem of incorporating defensive systems in satellites has been explored there.

Very important progress has also been made in the development of tactical reconnaissance systems for obtaining up-to-the-minute information on the disposition and movements of enemy troops and equipment in a theatre of war. Apart from the long-familiar use of manned aircraft for such purposes, there are now available or in development many different unmanned vehicles, ranging from miniature helicopters to quite large unmanned aircraft which can tour a battlefield under programmed or command control and either relay what they find directly or bring back a package of data for processing. This is likely to remain a major technological growth area for some time to come.

Mention should also be made here of developments in radar technology. Although there continue to be military requirements for the familiar types of surveillance and tracking radar in which one or more high-power transmitters and sensitive receivers are associated with a mechanically operated directional reflector, the much more versatile phased-array type of radar, using large planar assemblies of relatively low-power transmitting and receiving elements, is now widely used. Combined with elaborate data-processing facilities, such radars can make much more economical use of power and time; and a radar that can comfortably locate and track half-a-dozen targets simultaneously can now –

thanks largely to the advent of microminiature electronic devices – be accommodated in the nose of a fighter aircraft. At the other extreme the enormous radars built for the American ABM system also used planar arrays.

ELECTRONIC WARFARE

Battlefield sensors

Closely related to the foregoing is the development of techniques of identifying particular activities on a battlefield or in the vicinity of a defended area. The night vision and thermal sensing techniques mentioned in connection with anti-tank warfare are of obvious relevance here as also are special types of radar for the detection and identification of particular types of movement (men or vehicles, one man or many, men crawling, walking or running and so forth) seismic sensors for similar purposes and chemical devices to detect body odours. One system currently being developed in the USA is intended to scatter a variety of passive sensors from an aircraft or missile and subsequently receive reports from them for remote analysis.

Countermeasures

As each idea for a new or improved weapon, system or counter-weapon becomes known, someone starts thinking of ways to defeat it. The tank breeds the anti-tank weapon and the anti-tank weapon breeds improved armor and ideas on anti-anti-tank weapons; the ICBM generates the search for an ABM system; the reconnaissance satellite invites the means of its own destruction. With the great increase in the use of electronic systems in warfare the development of electronic countermeasures (ECM), electronic counter-countermeasures (ECCM), and electronic data gathering (ELINT – electronic intelligence) has become a major sub-discipline of military electronics. Simple jamming techniques were often sufficient to neutralize early radar and radio control systems but in areas of major importance the battle is nowadays much more sophisticated. Fundamental to all such operations is ELINT, which is one of the missions of most reconnaissance systems and is, incidentally, the purpose of the Russian 'trawlers' which are to be found in the vicinity of NATO naval vessels during major exercises.

Mobility and navigation

Developments in propulsion systems for ships, land vehicles and aeroplanes are not in general peculiar to military applications; although these may often stretch current technology to its limits and the ability to propel a 50-ton tank at 35 miles per hour or a submerged nuclear submarine at 30 knots may warrant comment. A high degree of mobility is essential to most operations in modern warfare; but of at least equal importance is the need for the mobile forces to know where they are. Among the more impressive achievements in this area are the systems that enable a supersonic strike aircraft to hedge-hop its way over hill and dale to a small target hundreds of miles away or a nuclear submarine to navigate under water for days on end and still know where it is with sufficient precision to program its ballistic missiles to reach targets more than 2000 miles away with acceptable accuracy.

One special military problem, however, is that of communicating with a submarine reliably and over long distances. The whole essence of the relative invulnerability of the nuclear ballistic missile-launching submarine is its ability to escape detection by remaining submerged for long periods out of sight of aircraft or reconnaissance satellites; unfortunately, however, radio waves at normal communication frequencies cannot be detected far below the surface of the sea. For some time now it has been the practice to communicate with remote submarines on station using LF and VLF signals which can be picked up on a trailing antenna which is paid out from the submarine and buoyed a few feet below the surface. This, however, limits both the depth and the speed at which the submarine can cruise to considerably less than its design capability: it also renders it liable to be detected by the improved airborne magnetic anomaly detection apparatus which is now becoming available.

To overcome this difficulty the US authorities have for some years been working on ELF (extremely low frequency) communications systems, the radiation from which can be received by a deeply submerged submarine and for which a single enormous transmitter can give global coverage. The system has been shown to work; the difficulty is the cost of the system and the environmentalist opposition that it has aroused. The frequency band concerned is in the 40–80 Hz region – the wavelength thus being between 3750 and 7500 kilometres (2330–4660 miles) – and the transmitter needs to be associated with an aerial system about 2000 miles long buried in the earth in a geologically suitable location covering about 500 square miles. This is obviously a formidable undertaking; but when it is remembered that without reliable communications a missile submarine would not even know that war had broken out it is evident that heroic measures may have to be contemplated.

Chemical and biological warfare

This has been left to the end of this introductory survey mainly because there is so little in the way of hard information on the subject. That a great variety of lethal and incapacitating chemicals may be found in the laboratories and probably the arsenals of many countries is a reasonable certainty; and much the same is true of virus and bacterial preparations which are capable of spreading disease and death far and wide. Some further comment on the implications of these supposed (and indeed probable) facts will be found in a later section of this study; but for the moment it will suffice to draw attention to the two types of chemical warfare that have actually been used in recent years. The first is the highly controversial use of defoliants by the Americans in Vietnam; the second is the use of lachrymatory gases in riot control or counter-insurgency operations. The first is probably not of long-term importance; but the second almost certainly is. It is also perhaps worth noting that the only country in which troops are regularly exercised in respirators and protective clothing is Russia.

The nature of military research and development

Concentrating as it does on the overt results of some (but by no means all) of the more important military research and development programs of recent years, this introductory survey may have given some indication of the diversity and breadth of the subject but does not convey its scale. Before an attempt is made to define the scale or to explore the reasons for the transition, in less than a lifetime, from near-total military disregard of scientists and engineers to near-total dependence on them – and it should be noted that the phenomenon has been the common experience of many countries – it may be helpful to draw attention to an important distinction between the scientific and technological activities sponsored by the military and some other areas of scientific endeavor. Although major scientific advances tend nowadays to be made more by teams with substantial resources at their disposal than by solitary scientists in private laboratories, there are still many disciplines and sub-disciplines in which important advances are made preponderantly by people who can be described as scientists rather than technologists, technicians, craftsmen or laborers. In most areas of military scientific activity, and certainly in all the very expensive ones, however, the end-product is a piece of military hardware, the effectiveness of which must be demonstrated satisfactorily before an advance can be said to have been made; and the costs of the basic laboratory work may be small when compared with those of the electrical, mechanical or structural engineering work involved in the construction of experimental or prototype equipment.

Engineering costs of these kinds, though seldom of comparable magnitude, are of course familiar features of some less warlike scientific occupations such as radio astronomy or high-energy physics; and in any large civil or military project to categorize the component activities under such headings as 'science' and 'engineering' (or even 'research' and 'development') is both difficult and unrewarding. Since it is a popular political pastime, in some quarters, to contrast expenditure on such socially undesirable objectives as the development of improved intercontinental ballistic missiles with that which is or should be spent on, say, finding a socially desirable cure for multiple sclerosis, however, it is important to realize that the comparison is false. As a simple piece of bookkeeping the money could be transferred from one side of a national budget to another, but a real transfer of the labor and materials represented by the money would be ludicrous. Even if all the nations of the world were today to decide to beat their swords into ploughshares, it would take certainly many years and probably decades to guide the footsteps of those currently involved in warlike activities usefully into the ways of peace.

The scale of military research and development

It is not intended that the foregoing should be regarded as a categorical defence of current levels of military spending, the global annual total of which has been recently estimated at $280 billion[2] and shows no sign of diminishing. That so many millions of people, even in an overpopulated world, should be thus employed must be a matter for regret, however arguably justifiable it may be in the present or any other circumstances.

That said, however, the present fact has to be recognized for what it is; and it is against the present fact and the recent history of military expenditure that the many proposals and programs for reducing the related international tensions must be judged. It would be foolish to take the pessimistic view that nothing can be done until the present arms race culminates in a hideous holocaust – foolish both because it is seldom true that nothing can be done and because the arms race does not necessarily lead to the holocaust – but those who propound simple solutions to this immensely complex problem may reasonably be asked to explain in considerable detail how their pet panaceas are to be applied in practice.[3]

A full analysis of the reasons for the uninterrupted growth of global military spending is beyond the scope of this study but a glimpse of its present distribution provides a useful background to a discussion of the research and development element. Well over half the total is accounted for by the USA and the USSR and roughly three-quarters by these two countries plus China, West Germany, France and the United Kingdom. Another fifteen countries cover the gap between three-quarters and nine-tenths, while the remaining tenth is distributed among more than a hundred countries.[4] A different analysis of the same global estimate shows that more than three-quarters is spent by the North Atlantic and Warsaw Treaty Organizations.

Examination of what is known of expenditure on research and development reveals an even greater concentration of effort. Total expenditure under this heading is very difficult to estimate, but it appears to be currently in the region of one-tenth of the global total of all military spending – almost certainly more than $25 billion but probably less than $30 billion annually.[5] Of this total between 70 and 90 per cent is accounted for by development programs in the USA and the USSR and about 10 per cent by France and the UK. China's contribution is virtually impossible to assess but is probably at least comparable with that of the UK; and the remainder is contributed almost entirely by some twenty other countries.[6]

It is extremely difficult to express expenditures of such magnitudes in terms that can be grasped by the human brain. Even in this relatively numerate age a billion is an unimaginably large number for most people. With considerable further loss of a precision that is already regrettably low, the figures can be brought more nearly into perspective by suggesting that there must be around three million people employed full-time on military research and development projects[7] and that about one in eight or ten of them are qualified scientists or engineers.[8] An estimate for 1974 put the total of scientists and engineers at 400,000.[9] The world has come a long way in the sixty years since the man from the

American Chemical Society called on the US Secretary of War.

REASONS AND OBJECTIVES

Although the scientist and the engineer have now penetrated almost every area of human activity from marriage guidance to space exploration, this immense concentration of skilled effort on warlike activities cries out for explanation. A simple answer is that there has been an escalating power struggle between the USA and the USSR in which each has found it necessary to demonstrate its ability to crush the other if sufficiently provoked; and that this demonstration is most effectively provided or countered by continuously inventing and deploying ever more devastating weapons and ever more effective countermeasures.

This explanation is too simple and must be qualified by some analysis, but it will serve as a starting point for a discussion of the trends in modern weapon development and the motives that lie behind them. In addition to the Russo–American power struggle – which also has Chinese overtones – it is possible to identify such subsidiary explanations as the different threat perceptions and reactions of the European countries, the attractions of the arms trade and the desire for military self-sufficiency in an unpredictable environment: nevertheless the dominance of the Russo–American contest makes it only sensible to deal with that first. The political and geopolitical motives that lie behind this struggle have been argued over for years, and they will doubtless continue to provoke discussion for many years to come; but it would serve no useful purpose to consider them in detail here.[10] Obviously, the international political situation which developed in the years immediately following the Second World War was of major importance; but the results would have been far less spectacular had it not been for a small number of vital technological developments during that war and in the years immediately preceding it.

First, the development of the atomic bomb gave to a single aircraft or other weapon carrier a capacity for destruction far in excess of anything that had previously been available. If reliable delivery could be reasonably assured, the nature of long-range warfare directed against an enemy's lines of communication or industrial base would be dramatically changed. Secondly, the British development of the jet engine and the German work on rockets indicated two possible ways in which the reliability with which the atomic weapon could be delivered might be improved : during the war the ability of ground forces to counter aerial bombardment had tended to decline as the speed, range, and operating altitude of bombers had increased and there was no known practical way of intercepting the German V2 missile. Thirdly, the great wartime advances in electronics technology, including radar, radar countermeasures and other forms of jamming, offered possibilities of superior operation in both attack and defence which could be grasped and exploited by a technologically advanced nation.

These and many related possibilities were clearly perceived by the Americans, who also had both the advantage and the responsibility of being temporarily the sole custodians of the atomic bomb. Moreover, their own wartime experience and their observation of the experiences of their allies and enemies had taught them many important military lessons. In the land battle they had become aware of the crucial importance of local air superiority and had witnessed the disruptive effect of a combination of armour and aircraft in a *blitzkrieg* thrust : at sea they had experienced the vulnerability of surface vessels to air attack and had become committed to the aircraft carrier as the principal element of a naval strike force : in submarine warfare they had perceived the effectiveness of the radar-aided anti-submarine aircraft patrol against submarines that must spend a significant proportion of their operational time on or near the ocean surface.

Each of these lessons and many others posed problems and suggested solutions that were grist to the mill of the wartime military infrastructure of academic and industrial development teams and government establishments that had been set up in the USA to solve the urgent problems of land, sea and air warfare against determined and highly-skilled enemies.[11] In the important area of rocketry, moreover, 'a further stimulus was provided by the discovery of the important advances made before and during the war by German scientists and engineers and by the transfer of Wernher von Braun and some of his team to the USA.[12] Even so, the exploration of these possibilities might have been conducted with far less vigour, if at all, had not the United States emerged from the war not only victorious and supremely powerful but also committed to the maintenance of overseas forces and organizations, for purposes of occupation and reconstruction, on an unprecedented scale. This fact, coupled with a great desire to exploit wartime technology for peacetime purposes (notably in civil aviation) delayed the rundown of both the military forces and the civilian infrastructure and enabled the technological momentum of the latter's activities to be maintained.

Events in Europe and the Far East gave added impetus to these endeavours. Although there were those in the West who, in the immediate aftermath of the war, were so mistrustful of the Russians that they were prepared to advocate a surgical strike to eliminate the Communist threat to the world, the general run of opinion was against them. Russian actions in Eastern Europe, however, coupled with Stalin's speech in January, 1946, in which he made it clear that top priority was to be given to enhancing Russian military power, were early steps in disillusionment as also were Communist involvement in the First Vietnam War and the Greek Civil War. Following these blows came the Chinese Civil War, the Berlin blockade and the announcement that the Russians had developed an atomic bomb.[13]

Few in authority in Western Europe by then doubted the desirability of American support to counter the threat posed by the evident and increasing power of the Russian military machine; and the world was brought face-to-face with the dangers of the international situation by the outbreak of the Korean War in 1950. Although American forces were deeply involved in that war, however, and although it was fought primarily with weapons that were essentially similar to those used in the Second World War[14] the principal world military role envisaged for the USA was essentially one involving strategic weapons. It was to these high-technology areas, therefore, that a large part of the American home-based defence effort was directed.[15]

In Russia, on the other hand, the resources to pursue high-technology projects on a broad front were simply not available in the late 1940s. Their war had been fought from a very narrow technological base and they had been heavily dependent on their allies for equipment of even quite modest sophistication. Although they put what effort they could – and with remarkable success – into the atomic bomb development; and although they too had acquired part of the German rocket development operation and set to work vigorously to exploit it, the main thrust of their initial post-war military reconstruction was directed towards the development and rapid mass-production of reliable low-technology weapons with which to improve the capabilities of their enormous land forces and supporting air forces. Intensive effort, again with German assistance, was also put into the reconstruction of Russian naval forces; but the lead-time here was necessarily such that significant results could not be seen until the 1950s.[16]

Differences in the method of approach

In this area of naval weapon and system development, however, the fundamental difference between the Russian and American approaches to similar problems was demonstrated in a way that is both interesting and important in that it is still perceptible in the much-changed circumstances of to-day. In both countries the potential of the submarine as a platform for weapons other than the traditional torpedo was recognized at an early stage.[17] Any submarine has the advantage of being able to move about the ocean more stealthily than a surface vessel; and with the prospect of the availability of potent guided missiles in view it was logical to con-

sider the potential of the submarine as a mobile launcher for
such weapons.

Both countries embarked on the development of cruise missiles for launching from either surface vessels or submarines
and ballistic missiles[18] which might be launched from surface
vessels but were seen as primarily submarine-launched
weapons. The cruise missile programs proceeded roughly
in parallel; but the Americans, starting with wider experience of most aspects of the technological problem, were able
to deploy their Regulus missile, in a form suitable for surface launching from submarines, in 1954, whereas the first
Russian missile, much inferior in performance, was suitable
only for launching from large surface vessels and was not
deployed until about 1958. They pressed on with the programme, however, and deployed a highly-successful missile
which could be launched from small fast patrol boats in 1959
and a much more powerful weapon for launching from sur-
faced submarines in 1960–61. The earlier missiles were intended for anti-ship or shore bombardment operations and
carried conventional high-explosive warheads:[19] the third
had a range of some 400 miles and could carry a nuclear
warhead. In that it flew supersonically and in its guidance
system it was somewhat superior to the Regulus missile;
but a further six or seven years of development in the USA
would have produced a much better weapon.

After Regulus, however, the US Navy abandoned the
development of naval cruise missiles[20] and concentrated instead on what subsequently became the Polaris submarine-
launched ballistic missile (SLBM) program. In this program they adopted a high-technology approach from the
outset : the weapon was to be launched from a submerged
submarine, and was to have the range and nuclear warhead
power and the guidance accuracy and invulnerability appropriate to a strategic missile. These requirements gave rise
to a natural association between the missile program and
the nuclear-powered submarine program which offered
the advantage of being able to transport the strategic missile
launchers to any ocean location without surfacing. A par-

ticular point of affinity between the two programs was the need for an inertial system[21] of navigation for the submarine and guidance for the missile. Not surprisingly the incorporation of so much sophistication into a single project took time : the first US nuclear submarine entered service in 1957 but the missile-launching type, equipped with Polaris SLBM, did not enter service until 1960.

In the USSR, on the other hand, the policy was evidently one of deploying operational systems first and attending to technological improvements later. The first ballistic missiles to go to sea were relatively unsophisticated, relatively short-range weapons which could be launched by a diesel-powered submarine (Z-class) on the surface – but they went to sea in 1955. By 1958 they had an improved missile[22] in service in both diesel-powered and nuclear-powered submarines. The missile was decidedly inferior to the first Polaris in range (about 300 or possibly 500 miles as against 1400 miles) and probably in guidance accuracy but it was operationally available with a nuclear warhead whereas Polaris was not.

These Russian policies of early operational deployment and of low-technology advance on a broad front coupled with higher technology on a much narrower front (which, however, broadened in response to a massive educational drive to produce scientists and engineers) continued beyond the death of Stalin and well into the Khruschev era. A major change of policy introduced in 1960, however, placed greater reliance on nuclear weapons and greater emphasis on the development of the hardware associated with them. Although this new policy was unpopular with the Russian military and probably contributed significantly to Khruschev's downfall, it was probably also a major factor in enabling the Soviet Union to catch up with the United States (and probably overtake them) in the field of strategic weapon development.[23] In more recent years conventional force improvements have again been given a place in the sun, but the technological momentum of the Khruschev policy has all too evidently not been lost.

In the USA there was also a change of policy. While the

country retained a clear superiority over all others in the quality and quantity of nuclear weapons at its disposal and was reasonably assured of its ability to deliver them effectively, an emphasis on strategic deterrence was at least arguably a suitable policy to adopt. Apart from a brief interlude during the Korean war, when the needs of an army in contact with the enemy forced an increase in the share of the budget allocated to conventional forces, this policy was maintained until in the 1960s it became apparent that American nuclear superiority was no longer overwhelming.

One inevitable consequence of this discovery was an upsurge of interest in the development of defensive and counterforce measures designed to protect the USA in the event of nuclear war. A less obvious consequence was the realization that a conventional war, or even conceivably a tactical nuclear war, might develop under the nuclear umbrella, so to speak; and that such a conflict would have to be resolved quickly in one way or another, using local or immediately available forces, if escalation to all-out nuclear war were to be avoided. This concept has formed the basis of a considerable military debate; but it has also involved a general recognition of the importance of fast-reaction weapons and systems for use at the tactical rather than the strategic level.

Further encouragement for the development of sophisticated conventional weapons resulted from the unhappy experiences of US forces in Vietnam. Early attempts to apply land and air warfare techniques that had been successful in the war against Japan and in Korea were conspicuously unsuccessful both because of the totally different nature of the war and because, in the later stages at least, the Americans came up against several Russian-supplied weapons with some of which they were unfamiliar and which were uncomfortably effective.[24] To these troubles the American defence establishments and industries reacted with characteristic vigour; a great deal of effort was put into the development of conventional weapons and associated sensors and countermeasure equipment and the results include some of the more important military hardware improvements of recent years.

One well-known one is the precision-guided 'smart' bomb : less well-known, perhaps, are the many devices developed for detecting the presence or movement of an enemy in thick undergrowth.

European viewpoint

In Europe[25] the main research and development emphasis in the countries outside the Warsaw Treaty Organization (WTO) has been fairly consistently on non-nuclear tactical weapons and systems. Both the British and the French have developed their own nuclear weapons and the French have developed both land-based and submarine-launched medium-range ballistic missiles and a short-range mobile battlefield support missile (Pluton) as delivery vehicles. All three are operational and there is a continuing program of development work. The British had programs for the development of a land-based medium-range missile, a short-range battlefield missile and an air-launched missile[26] but the first two programs were cancelled before completion and the air-launched missile is no longer in service. The British nuclear ballistic submarines are equipped with Polaris missiles with British warheads and support systems and the mobile battlefield support missiles deployed with the British and some other NATO armies are American.

Great efforts have been made in many European countries, however, in the development of anti-aircraft, anti-tank and anti-ship missiles and of all the reconnaissance, surveillance, countermeasures and communications equipment that goes therewith; and some of the resultant systems are as good as or better than anything that has come out of the USA or Russia. The British Seawolf missile, for example, is a short-range ship defence weapon, currently in the later stages of development, which will give what is probably a better chance of reliably intercepting an incoming anti-ship missile than is offered by any other system known to be deployed or in development.

In part, no doubt, the European concentration on tactical rather than strategic weapons has been encouraged by what is

now the enormous cost of producing a strategic weapon that
is likely to be effective in a modern war environment. Geographical proximity to Russian tanks, aircraft and missile
launching ships has probably been at least as persuasive,
however, if not more so : also some cautious mention should
be made of the commercial aspects – cautious because so
many people are implacably opposed to the idea of making
money by selling arms.[27]

The arms market

What has to be recognized is, first, that there is a substantial
and currently still growing demand for arms among the
newly-independent countries of the world; secondly, that
whoever supplies such countries with arms is often able to
derive political as well as financial benefit from the trans-
action, so that the commercial competition becomes part of
the international power struggle; thirdly, that the adoption,
by NATO, of a particular weapon or system usually means
very big business indeed;[28] and fourthly that a country's
ability to afford to procure a system even by internal manufacture may be dependent upon its ability to persuade others
to share the capital cost by buying some of the equipment.[29]

Although it represents only a small fraction of total military expenditure, the value of the international trade in
arms has been rising rapidly in recent years and shows no
sign of levelling off : a recent estimate[30] put the current annual value at $10–12 billion. The USA takes the lion's share,
followed by the USSR, France and the UK. Another important feature of the trade is the change in the quality as
well as the quantity of arms supplied to 'third world' countries. It was once customary for such supplies to be made up
mainly of surplus or obsolete equipment which the vendor
country was prepared to dispose of cheaply : nowadays,
however, it is often the very latest type of equipment that is
supplied and quite often the supply is a prelude to licensed
manufacture. Arms industries are proliferating.

The quest for self-sufficiency

One reason for this proliferation is the desire of some relatively uncommitted nations for a degree of military self-sufficiency in what they regard as an unreliable international environment. A near-perfect example is Israel who over the years has at various times been armed and denied armaments by Russia, France, Britain and the USA and who has at different times seen her hostile neighbors armed and denied armaments by the same four suppliers. No other country has had quite such a bizarre experience, but South Africa's drive for self-sufficiency has almost certainly been similarly motivated and considerations of a similar type have probably entered into the plans of such countries as Brazil, India and Pakistan and probably many others.

Compared with the activities of the major powers the investment by such countries in military technology is small: nevertheless it is worth remembering that at least two of them (India and Israel) are not far from being able to construct their own atomic bombs – if indeed they have not already done so.

CURRENT TRENDS

From what has already been said it will be apparent that the scale of effort applied to military research and development and the pace and variety of its progress precludes anything resembling a comprehensive survey of the present position or a prediction of the future. The annual output of even non-secret information far exceeds the absorption capacity of any individual; and the 'defence expert' is a mythical creature.[31]

With this important reservation, however, it is possible to identify what appear to be some important trends in what is currently known to be afoot. They are certainly not the only trends; and they might well prove not to be the most important ones if all the facts were known; but at least they appear to be interesting.

First there is the trend towards increasing precision in

weapon aiming and control. Manifestations of this range
from improved methods of aiming a rifle[32] through the use
of laser designation and homing for air-to-ground missiles,
'smart' bombs and laser-guided shells and of remote television aiming using a camera mounted in the nose of a missile,[33]
to the need to score a near-direct hit with one ICBM on
the launcher of another if the latter is to be knocked out.[34]
The military mind has not yet fully detached itself from a
belief in the efficacy of suppressive fire, saturation bombardment or 'massive retaliation'; but there is a growing awareness of the wastefulness of the use of massive firepower and
of the problems of supply which it creates – and which might
well be intolerable in a major war. Moreover, the destructive
capabilities of modern weapons are so great that their excessive use might create more problems for the user than it
solved. Towns are more attractive to occupy than radioactive deserts.

Closely associated with this trend is one towards the more
extensive use of optics in defence systems. Optical techniques were for many years overshadowed by radar techniques because radiation at radar frequencies was more
amenable to manipulation than were light waves. This gap
has now been narrowed, and the relative invulnerability of
some optical systems to jamming is one important advantage; another is that optical systems provide ready access to
what is still the best and most easily produced computer in
the world – the human brain. It is not suggested that the
radar era is ending; but the advent of the laser and improvements in night vision systems have gone a long way towards
reducing its importance.

One recent use of radar for precision guidance which is
certainly worth mentioning, however, is in terminal homing
for long-range missiles. Radar has been used for short-range
anti-aircraft and anti-ship missile homing for many years
but the more recent system – known as TERCOM (for Terrain Contour Matching) – is a self-contained precision
navigation system which scans the ground under the missile
with a narrow radar beam and, by measuring the distance

of the ground as it scans, builds up a contour map of the area over which it is travelling. This it compares with a 'map' which has been prepared from reconnaissance data and is stored in a computer, and correlates the two to determine its position and track. This system is being installed in the Tomahawk long-range strategic cruise missile which is being developed in the USA. It works well over land which has prominent contour features; but it cannot be used over the sea or very flat terrain. An alternative possibility for these difficult conditions is a system (MAGCOM) which detects and maps the variations in the earth's magnetic field; but many problems remain to be solved before this can become an operational system. In the Tomahawk the TERCOM system is supplemented by inertial navigation equipment to form a combined system known as TAINS.

As already noted, a lot of effort is being put into the development of laser weapons. This may turn out to be a blind alley but results so far seem to be encouraging. The big difficulty at present is the provision of adequate power within the compass of a system of adequate mobility : unless that can be done the weapon is likely to be so cumbersome that its usefulness will be severely restricted. One obvious power source is nuclear fission (or fusion); and if the energy conversion problem could be solved the high-power laser could transform the nature of warfare.

Another trend is in the direction of improved reconnaissance techniques. This, too, is associated with the quest for precision : precise information on enemy positions and movements permits the economical use of force. The use of manned aircraft for battlefield reconnaissance missions is still important but the future here seems to lie with the unmanned vehicle and with the passive reporting sensor. A major reconnaissance problem remains to be solved at sea, however, and here it may well be that the manned aircraft has a continuing role to play. The detection of surface vessel movements is fairly readily achieved by satellite reconnaissance but much progress needs to be made in the accurate

tracking of submarines at great depths. Available techniques
are sonar, magnetic anomaly detection and sea-bed sensors;
but all have their limitations and can scarcely yet be handed
over to robot operation. Thermal detection systems can be
used to detect submarines in shallow water; but detectable
symptoms of the passage of a submarine at great depth tend
to be random in position and considerably delayed.

Probably the most promising line of development at
present is in magnetic anomaly detection using cryogenic
techniques for greater sensitivity. Unless the submarine is
very deep and stationary the presence of so large a mass of
moving metal causes a disturbance in the earth's magnetic
field which can be detected by sensitive airborne apparatus :
it is the improvement in this technique that makes the
ability to communicate with a submarine in deep water so
important.

Sonar has already been developed to near the limit of
practical possibility : the most modern transmitters and receivers are probably about as powerful and as sensitive as
can reasonably be expected : the major limitation lies in the
nature of the medium and the effects of turbulence and particularly changes of density with temperature which either
reduce accuracy or in extreme cases make detection impossible. Sea-bed sensors can be used to monitor straits and
other necessary channels of passage; and associated with
'smart' mines (such as the American CAPTOR which is
designed to launch a homing torpedo when a hostile submarine comes within range) they can form a valuable barrier to clandestine operations : the drawback of such
arrangements is that they are obvious early targets in the
event of hostilities and less obvious systems need to be developed.

One final trend that has profound implications for the
future of warfare is that towards the development of more
and more powerful man-portable weapons. A single soldier
can now readily carry a weapon with which he can disable
a tank or another with which he can shoot down an aircraft,
and it is likely that these weapons will be improved and others

of equal or greater potency developed. Within the framework of an organized army such weapons can provide valuable capabilities for front-line troops; but it must not be forgotten that some of the same capabilities are already available to irregular forces – guerrillas, 'freedom fighters' and the like – whose increasing prominence in the world is almost certainly receiving far less official attention from the major powers than it merits.

Debating points

There is little unanimity among military commentators. Although enormous quantities of information are available the only thing that can be said with certainty about it is that while some of it may be the truth it is certainly not the whole truth and even more certainly not nothing but the truth. When to this difficulty are added national rivalries, inter-service rivalries and conflicting political ideologies and moral viewpoints it is scarcely surprising that the result is a large measure of confusion, or that the pages of the world's military journals contain more argument than fact.

This point is made largely because it is important to realize that the selection of material for this study and the comments made on it are essentially personal : there is no received view of the world military situation or of any significant aspect of it. A subsidiary reason is to draw attention to the general problem of communication in defence matters : the requirements of official secrecy generally ensure that the parties to a discussion do not have access to the same body of data (factual or otherwise) so that the ordinary difficulties of human argument are artificially increased.

An extremely important example of a major difficulty in this area concerns Russian strategic missile developments. Whereas some information on less potent Russian weapons filters out from uncommitted countries who have purchased them or as a result of their capture by, for example, the Israelis, virtually all that is known about the Russian strategic missiles has been discovered by American reconnaissance. This puts the American intelligence authorities in the difficult position of having both to interpret the results of their work and to decide to whom they are going to release what information, knowing that the recipients will be unable to cross-check it and will have to make similar decisions. In such circumstances it is no more than natural that personal bias should often enter into the communication equation. Such bias could be extremely important where, for example, the relative merits of deterrence or détente are being debated.[35]

Another debate that suffers from a shortage of hard fact is that concerning the relative merits of quality and quantity in major international confrontations. As pointed out earlier the general tendency in Russia has been to compensate for technological inadequacy by investing massively in relatively simple weapons whereas the Americans have generally selected the high-technology solution to a problem – which has involved heavy capital investment and long development programs. It is generally maintained by the US authorities that they still have a substantial technological lead over the Russians (which is probably true although the way in which it is quantified may not be) and that the way to maintain a military advantage is to ensure that the weapons and systems that it deploys take maximum advantage of this technological lead. In major strategic weapon programmes this may well be a valid argument; but it is certainly arguable that in many other areas the level of technological sophistication aimed at by the Americans is too high and that the resulting weapons and systems are unnecessarily complicated and far too expensive.[36] The fact that US dominance in the NATO alliance tends to draw its allies into the same high-technology net spreads anxiety on this topic beyond the shores of North America.

On the evidence available it would be unwise to say that either the American or the Russian approach to the arms race is the right one. What is certainly true is that, apart from the experience of the Korean and Vietnamese wars, noted earlier, which had an influence on the conventional weapon development programs in the USA, the bulk of the American and Russian military research and development effort has for more than thirty years been devoted to finding solutions for theoretical problems. As one commentator has put it 'though it is customary to speak of the arms race between the United States and the Soviet Union, it is worth considering the possibility that there is no race – just two military bureaucracies, each doing its own thing'.[37]

THE TECHNOLOGICAL FUTURE

Some indication of the likely shape of military developments may perhaps be obtained by extrapolation of the trends discussed earlier in this study. Most of these extrapolations may be left as an exercise for the reader, but it should be said here that if – and it is by no means certain – the laser weapon should be developed as a practical weapon for use over ranges of a few miles, and if it is compact and inexpensive enough to be deployed in reasonably large numbers, it will offer a very serious threat to both armoured vehicles and manned aircraft. It may well not eliminate them from the battlefield – both have been 'eliminated' by other weapons but somehow continue to remain in service – but it must necessarily significantly modify the way in which they are used.

If the laser weapon can be used over much greater ranges, even in a more cumbersome form, it may provide the answer to the ballistic missile, which otherwise seems as likely to be unanswerable in the 1990s as it is today. It seems probable, however, that if the laser weapon is to provide such an answer it will have to do so from earth orbit : it seems unlikely that a beam of sufficient power could be radiated through a damp or dust-laden atmosphere without an unacceptable degree of dispersion.

Laser target marking, on the other hand, may not survive the test of time, not because the system does not work – which it certainly does – but because the requirement for the target to be continuously illuminated by the laser while the precision bomb or shell homes in on it may prove to be unacceptable. What is required is a marking system which will enable the forward observer to mark his target and then withdraw before the shell or bomb arrives, and this may well be found.

Both in this context and in the wider sphere of weapon control development an increased use of television techniques may be expected. The use of optical sensors in conjunction with scanning rasters and sub-miniature computers offers a range of automatic target designation and homing possibilities which has certainly not yet been fully explored.

Major advances in anti-submarine warfare are likely. The physical barrier to sonar detection systems caused by temperature inversion in the ocean cannot be tolerated when as deadly a system as a nuclear submarine may be hiding beneath it. Ocean-bed sensors will probably be widely deployed to detect such targets and long-range unmanned reconnaissance submarines will probably be used both to monitor sonar reports and to seek out other submarines. For combat purposes deep-diving target-seeking torpedoes will probably be used : the best existing weapons are already highly educated and the improvement required will be relatively small.

On the surface of the ocean it seems reasonable to expect the high-speed surface-effect ship gradually to oust the conventional warship and it is difficult to believe that the fixed-wing aircraft carrier has a significant future. Ships have long lives, however, and changes will take time.

Perhaps the biggest problem in this region is the adequate protection of merchant shipping. In the very distant future it may be that the ideas of sub-surface transport that have been discussed for many years will gain acceptance; but for a long time to come European countries at least will have to find ways of protecting large surface transports carrying essential supplies against air, missile and submarine attack. Conceivably this could be done by improving the defensive capabilities of surface vessels – and this is certainly the underlying philosophy of much current development, notably in the USA – but the improvements in anti-ship missile systems that may be expected would seem to make this a hazardous line of approach. In home waters it may be possible to use aircraft, small fast surface vessels and shore and ocean-bed installations to good effect; but for long-range operations the nuclear-powered submarine seems to offer the best platform for convoy protection.

One other problem that may be mentioned briefly concerns the protection of oil rigs at sea. Apart from taking such obvious measures as the provision of local defences, minefields and the like there would appear to be no way of guaranteeing their safety. This is certainly the view currently taken by the Norwegian defence authorities.

Laser weapons apart, it is much more difficult to predict the future of the military aircraft. For as long as the low-level strike aircraft can be so equipped as to survive in the battlefield environment, the special abilities of the human pilot are likely to continue to earn him a place in the military arena; but it is difficult to see a role for the high-flying manned aircraft in operations over land : the number of threats to which it could be exposed is so great that the provision of adequate defences would price it out of any rational budget. In passing it may be noted that this is a current criticism of the extremely expensive Airborne Warning and Control System (AWACS) whose purchase, for air-raid warning duties in Europe, is currently being urged on the European NATO nations by the US authorities.[38]

It seems reasonable to expect further developments in chemical warfare but probably not in biological warfare; the latter is essentially associated with ideas of mass destruction and uncontrollable warfare which seem to be opposed to the more significant modern trends. The use of chemical agents for temporary incapacitation is already a feature of some para-military operations, however, and it could well be that the wider use of such techniques will be contemplated.

From a strategic point of view the use of poison gas is unattractive, quite apart from its illegality. The difficulty of delivering any poison gas in sufficient quantity to have a major impact on an enemy and the uncertainty regarding its probable effect, coupled with the reasonable certainty that its use would provoke severe retaliatory measures, virtually rule it out of any conflict between well-armed nations: apart from its use in the First World War the only known examples are the Italian campaign in Abyssinia, the Japanese war against China and the UAR campaign in the Yemen.

It is, however, possible that some poison gases might be used as tactical weapons; the most likely group being the nerve gases because a respirator alone does not provide adequate defence against them and a lethal or incapacitating dose is small : protective clothing is required to prevent ingress through the skin. Psychic weapons such as LSD are among other possible candidates. In general, however, since all the major powers certainly have the means to bring such weapons into use, and since most of them have agreed not to be the first to do so, it would seem that the near-certainty of retaliation is likely to deter the would-be user in any but the most exceptional circumstances. Nevertheless, such considerations are unlikely to inhibit the further development of either chemical or biological weapons.

Similarly, there is no sign of an abatement in the development of improved strategic ballistic missiles. The Americans are working towards a twofold increase in accuracy for the Minuteman missile and have new land-based and submarine-launched missiles in development for which accuracies are likely to be redoubled and perhaps doubled again : the Russians have maintained a steady flow of new types of missile; the Chinese programs have slowed and speeded up more than once but it seems unlikely that they will stop. Having regard to the real and supposed discrepancies in the capabilities of the three powers and to the existence of actual and potential nuclear threats from other directions, it seems very unlikely that the development work, at least, will stop – even if procurement and deployment are checked.

Developments in the application of new materials to the solution of military problems of mobility and protection may be confidently expected. The development of lighter and stronger materials is, of course, essentially one of general rather than specifically military technology, although the initial funding of such developments is often provided from defence budget sources. Two recent examples of special developments in the UK will serve as pointers to the future, however : one is the new type of vehicle armour which is said to be highly resistive to all present anti-tank weapons; the other is a photochromic glass which reacts to intense radiation by becoming opaque so quickly that it can protect the eyes from the flash of a nuclear explosion.

Another important line of development which is likely to continue is in the provision of personal armour. Jackets and trunks which give a large measure of body protection against bullets and fragmentation weapons are already in widespread use and visors and shields are familiar features of civil disorders. The garments now in use interfere very little with the wearer's freedom of movement and more comprehensive protection with no added inconvenience may be expected in future.

Finally, some mention should be made of the paranormal. Although this subject is dealt with elsewhere in this book, it is worth noting here that the possibility of using extrasensory perception for military purposes has not been neglected – although it has probably received less attention than it deserves. Radar, sonar and optical techniques can at best tell the user where something was a short time ago : to know where it will be would be of great benefit to a defender; and no reasonably open-minded observer can deny the possibility that ESP techniques might be developed to provide such information.

The problematical future

Whatever means of destruction or defence the scientist and the engineer may provide now or in the future, what actually happens will be determined by people rather than hardware. Attitudes of mind, personal and collective hopes and fears, understanding and misunderstanding and all other human emotions and thought processes are ultimately what determine the conduct of human affairs. Push-button war depends on the decision to push the button and the finest and most accurate weapon in the world cannot be expected to work well in the hands of a man who is terrified out of his wits.

To expect a spontaneous change in human nature to usher in an age of peace is futile : left to its own devices the collective and individual human mind will continue to harbor aggressive tendencies and the competitive instinct. It might be possible to eliminate such innate characteristics artificially; but such a cure would be worse than the disease.

It is not a counsel of despair to postulate a continuing human tendency to engage in occasional wars or to expect that a large part of the world's manpower and materials will continue to be devoted to preparations for war : the human race has on the whole managed to contain this problem with remarkable success and will probably continue to do so. Nevertheless, those at least who value the qualities of the general run of human existence owe it to themselves to discourage some of the more extravagant possible manifestations of the aggressive instinct such as all-out nuclear or biological warfare.

It is unlikely, however, that sufficient discouragement will be provided by formal agreements such as those between Russia and America on strategic arms limitation : no Russian leader dare abandon a potential advantage in such circumstances and American leaders must also look over their shoulders from time to time. A disturbing feature of the present confrontation is the resemblance of the NATO attitude to the Maginot concept of the inter-war years : a more flexible series of alliances and accommodations covering a much wider area might expose the irrelevance of the nuclear arsenals to any rational redistribution of world power.

For all that, nuclear war may yet come. Those who are given to apocalyptic pronouncements are fond of pointing out that the world's nuclear arsenals have the potential to exterminate the human race, and no doubt they are right –

provided the human race obligingly lines itself up in appropriate extermination patterns and provided all the
weapons work. It is probably equally true to say that there
are enough rifle and machine-gun bullets, mortar bombs and
artillery shells currently stored around the globe to achieve
the same result – with the last gunner committing suicide.
That this will not happen is obvious : it is perhaps less obvious, but it is equally true, that nuclear war would not exterminate the human race even if – as is most unlikely – it
were continued until the last missile had been fired.

Nuclear war is terrible to contemplate and the prospect
of all-out strategic nuclear war is such stuff as nightmares
are made on; but the human race has already shrugged off
some 60 million war dead and millions more in purges already in this century. It will survive.

REFERENCES

1. John E. Dawson, *An American View of Defence Management*,
   in *The Management of Defence* ed. Laurence Martin (Macmillan, 1976), p. 50.
2. Estimate for 1975 published in *World Armaments and Dis-
   armament: SIPRI Yearbook 1976* (annual publication of the
   Stockholm International Peace Research Institute), p. 127.
3. One such panacea, the more interesting because it was propounded by a very distinguished scientist, was put forward by
   Professor Joseph Rotblat (Secretary-General, Pugwash Continuing Committee) as one of a series of twenty-year forecasts
   published by *New Scientist* in 1964. He advocated, in essence,
   the formation of international advisory councils of independent scientists to whom international political problems
   would be referred for the objective formulation of solutions
   based on computer assessments of popular opinion which
   politicians would disregard at their electoral peril. What
   persuasive force such recommendations would have on the
   leaders of one-party states was not explained. The series was
   republished by Penguin Books in 1965. *The World in 1984*
   ed. Nigel Calder, vol. 2, p. 118.
4. Supporting data for the estimate in *SIPRI Yearbook 1976*
   (q.v.) yield the following percentages of the global total for the
   six biggest spenders in 1975 :

	%
USA	30·01
USSR	28·57
China	6·13
W. Germany	3·62
France	3·14
UK	3·08
	74·55

It is reasonably certain that this ranking order is correct for the year in question and the figures for the four NATO countries are based on well-documented information. The Russian contribution is less certainly known, however, and that for China is of very uncertain reliability.

5. In his *Annual Defense Department Report: FY 1976 and FY197T* (February 1975) the US Secretary of Defense, James R. Schlesinger, introducing a budget proposal containing $10·3 billion for US military research and development in FY 1976, claimed (pp. 1–5) that the USSR was spending 20 per cent more under this heading than the USA. This estimate gives a datum of some $22 billion for the two superpowers to which can confidently be added at least $2 billion for France, West Germany and the UK, and a large number of smaller contributions must bring the total to at least $25 billion, without making any allowance for Chinese expenditure which cannot be negligible. An estimate published, without supporting analysis, in *SIPRI Yearbook 1975* (p. 102) put the total for 1974 at between $20 and $25 billion.

6. *Resources Devoted to Military Research and Development* (Stockholm International Peace Research Institute, 1972) provides a valuable analysis of expenditure in some twenty countries from the mid-1950s to 1971.

7. This estimate is little more than an educated guess based on what is known of the military expenditure, gross national product and working population in five of the most important countries. Some confirmation is provided by an extrapolation from what is known of defence employment and expenditure in the UK.

8. *Research and Development and the Prospects for International Security* by Frederick Seitz and Rodney W. Nichols (Crane Russak, 1973) contains estimates of the numbers of

scientists and engineers employed on defence-related research and development projects in the USA (p. 23). They range from 114,000 in 1954 through 209,000 in 1961 to 149,000 in 1971. See also note 9 below.

9. *SIPRI Yearbook 1975* (p. 102). This figure is quoted without supporting analysis and should perhaps be viewed with some caution, but it cannot be far wrong.

10. Dawson (op. cit., Note 1) provides an interesting brief analysis of the change in the US attitude to external military involvement during this century.

11. The similar, though much smaller, infrastructure created in the UK reacted in much the same general way. Here and elsewhere in this study concentration on the activities of the USA and USSR is not intended as a belittlement of the importance of the activities of other nations: it is simply that the scale of the superpower activity in terms of recognizable products dwarfs that of all other countries. Some of the most important fundamental technological advances have been made in the smaller countries: it would be difficult, for example, to overstate the importance of the British development of the cavity magnetron to the success of wartime radar.

12. Von Braun and most of his associates decided in January 1945, when the imminent collapse of the Third Reich was apparent, that they would move south from Peenemünde so that they could surrender to the American forces rather than to the Russians. A good first-hand account of the touch-and-go operation that secured much of the German rocket equipment for the USA (subsequently christened Operation Paperclip) appears in *History of Rocketry and Space Travel* by Wernher von Braun and Frederick I. Ordway III (Thomas Nelson, 1967).

13. In September, 1949, a month after the ratification of the North Atlantic Treaty. It had been recognized in 1946 that the US monopoly was only temporary, but it had been thought that it would take the USSR about ten years to develop their weapon. The subsequent development of the Russian hydrogen bomb only a year behind the Americans came as an even greater surprise.

14. It was, however, the scene of the first combat between jet aircraft. In November, 1950, a MIG–15 was shot down by a Lockheed P–80 Shooting Star.

15. An account of the growth of the American strategic missile

programs is given by von Braun and Ordway (op. cit., Note 12), p. 120 et seq.

16. The pattern of reconstruction is surveyed in some detail in *The Soviet Navy Today* by Captain John Moore (Macdonald and Jane's, 1975), p. 19 et seq.

17. It had been recognized even earlier in Germany where the first rocket launches from a submerged submarine were made as early as the summer of 1942. See Ernst Klee and Otto Merk, *The Birth of the Missile* (Harrap, 1965), p. 92.

18. A cruise missile is one that travels under power – commonly propelled by a turbojet or ramjet engine – all the way to its destination. A ballistic missile is under full power only in the early part of its journey after which it travels under the influence of external (mainly gravitational) forces only, except that it may be provided with small motors for trajectory correction or terminal guidance.

19. The second missile, known to the West as the SSN-2 or *Styx*, was the first such weapon to be used operationally (by the Egyptians against the Israelis in 1967).

20. With it they also abandoned ship-to-ship guided missile developments until the beginning of the present decade, by which time many such devices had been developed in other countries, mainly in Europe.

21. An inertial navigation or guidance system is one which depends for its operation on the measurement, in three dimensions, of the accelerations to which the vessel or vehicle is subjected and hence computes its velocity and thence the distance travelled in any direction. To make these measurements and calculations with the order of accuracy required for the Polaris program a highly sophisticated piece of apparatus is required: there is also, for a SLBM, the problem of transferring navigation information from the ship's system to the missile system right up to the moment of launch – so that the missile 'knows' not only where it is to go but also where it is when it starts. Being entirely self-contained, inertial systems are not susceptible to jamming in the way that radio and radar systems are: like the sailor's compass and log, however, they are dead-reckoning systems and their errors are cumulative. To maintain accuracy, therefore, an inertial navigation system must be corrected periodically by reference to some absolute position-fixing system.

22. Known in the West as the SSN-4 or *Sark*. Hardly any Russian

missiles are known by their Russian names: names such as *Sark* are NATO inventions and the SSN–4 designation comes from an American alphanumeric code which is widely accepted outside Russia.

23. See, for example, Ken Booth, *Soviet Defence Policy in Contemporary Strategy, Theories and Policies* (Croom Helm, 1975), p. 210 et seq.

24. The Israelis had a similar experience in the 1967 war and a more acute one in 1973.

25. It should perhaps be made clear that 'Europe' is to be taken to include Scandinavia. Sweden has a substantial and broad-based defence industry and the country maintains a high level of sophisticated non-nuclear armament. The Norwegian industry is much smaller but a good deal of attention has been paid to naval defence systems.

26. Code-named Blue Streak, Blue Water and Blue Steel respectively. Few argue nowadays about the cancellation of Blue Streak but many would contend that the Blue Water cancellation (American missiles were purchased instead) was a mistake.

27. Above a basic level of moral hostility attitudes seem to vary cyclically. Before 1914 the arms trade was as respectable as any other. Between the First and Second World Wars the successful dealer in armaments was the archetypal villain of popular adventure fiction. In comparatively recent times he became, in Britain at least, something of a hero but has since gone into a decline again. Most countries place some restrictions on the export of arms; but it is worth noting that two countries, Sweden and Switzerland, who not only operate severe restrictions in this respect but also are high in the international peace-making league, nevertheless derive considerable income from licensing weapon manufacture to their designs in other countries.'

28. The decision, in 1975, by four European NATO countries to purchase the American F–16 fighter instead of the competing French and Swedish aircraft was described, without excessive hyperbole, as the 'arms sale of the century'.

29. A spectacular, but by no means isolated, example is AWACS, the enormously and quite possibly excessively expensive American airborne (radar) warning and (interception) control system which is on high-pressure offer to European NATO countries at the time of writing.

30. *SIPRI Yearbook 1976*, p. 16.
31. The most comprehensive generally-available reference book on the larger weapons and systems in use or under development is *Jane's Weapon Systems* published annually by Macdonald and Jane's, London. It does not deal with small-arms and certain other infantry equipment, however, nor significantly with ships or aircraft as such. For these the sister annual publications, *Jane's Infantry Weapons, Jane's Fighting Ships* and *Jane's All The World's Aircraft* may be consulted. The last two have been in continuous publication for many years and their earlier issues are useful source books for historical data. All these books, however, derive their information from 'open' (non-secret) sources and thus can only guess at the facts relating to some of the systems that they describe.
32. Several systems, based on a Swedish invention called 'single-point', are available. See, for example, *Jane's Infantry Weapons, 1976*, p. 619.
33. A useful survey of precision guidance and some related topics can be found in *New Weapons Technologies – Debate and Directions* by Richard Burt. Adelphi Paper No. 126 (International Institute of Strategic Studies, London, 1976).
34. The problem and the recent state of the art are discussed and some useful data provided in *Strategic Survey, 1974* (International Institute of Strategic Studies), p. 46 et seq.
35. Information on Russian and Chinese strategic missiles was also a major input to the great debate on the desirability of installing a comprehensive ABM system in the USA. See, for example, the extensive series of papers in *Implications of Anti-Ballistic Missile Systems*, Pugwash Monograph II (Souvenir Press, 1969).
36. For a vigorous criticism of many US weapon programs see David T. Johnson, *Overview of the Military Budget* in *Current Issues in US Defense Policy*, ed. Johnson and Barry R. Schneider (Praeger, 1976), p. 115 et seq.
37. John C. Garnett, *Some Constraints on Defence Policy Makers* in *The Management of Defence* (Note 1), p. 45.
38. See Note 29.

ASTRONOMY AND SPACE SCIENCE

Peter Beer

Peter Beer is a Physics graduate of the University of Manchester. After holding research and teaching posts he is now a full-time writer and broadcaster on science subjects.

For the last 10 years he has produced weekly science magazines – *Science in Action* and *Discovery* for the BBC World Service. He is also joint Editor of the international review journal *Vistas in Astronomy*.

Turning points

No field of science and technology has seen more remarkable changes within the span of our lifetime than astronomy and the exploration of space. We are living through not one, but three revolutions – turning points at which Man's knowledge of the Universe acquires a new dimension.

The best known is also the most recent – July 21st, 1969. On that day two men stood for the first time in history on the surface of another planet – the Moon. In itself, a tiny step into the vastness of space – but Neil Armstrong was not exaggerating when he called it a great leap for Mankind, as his feet touched the surface.

But what of the other two revolutions within our life-time? One at least can be dated with almost equal precision – and it began with one of those chance observations which, seized on by an alert mind, have so often led to a great stride forward in science. It was in 1932 that Karl Jansky, a radio engineer at the Bell Telephone laboratories in the United States, was studying the interference to radio communications caused by thunderstorms. He noticed that the level of interference increased regularly every day. And by careful observation he was able to show that the radio signals were coming from the Milky Way – that great belt of stars that stretches conspicuously across the sky. It was the first-ever detection of radio waves from space, and it was to lead to a profound advance in our knowledge of the Universe.

For all we can ever see of it, from Earth, is with the aid of whatever radiations can penetrate our atmosphere; and in the whole vast spectrum of electromagnetic waves, there are only two relatively narrow bands which can pass through it without being heavily absorbed. Visible light, and a part of the radio spectrum, are our only two windows into the world beyond the atmosphere – and until A.D. 1932, only one of those windows was open.

Astronomy has been practiced since the times of the Druids and the ancient Greeks and Egyptians – certainly for a few thousand years – but the new science of radio astronomy is little more than 40 years old. And already it has transformed our understanding and revealed whole classes of objects whose existence was never before suspected.

A new window into space

The second astronomical revolution in our times can now be seen as a logical next step. If we can 'see' only some light and radio waves through the atmosphere, and yet we know that there is a whole continuous range of radiations, from the very shortest X-rays and gamma-rays up to the longest radio waves, then the only way left to widen our knowledge is by getting above the atmosphere, or at least sending suitable instruments there. And this is just what became possible in the 1940s with the development – largely for military purposes, by German scientists – of rockets that could reach heights of 50 miles or more. Simple sounding-instruments were carried aloft on rockets, in fact, soon after the war, when the United States began a research program at its White Sands proving grounds, based at first on captured German V–2 rockets, but soon developing more powerful versions of their own under the guidance of Werner von Braun and others.

The idea, at least, of getting above the atmosphere is not new : astronomers have been placing their instruments on high mountain sites since the last century, at such places as the Pic du Midi in France and the Lick and Mount Wilson observatories in California.

In fact, pioneering balloon flights in the late 1930s by physicists trying to measure cosmic rays were the first real attempt to rise above at least the bulk of the atmosphere. But the real stride forward came with the successful launching of artificial satellites. It was the Soviet Sputniks 1 and 2, in October and November 1957, and the American Explorer 1 in January 1958, that really heralded the dawn of the Space Age. Explorer 1, incidentally, though much smaller than the Sputniks, was instrumented and planned from the outset as a science satellite, and in fact made at least one vital discovery – the existence of the Van Allen radiation belts in the upper atmosphere, belts of charged particles that greatly affect radio observations. From these modest beginnings, the new sciences of X-ray and ultra-violet astronomy have developed extremely rapidly, as will be seen.

Sun and planets

Man's first cautious ventures beyond the surface of this planet, to which he has been confined throughout time like some wingless insect imprisoned on a lamp globe, will be described later in this chapter. But before we look at the way in which men in our generation were the first of all to break that barrier, what *is* this Universe that so beckons the imagination that we must at all costs try to explore it?

The first revolution in our ideas about the cosmos did not begin in our own times but some 500 years earlier – admittedly only an instant in the great stretch of time, but long enough in human terms. It began with an obscure Polish cleric who served his time as a Church and civic official in Frombork, a small town on the Baltic, not far from Danzig. In his leisure hours he, like many before him, grappled with the problems presented nightly to anyone who looks at the starry sky : what are they, where are they, what holds it all in place, what are the laws that govern the movements of time and season?

The man was Nicolaus Copernicus, and the great book setting down his thoughts was published in 1543, in the last months of his life.[1]

The decisive step he took was a great leap of imagination, for he challenged what had till then been the most natural – if also the most conceited – of ideas : that our Earth is the center of the Universe, the center of God's Creation. It was an idea that, naturally enough, found favor with the Church – so much so that, even a century later, when Galileo Galilei insisted that the Sun stood still while the Earth revolved about it, only his age and dignity, and a forced recantation, saved him from being burnt at the stake. But, like it or not, the new picture of the Universe had arrived. Galileo turned the newly-invented telescope on the heavens for the first time in 1609, and what he saw and described at once provided telling evidence for the new order of things. And the genius of Kepler and of Newton soon clothed these ideas in a mathematical structure of such firmness and clarity that its essential truth could no longer be denied. The universal law of gravity, laid clear by Newton, provided a master plan to explain the motions of the Sun and planets that has stood virtually unshaken to the present day.

Stars and galaxies

In the three centuries since then, our horizons have continually widened. Galileo's observations, Newton's and Kepler's calculations, had provided a valid explanation of the complex movements of the Earth and other planets as they travel round the Sun in elliptical orbits of successively greater diameters, at precisely determined rates of revolution – and in this way built up the clear and easily grasped picture of the Solar system that we are now all familiar with.

They also explained the motion of the satellites, such as our own moon and the four principal moons of Jupiter, and showed that all satellite systems obey the dynamical laws of Newton and Kepler – the same laws, indeed, that are used today with the help of great computers to calculate in a fraction of a second the trajectories and orbits of spacecraft and artificial satellites.

But there was no explanation of the role of the fixed stars, so called because – apart from their nightly journey across the sky due to the Earth's rotation – they do not appear to move at all. It was appreciated quite early, by a mixture of logic and intuition, from the very fact of their immobility, that these fixed stars were much more distant than the planets. The ancient Greeks imagined them as carried on some great crystal sphere, surrounding the whole planetary system like a curved backdrop in a theatre. It was indeed not until the last century that it became possible to measure the distances of some of the nearer stars, and thus show that they were not all at the same distance from us, as the sphere idea demanded.

This was first achieved by Friedrich Bessel in 1838 using the principle of parallax, which simply means the movement of nearer objects relative to more distant ones, in just the same way as when we look out of a moving train the nearby hedges and trees pass across in front of the distant landscape. In our case, the movement of the train is represented by the earth's movement from one end of its orbit to the other. The measurement is easy in theory and very hard in practice, simply because all the stars are so distant that even for the nearest of them, the parallax movement is so very small – never more than one second of arc, or 1/3600 of a degree, in a whole 6 months' movement. And this is only a two-thousandth part of the full moon's diameter. It can, however, be detected by carefully comparing two photographs of a region of the sky, taken 6 months apart, when the tiny movement of a few of the nearer stars becomes just measurable.

Island universes

The most urgent and difficult task confronting the 18th- and 19th-century astronomers was to try to detect a pattern in the arrangement of the stars – to map the Universe, as it were. As Galileo had shown, the Milky Way in fact forms a vast belt of stars stretching right across the sky; and slowly the idea grew that our Sun, and we with it, form part of this vast association of stars, which became known as the Galaxy. Plotting its shape, however, was extremely difficult, working – as we have to – from a position right inside it, but finally it was pictured as a flattish disc, rather like two soup-plates placed face-to-face.

A clearer idea of the Galaxy's possible shape came eventually from an unexpected source. A few of the 'fixed stars' had long been recognized as rather peculiar, fuzzy objects quite different in appearance from normal stars, and they were given the name nebulae, meaning clouds. The 103 brightest of them – some, like the great nebula in Andromeda or that in Orion, visible to the naked eye – were first catalogued in the 18th century by the French astronomer Messier. In the years around 1800, William Herschel and his sister Caroline observed and described about 2000 of these objects. But their true nature remained obscure. In 1845 the Earl of Rosse built himself a giant telescope; with a 6-foot mirror, it was far and away the largest in the world, and indeed remained so for almost half a century.

Rosse looked at many of the nebulae and made striking drawings of what he saw. Fifty years later, when the first photographs of nebulae were taken, they showed far more detail of the faint outer spiral arms than could ever be seen by the human eye. Modern photographs of the nebulae first drawn by Rosse reveal them in all their splendour.

So the idea took root that our own Galaxy, too, might have a spiral shape, though it was not until this century that this was finally shown to be the case. More importantly, however, the work of Rosse and the Herschels set off fierce arguments about what and where the nebulae were. Slowly the view gained strength that many – though not all – of them were not part of our own Galaxy at all, but were 'island universes' – other galaxies, far off in the depths of space.[2]

Lighthouses in space

The problem of the galaxies turned out to be a tough one, not finally settled until well into the present century. The great problem was to try and estimate the distances of these objects, which (as we now know) are very remote indeed. The parallax method that worked for nearby stars was of no help – the galaxies *are* in fact the 'distant landscape' which serves as the truly fixed background for the nearer stars – but other methods were gradually devised. In these methods, what is done is to look for some property – some special characteristic - of a recognizable class of objects. A typical example, discovered in 1912, is the class of variable-brightness stars called the Cepheid variables : it turns out, remarkably, that the rate at which their brightness fluctuates is an exact

measure of their actual brightness. These objects, whose real
brightness is thus known with fair accuracy, are scattered
through space like a series of 'standard candles' or light-
houses.

The further away they are, the fainter they appear to us,
and so by measuring their *apparent* brightness it is possible
to estimate their distance. In this way, a series of different
methods was evolved, each with its own limitations but each
in turn pushing the limit of measurable distance further and
further out.

It was one of the great triumphs of modern astronomy
when, in 1929, Edwin Hubble at the Mount Wilson Observatory finally succeeded in locating some of these light-
houses in space within a few of the nearer nebulae, and thus
proved conclusively that they really are separate galaxies,
far beyond the confines of our own. But he did far more than
this, as we will see. It had been possible for quite a long
time, curiously enough, to measure the speed with which a
star or a galaxy is moving towards or away from us by the
amount of reddening of the light reaching us from the star –
an effect discovered in 1842 by Christian Doppler. The star-
light collected by a telescope can be split into its component
colors by a prism, and the spectrum so formed is found to
show a number of sharp lines due to the presence of particular chemical elements in the light source. The lines are
shifted towards the red end of the spectrum by an amount
proportional to the speed with which the star is receding
from us. This is the famous red-shift which we will come
across again; and with its help, it became clear that many of
the stars, and all the distant galaxies, are indeed moving
rapidly away from us.

It was this that led to Hubble's second great discovery ...
he compared the distance from Earth of all those galaxies
where this could be reasonably estimated, with their speed
of recession, as measured by the red-shift – and found that
the two quantities were in strict proportion. The further a
galaxy was from us, the faster it seemed to be receding.
Now this simple and beautiful result has been subject to a

vast amount of discussion and research since Hubble's time; it has been modified in detail, but fundamentally is still thought to be true. And the beauty of the result is that it is exactly what we would expect to observe in a *uniformly expanding* universe – in other words, one where every distance, between every pair of objects, is steadily increasing with time – just like a giant balloon being gradually inflated. The more distant regions of space, being already more widely separated, clearly have to move apart faster to maintain that uniform expansion.

Origins

This idea gave rise to a ferment of discussion and thought about the real structure and the origins of the universe – the study of which we term cosmology. Prominent among the astronomers who tried to explain these observations were the Belgian Abbé Lemaitre and the Cambridge mathematician Sir Arthur Eddington and Sir James Jeans. Now the obvious conclusion, if the Universe is steadily expanding now, is that once upon a time it was highly concentrated and occupied only a very small space. Then, something – perhaps a colossal explosion – happened, to start the galaxies on their long journey across space. This theory – called, for obvious reasons, the Big Bang theory – is still very much in favor, and as we shall see new evidence from radio astronomy lends it strong support – so much so that despite many other attempts and theories, most astronomers today believe that the Big Bang comes closest to explaining all the observed facts.[3]

An intriguing variant of the theory was first put forward by Lemaitre. He suggested that, at some time in the distant future, the expansion will stop, to be followed by a phase of contraction as all the matter in the Universe falls back again under its own gravitational attraction, until it occupies once more the very compressed small region where it all began. Then there could be another Big Bang and the process would start all over again. This at least has the philosophical merit of avoiding such awkward questions as 'what came before the beginning ... and what happens in the end?'

At present, there is simply not enough evidence to confirm or deny the oscillating universe theory – but one vital piece of evidence now being provided by radio and optical astronomers is just how much invisible matter there is between the stars – clearly a vital factor in working out if the gravitational forces will be enough to halt the Universe's expansion. So cosmologists are hopeful that, with this evidence and the help of nuclear physicists, they can study the likely events in the first few seconds after the explosion, and so come to take a position on the theory within the next decade or two.

A new science: radio takes a hand

By 1930, with Hubble's twofold triumph to crown the intensive study of the galaxies that the new instruments had made possible, astronomers had good reason to feel a great sense of achievement; order had at last been created out of chaos, though much of course still remained to be done. The report that an obscure radio engineer had detected an unusual source of static while working on thunderstorms passed virtually unnoticed. Even when a second and more powerful radio telescope was built – entirely by the efforts of one man, the American Grote Reber – its results drew little attention. The more urgent needs of war overshadowed all else – yet, paradoxically, in the end they gave the new science its greatest boost. The techniques first developed for radar proved invaluable; and even in the darkest days, some important discoveries were made – again by accident – and noted by the alert minds of British radar scientists like J. S. Hey and his collaborators. They were the first to pick up radio emission from the Sun, and to observe how it increased dramatically whenever a solar flare disturbed the surface of the Sun; and soon, they also detected echoes from meteor trails in the upper atmosphere.

It was not long before the world's first radio astronomy observatory was set up, in a muddy field in Cheshire near the village of Jodrell Bank, where the University of Manchester had a botanical research station. A young lecturer named Bernard Lovell took two ex-army trailers full of surplus radar equipment there – and radio astronomy got its first real home. As a young student I vividly remember looking up at the first large aerial that they built, and thinking what a fearsomely unstable arrangement of masts and guy-ropes it looked. But this was only the prelude to greater things. Lovell's ambition was to build a really large aerial that was fully steerable so that it could be aimed at any part of the sky. It was to be a giant bowl with a diameter of 250 feet. Such an engineering project had never been attempted before. The trials and tribulations that took place before it was finally completed have been brilliantly described by Sir Bernard himself;[4] enough to say that in the end, he and his consulting engineer, H. C. Husband, triumphed over all the odds.

Almost before the great instrument was working, it had an unexpected chance to show its mettle. By one of those fortunate coincidences that mark the story of science, there came a curious link with the second part of this story – the exploration of space. For several years, Lovell's dish was the biggest and most sensitive steerable aerial in the world; and it was just what the Americans, and the Russians, needed to track their first rockets and satellites sent into space. With a fine impartiality Lovell put his telescope at their disposal, with brilliant success.

Observing radio stars

Of course, Jodrell Bank did not long remain the only radio observatory; others were started at Cambridge, in Australia, Holland and the United States, and now there are dozens of active research centers for this new branch of astronomy. Not all are equipped with the familiar great dishes. Valuable as these are, they have one fundamental limitation, they cannot restrict the signals they receive to as narrow a beam as astronomers would like in order to define the position of a radio source to a very small fraction of a degree. To do this would require a dish of several miles in diameter, an obvious impossibility. Various tricks have been used to get round this problem. One of the most ingenious was devised by

Britain's Astronomer Royal, Sir Martin Ryle. He uses two or more quite small aerials suitably arranged on a long straight baseline to imitate the effect of a huge aerial as wide as the baseline – which in the case of the Cambridge instrument, is three miles long. Although the aerials only cover a very small part of the baseline, they can be moved along it to a number of different positions in turn. At the same time, the earth's rotation causes the aerial beams to sweep out great circles in the sky. Thus, slowly but surely, the same effect is built up as if the telescope was a giant dish three miles across.

It was not long before the new radio observatories springing up round the world began to produce results – and results that were more spectacular than anyone could have imagined.[5]

Remarkably strong radio signals were picked up from Jupiter, and weaker ones from the Moon and the other planets. Probably it is Jupiter's powerful magnetic field that accounts for this difference. But much more significant was the discovery of a small, sharply defined radio source in the Milky Way, in the constellation Taurus. This was one of the strongest radio sources in the sky, and it was soon identified with the Crab Nebula.

We recognize this as the remnant of a supernova, a colossal explosion of a star which grew to such astonishing daytime brilliance that it was recorded in detail in the Chinese chronicles for the year 1054 A.D. This nebula has now been studied in great detail and has become a fruitful source of data for astronomers trying to understand the mechanism of these exploding stars.

The riddle of the quasars

The Crab Supernova, fascinating though it is, is a relatively near object in our own Galaxy. Perhaps one of the most remarkable aspects of radio astronomy, however, has been its ability to reveal very distant objects – the most remote in the Universe, as it turns out. It was not long before the observers found clear signs that some of the signals being received were coming from far beyond our Galaxy. When their positions were compared with the very detailed sky maps or surveys taken with the great optical telescopes in America and Australia, it became certain that many of these sources were in fact located in distant galaxies. We now believe that all normal galaxies emit radio signals, but rather weak ones – with much less radio energy than is contained in the visible light from the galaxies. But it also soon became clear that there were some non-conformists that emitted vastly more radio energy, ten to a million times as much as an average galaxy. These, naturally enough, were christened radio galaxies; and some of them are so powerful that we can detect them across unimaginably vast distances.

But then, in the early 1960s, came another and even more baffling find. Some of the most intense sources of all could not be identified on the sky survey photographs with any visible galaxy; at best, their position seemed to coincide with some quite insignificant-looking starlike objects, certainly very much more compact than any ordinary galaxy. Consequently, these sources were named 'quasi-stellar objects' – quasars for short. And they posed enormous puzzles. For a start, their red shifts showed them to be moving away from us at almost unbelievable speeds. Some of the most distant are travelling at over 90% of the speed of light. According to Hubbble's law, this also makes them the most distant objects in the Universe. And this in turn means that they must be pouring out simply incredible amounts of energy. The riddle of where all this energy comes from is one of the main questions in astronomy today.

The quasars are of great interest for a second reason. The light and radio waves that reach us from the most distant quasars have been travelling across space for over 20,000 *million years* – so what we are in fact seeing today is what they looked like, all that time ago. For all we know, the quasars have long since ceased to exist. In other words, *we are looking now* at objects as they were soon after the creation of the Universe – that is, if we follow the now generally accepted theory of the Big Bang. And it is this fact that gives the quasars their fascination for cosmology. We will undoubtedly learn a great deal more about the quasars, and hence about the very beginnings of our Universe, in the next 10 to 20 years.

Big Bang or continuous creation?

In 1948 Fred Hoyle, one of the most brilliant – and most controversial – astronomers of our time, was searching for a viable alternative to the Big Bang theory : a theory that seemed logical enough, but also left unanswered some key questions, both mathematical and philosophical. It seemed to Hoyle and two of his closest colleagues, Herman Bondi and Tommy Gold, that there *was* an alternative that would also account for the observed fact of the expansion of the Universe. Supposing, said Hoyle, *new* matter is continuously being created in the vast depths of space? If this is so then it could form new galaxies, in the end, to take the place of those that, driven by the relentless expansion of space, were actually disappearing from our range of sight – vanishing forever from our view.

A pretty idea and no more, if it had come from anyone else : but it was backed by the detailed and quite convincing calculations of three of the finest mathematicians alive. It was to be nearly 20 years before Hoyle, the stubborn York-shireman, finally had to agree that the new evidence gathered by the radio astronomers was too powerful to be withstood. Today, as we have seen, almost all cosmologists accept the Big Bang as the nearest to the truth that we can get at present.

What was this evidence that finally settled the great debate? The first clue came when the radio astronomers working with Sir Martin Ryle at Cambridge made systematic counts of the numbers of radio galaxies at various distances in space. There seemed to be relatively far too many of them at great distances to fit in with Hoyle's theory. But the final blow came – once more – almost by accident. Two American researchers, Penzias and Wilson, detected in 1965 a faint but unmistakable background of radio noise, present everywhere in space. Cosmologists thought they knew what this was : a sort of remnant from the Big Bang, when of course enormous amounts of energy were radiated in all directions. Now, more than 20,000 million years later, the primeval fireball radiation would have cooled down almost – but not quite – to zero : in fact, according to theory, to just 3° above the absolute zero of temperature. And that is precisely what Penzias and Wilson had picked up. The proof was convincing to all but a very few astronomers, and the Steady State theory was set aside. But in passing, it is as well to recall the remark by another distinguished cosmologist, Martin Rees of Cambridge : 'Being stimulating – as Hoyle has been – can often be much more important than being right.'

Cosmic clocks: the pulsars

The strangest, almost certainly, of all the discoveries of the radio astronomers happened nearly ten years ago. It was in 1967 that Dr. Anthony Hewish and a young researcher, Jocelyn Bell, working at Cambridge, picked up signals quite unlike anything that had been found before. They were pulses – as regular as the ticking of a clock – so regular that Hewish's first thought was that they were some kind of man-made interference, or possibly secret military transmissions. Even when all this was ruled out, there remained one bizarre possibility : that they were signals deliberately beamed out by another civilization far off in space – 'the little green men', as Hewish jokingly called them. It was not until they were quite sure that these *were* natural signals, from some quite new kind of star, that Hewish released the news to the scientific world. Then the hunt was on, and within a few months several more pulsars – short for pulsating stars – were detected. At the last count over 100 had been found, with pulse frequencies varying from about one to thirty pulses per second – but with each individual pulsar keeping its rate steady with rock-like precision.

How can we account for this strange behavior? It now seems clear that the pulsars are very tiny stars, only a few miles across; this is inevitable if they are rotating rapidly,

as we believe they must be to account for the pulsating signals. But although small, they contain about as much matter
as a normal star like the Sun. The matter is incredibly dense
– a piece the size of a sugar-cube would weigh several million
tons. They are 'neutron stars' – composed of densely-packed
nuclear particles. Normal atoms are largely empty space,
their central nuclei kept well apart by their outer shells of
circling electrons; but in a neutron star, intense gravitational
pressure has caused the star to collapse, overcoming the
electrical forces, and sweeping away the electron shells so
that only tightly-packed nuclei remain.

As to the pulsation itself, here argument still rages, but one
popular theory is that a beam of radio signals is being
emitted from one region on the star, and as it rotates the
beam swings round with it like a lighthouse beam which we
can see only as it sweeps past us. Whatever the ultimate
answer to the pulsar puzzle turns out to be, it is certain that
pulsars will soon tell astronomers a great deal about the life
cycle of the stars, for they may well turn out to represent
one of the last stages in the death of a star.

So, are there little green men?

This seems an appropriate point to put this question. The
pulsars have been seen to be natural if bizarre in origin; but,
in all seriousness, what *are* the chances of other intelligent
beings in the Universe – and perhaps, ultimately, of our communicating with them? These are among the most hotly debated questions in science. The conditions that have to be
met to make life possible – a stable star with a system of one
or more planets, some of them at the right distance to avoid
the worst extremes of temperature – will rule out the large
majority of stars. But such is the number of stars in our
Galaxy that most astronomers now agree the odds are very
high that a number of them will have planets with conditions not unlike our own. And if life *can* start, why *should*
it not? Strong support for this idea has come from recent
observations of radio and infra-red radiations from the inter-

A BLIND MAN SEES

This San Francisco student has been blind since birth. A miniature TV camera sees images which are transmitted to *stomach* nerves through the waistband instrument. These nerves transmit the picture pattern to the brain which activates the visual center of the brain and the patient sees!

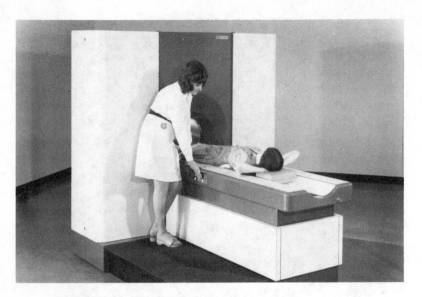

EMI BODY SCANNER

The top picture shows the Scanner in use which can scan an exact 'slice' through the patient in 20 seconds. Details invisible to normal X-ray – such as the brain, lung, pancreas, etc., can be seen.

The bottom picture shows a 'slice' through the chest and details of bone, fat and the heart (central).　　　　© *EMI*

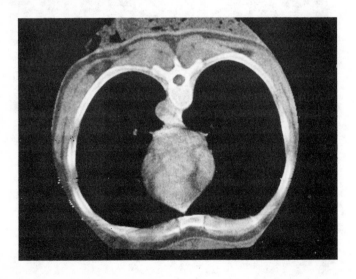

SPACE SHUTTLE TO BE LAUNCHED FROM 747 BACK
This photograph shows the launch method of the Space Shuttle
which was being tested in 1977.

The first operational flight will be 1979 and the re-usable space
ship is forecast to ferry up to 2000 people to orbiting platforms
and stations. © *NASA*

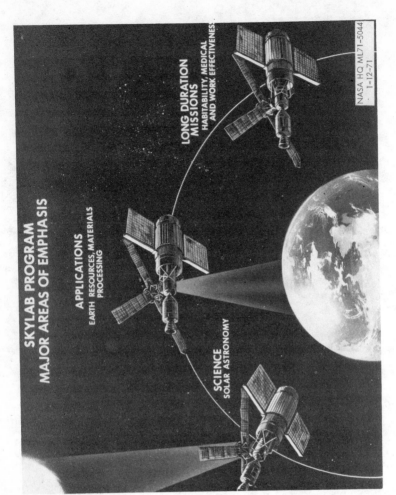

SKYLAB PROGRAM
MAJOR AREAS OF EMPHASIS

APPLICATIONS
EARTH RESOURCES MATERIALS
PROCESSING

LONG DURATION MISSIONS
HABITABILITY, MEDICAL
AND WORK EFFECTIVENESS

SCIENCE
SOLAR ASTRONOMY

NASA HQ ML71-5044
1-12-71

SATELLITE USAGE
The three major uses of satellites. Solar astronomy has been given an impetus with the realization that in 1983 a conjunction of planets will occur that is almost unique and is likely to have an enormous 'pulling' effect on the sun's flares. It is believed this could have a very important effect on the world's weather since the only previous similar conjunction was approximately 10,000 years ago – which is picturesquely the suggested timing for the Biblical Flood story. © *NASA*

D

GIANT SPACE STATION PLANNED FOR 1980's LAUNCH
The Grumman Aircraft Company has designed the station to carry
50–100 men in the outer modules.　　© *Keystone Press Agency*

E

FIREPOWER AND GADGETRY

The top photograph illustrates the weapon capacity of the RAF Phantom jets – carrying 4 short range 'Sidewinder' and 5 long range 'Sparrow' air to air missiles plus 5 cluster bombs – each of which ejects over 100 bomblets against troop concentrates.

The bottom depicts a remotely piloted WISP helicopter developed by Westland for battlefield surveillance but civilian usage is possible.

F

FIVE-MAN RESEARCH SUBMARINE
A 24-ft 'Beaver' submarine built to explore at 2000-ft depths –
note the remote grabs and arms.   © *Keystone*

UNDERWATER HOMES FOR MID-ATLANTIC
The idea of crews living for weeks in the 12-ft linked glass-
ceramic and titanium spheres shown in the artist's impression
would have looked futuristic even months ago. The launch of
'Beaver' – previous page – puts the probable date of test for this
permanent structure as the early '80s. A service submarine will
rotate crews and ferry supplies.   © *Keystone*

H

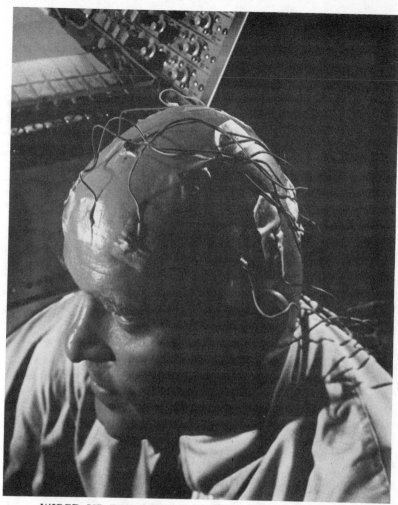

WIRED UP FOR SYNTHETIC EXPERIENCE

Electrodes can both monitor brain activity and alpha, beta, theta waves to measure altered states of consciousness. They can also be used to directly and accurately stimulate specific points of the brain to create the illusion of experiences – visual, sound and even sexual.

HEALER   PATIENT

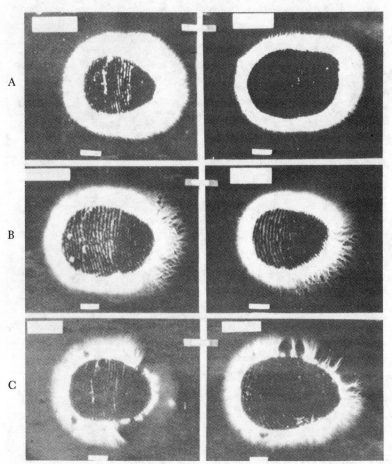

A

B

C

THE HUMAN AURA – PHOTOGRAPHED!
A, B and C show three separate series of Kirlian photographs of
spiritual healer Yehuda Isk's fingers and the effect on a patient.

Column 1 shows Isk's fingers *before* the healing session,
Column 2 the patient. Column 3 shows the patient's corona

J

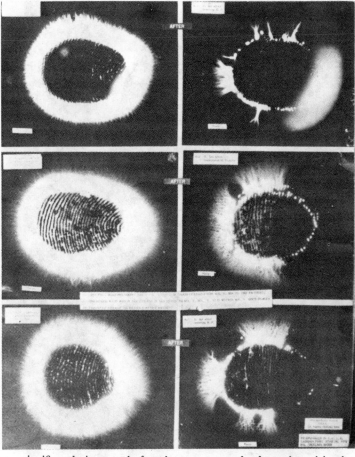

PATIENT   HEALER

significantly increased after the energy transfer from the spiritual
healer whose emanations of bio-energy have almost vanished –
Column 4.

The tests are from a large number conducted under stringent
conditions by US researcher Dr. Thelma Moss.

© *Dr. Thelma Moss*

K

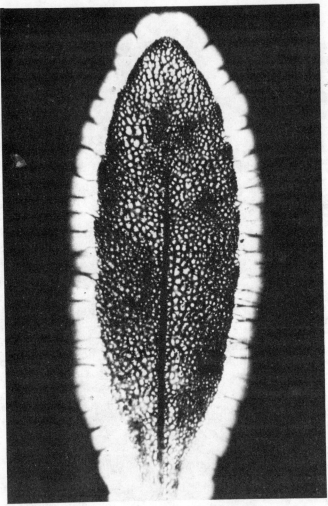

PLANT ENERGY BODY PHOTOGRAPHED
The 'Aura' of two leaves as photographed by the Kirlian process.
This dramatically shows the separate 'energy body' which exists
in plants as well as humans.   © *Novosti Press Agency*

M

ELECTRONIC AUTOMOBILE DIRECTION FINDING AND AUTOMATIC CONTROLS
The Marconi installation in a police car (the MADE system) showing the touch map and alpha-numeric keyboard, display and printer for instantaneous information from the mobile computer terminal.
© *Marconi Research Laboratory*

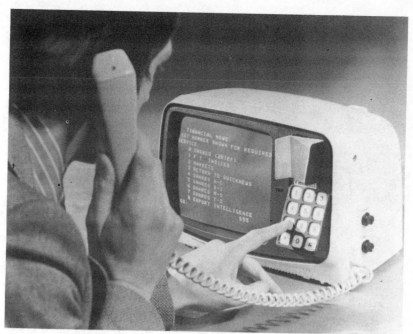

VIDEOFILE - *Unlimited information at the end of your phone*
By 1978 Videofile will be a public service. Harnessing the telephone and an ordinary TV set or a purpose built set as shown on these two pictures. You will dial a telephone number which links your TV set with constantly updated information. By pressing a keyboard to call up the information required, the data is displayed on the screen.

Information will range from news, sport, hobbies, train times, classified advertisements, recipes, personal messages and later to answer specific questions and problems. Whole encyclopaedias and reference libraries will ultimately be 'accessed' by Videofile.

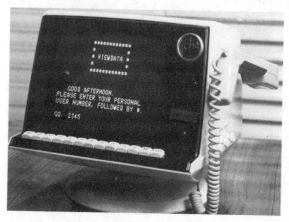

RAPIER ULTRA-LOW-LEVEL DEFENCE MISSILE
Deployed by NATO, the picture also shows the comparatively
small size of today's missiles – similar size or even smaller missiles carry tactical *nuclear* warheads and can be launched from
trucks!
© *BAC*

stellar gas. It turns out that this is not pure hydrogen, but also contains some of the key molecules containing oxygen, carbon and nitrogen that we believe gave rise to life in our own solar system.

So the general view is that life – and therefore in all probability intelligent life – does exist, probably in many places. In that case, what chance do we have of getting in touch with it (if that is what we want), or even of learning of its existence? Here one has to make an assumption, which may or may not be justified : that other civilizations, or some of them, may be advertising their existence, so to speak, by sending out radio signals. And we also need to assume that they (being highly intelligent) have worked out at least a short-list of stars that may support life, and included our Sun on that list. For in the vastness of space, it is absolutely essential for any artificial radio signal to be beamed towards its target. All this will have shown, no doubt, that the odds are in fact stacked pretty high against our being able to pick up signals of this kind – but that need not stop us from looking, or rather listening, for them, if we can do so at reasonable cost.

The first serious attempt was by a well-known radio astronomer, Frank Drake, back in 1959. In his Project Ozma he turned a radio-telescope on two of the nearer stars, and – perhaps not surprisingly – found nothing. Now, however, both NASA and the Soviet radio astronomers are planning new and more ambitious programs. The NASA one, called SETI for short – Search for Extraterrestrial Intelligence – involves using existing radio telescope aerials with a new type of receiver, designed to scan through up to a *million* different frequencies in what is thought to be the most promising part of the radio spectrum.[6] Even this has only a modest chance of success – but, as Frank Drake says, 'The trend for SETI is upwards.'

In passing, one should add that many people – including some very eminent scientists – are happy enough about *listening* for signals, but rather worried about the suggestions that have also been made for CETI – *Communication* with

Extraterrestrial Intelligence. Typical of this concern is the comment by Sir Martin Ryle, one of the world's leading radio-astronomers. The chances of success are remote, he says, but if there *is* such contact, then just possibly another civilization, much more advanced in its technology than we are, might see in us a tempting site for colonization. So perhaps we should think twice before we advertise our presence!

The invisible Universe

The story of how astronomers in the last two decades have been able to open a new window on the Universe through the use of invisible radiations is a striking illustration of the effects of space exploration – with which the rest of this chapter will be concerned. So before we turn to space science as such, let us look very briefly at this remarkable widening of the astronomers' horizons.[7, 8] The first clue came quite early: in 1949, when an experimenter put a piece of photographic film, wrapped in opaque foil, in the nose-cone of a V–2 rocket. The film, when recovered, was blackened – and the cause eventually turned out to be X-rays emitted from the Sun. It was another dozen years before further rocket flights led to the discovery of an X-ray star, in the constellation Scorpio. This was totally unexpected; the Sun's output is relatively weak, and it should have been quite impossible to detect a similar source at any great distance. So evidently the new X-ray star was very different from the Sun.

The next great step forward came with the advent of a satellite called Uhuru – the Kenyan word for Freedom, for it was launched on the new Republic's Independence day, December 12th, 1970, from the coast of Kenya. Uhuru started a new era in astronomy, for it detected dozens more X-ray sources. With those found by the later satellites, Copernicus and the highly successful British Ariel 5, we now know of over 100 X-ray stars – and a peculiar bunch of objects they seem to be. All of them must be extremely hot stars, giving out great amounts of radiation; some of them fluctuate with time, posing more puzzles for the astronomers trying to explain them. But already it is clear that many of the strongest are binary systems – two stars in close company, so close in fact that matter is being pulled from one star to the other as they rotate in their orbits. Moreover, one star in the pair is almost certainly a very small, highly condensed star – a 'white dwarf', as it is called, or maybe the even more compressed neutron star we have already met as the source of the pulsar's radio waves. There is even the possibility – though it is too soon to say for certain yet – that we may actually be seeing evidence of one of the weirdest of all astronomical ideas : the Black Hole.

Point of no return

Of all the concepts of modern cosmology none, surely, is more bizarre than that of the Black Hole. This is the point at which gravitation – the attracting force that keeps the planets in their orbits – finally overwhelms all the other forces that normally keep it in check : even the immensely powerful electrical repulsion that exists between the fundamental particles of matter, and which give the atom its immense stability in normal circumstances. When, and how, does this happen? As the nuclear fuel of a star is exhausted, it collapses in upon itself more and more, under the influence of its own weight. It may end up as a white dwarf, with its atoms tightly packed together, or as a neutron star, where the condensation is so great that the outer electron shells of the atoms are broken down, and the nuclei are pushed tightly together. This is what we believe makes up a pulsar – the neutron star's rotation, and its intense magnetic field, combining to emit tremendously strong regular bursts of radio waves.

But sometimes – not always – the collapse can go even further. The conditions – the size and mass of the star – have to be just right, not too little, not too much, as the advertisements used to say; but then the gravitational field is so strong that literally nothing can stop it, and the collapse of the star continues inexorably right down to a point – a point from which nothing, neither matter nor radiation, nor light, can ever escape. And since radiation cannot get out, the star is quite literally invisible, though it contains billions of tons of incredibly hot matter – hence its name, a Black Hole. What is more, so strong is the gravitational field that anything material – a cloud of gas molecules, an asteroid, even perhaps an unfortunate space traveller – that wanders too near to it will be sucked remorselessly in, to disappear for ever from our Universe. Time and space would play strange tricks on our unhappy spaceman, too; as he approached, he would be stretched out like a string of spaghetti – before merciful oblivion in the Black Hole overtook his component atoms.

Are these weird objects any more than a mathematician's flight of fancy? Is there *any* evidence that they really exist? Well – yes; cosmologists believe that the Black Hole not only follows inevitably from their calculations, but is also the best way to explain some of the latest observational evidence – such as the pulsating X-rays that satellites have picked up. So scientists are indeed coming more and more to believe in the existence of this strangest of invisible universes.[9]

Heat rays from space

We think of infra-red or heat radiation here on earth as emitted by hot bodies; and indeed it is used in this sense to determine the effective temperature of growing crops in surveys by satellite. But in astronomical terms it is connected with the coolest objects. We have all noticed how the color of a gas fire or an electric heating element changes as it heats up, from dull red through orange to yellow, almost white heat; and in fact a typical star like our sun is at about six thousand degrees C, and there are much hotter, bluish stars. For these, their main energy is radiated in the blue and even ultra-violet. But the infra-red radiation comes from the coolest objects in the heavens. With it we are literally looking at an invisible universe – and a very exciting one, because it is out of the cool gas between the stars that we believe new stars are created.

This explains the fascination of making observations in the infra-red, observations which have only recently become effectively possible. Much of this success has been due to the pioneer development of flux collectors, as infra-red telescopes are often called, by Prof. James Ring of London. He designed and operated the very successful sixty-inch telescope on Tenerife and is now leading the team finishing the construction of the world's biggest flux collector, forty times as sensitive as any other. It has a hundred and fifty-inch mirror and is being installed on a fourteen-thousand-foot mountain top in Hawaii. Looking further ahead, balloon- and satellite-mounted instruments are already being planned to get above even the very tenuous atmosphere of the mountain observatories. There's no doubt that this will be one of the fastest-growing fields of research in the next twenty years – as also will be observations in the almost completely un-explored ultra-violet region. These wavelengths are practically all absorbed by our atmosphere, so it is only from sounding rockets and satellites that worth-while observations are becoming possible.

What way ahead?

Astronomy, as has already become clear, will benefit immensely from the coming advances in space satellites and space platforms, shortly to be described. But it would be a mistake to think earthbound astronomy is at a standstill. Indeed, it has seldom been as active. There have been striking developments in observational techniques which have virtually transformed the scope of the subject. First there has been a surprisingly large improvement in the optics of telescope systems. Even without launching out into space, telescopes today have gained both in light-gathering power and in their precision – the fineness of detail they can resolve. And this has been accompanied by a parallel development in the light-sensitivity of the photographic materials used. It has even become possible to take color photographs of many quite faint objects in the sky – some of these beautiful pictures are reproduced in a book recently edited by a Swiss astronomer, H. Rohr.[10]

Photography is no longer the only method now available

for studying faint objects. Very sensitive image-detectors have been developed from the photomultiplier tubes used in various light-sensing applications; some of these devices are so sensitive that they will respond to only one or two 'photons', as the basic packets of light-energy are called, so they are in fact approaching the theoretical maximum of sensitivity.

Finally, the computer is now beginning to play a major part in astronomy, both in making possible calculations which would have been completely out of the question without its help; and in helping the astronomer to cope with the great mass of observational data the new telescopes are producing. To take just one example : a Schmidt-type telescope as used in a sky survey will record tens of thousands of star images on a single photographic plate. To measure them all individually would take literally months of effort. But automatic star-measuring machines linked to computers, like the GALAXY machine developed by the physicist Peter Fellgett, and its successor, COSMOS, have made the task possible. In fact it is said that COSMOS can measure more stars in one year than all that had ever previously been observed and recorded.

The leap into space

Let us now turn to the physical exploration of space – a science entirely of the 20th century (if we except the visions of such far-sighted writers as Jules Verne and H. G. Wells).

There is no room here to give a detailed history; in any case, the main facts are well known. The German rocket pioneers of the 1920s and 30s, whose work was soon to be sadly perverted in the interests of the Nazi war machine, showed that it would soon be possible to give a rocket sufficient acceleration to break clear altogether from the Earth's gravity. Simple mechanics shows that this will happen as soon as a certain velocity, the so-called escape velocity, is attained; and for the Earth this is about 11 kilometres per second (in more familiar terms, 25,000 miles per hour). Fast, but not impossibly so, even for simple chemically-fuelled rockets. The first major successes – the Sputniks and Explorers – have already been noted – so the end of 1957 marks the real start of the Space Age. In the mere twenty years since then, not only have literally hundreds of unmanned satellites been put into orbit, and probes sent as far into space as Jupiter – 500 million miles from Earth; but also, several dozen men – and a couple of women – have ventured into space, some as far as the Moon. What are the possibilities for further exploration?

Destination – the Galaxy?

'Space Travel is Bunk!' That off-the-cuff remark was uttered only twenty years ago by the then Astronomer Royal – Sir Richard Woolley. And he was not allowed to forget it when only a dozen years later three men boarded a rocket at Cape Canaveral and next set foot outside it on the surface of the Moon. Yet, was Sir Richard all that wrong? Compared with the vast distances of space – even no further than the nearest of all the billions of stars that might have its accompanying planets – the journey to the Moon is just an almost insignificant first step. A first small step, indeed, in Neil Armstrong's words as he jumped down onto the lunar surface. That was July 20th, 1969. How have things changed since then? A handful of men have ventured as far as the Moon – and twelve of them have stood on its surface, but beyond that, only instruments and unmanned rockets. Nor are there any plans at present for further manned flights beyond earth orbit. At present, if we think of space travel as the exploration of deep space beyond the confines of our solar system, then it is indeed bunk – and it's now clear that it will remain so for our lifetime, if not for much longer. Consider the facts for a moment. Suppose the target was just one of the very nearest stars. Barnard's Star, for instance, might make a good target: it is, in fact, the second nearest of all, and most significantly, the painstaking research over many years of Peter van de Kamp at the Sproul Observatory, near Philadelphia, has indicated strongly that it has one or more small, invisible companions – planets, in other words. Now Barnard's Star is 6 light-years from the Earth. That doesn't sound much; but a light-year is the distance light travels in a year, and at 186,000 miles per second, that amounts to almost six million million miles.

Put that against the 240,000 miles to the Moon – or even the 200 million miles or so to Mars – and the figure acquires more meaning. In short, Barnard's Star is about 30 thousand times as far from us as is Mars – and it took the Viking spacecraft almost a year to get that far.

So, with present chemically-fuelled rockets, it's not even a remote possibility. Sir Richard was undoubtedly right. But what if new, vastly more powerful rocket systems can be devised?

New power for new tasks

The first step is to apply nuclear technology to the propulsion problem. NASA has been experimenting for many years with Project NERVA – Nuclear Engine for Rocket Vehicle Application – and has shown that it is possible to generate a powerful thrust by exhausting at high velocity hydrogen which has been heated up in a nuclear reactor. Although this design of rocket still calls for the propellant supply of hydrogen to be carried with it, the quantity required is only about half that needed by present-day rockets. The NERVA motors have not yet been flown, but ground tests some years ago were successful. It has been calculated that this type of power unit would greatly reduce journey times for missions within the solar system, enable larger payloads to be carried and make easily possible sample-return missions to Mars.

Looking further ahead, however, space scientists see the best hope in the so-called ion motor, in which the propellant is a stream of charged particles – ions – accelerated to enormous speeds by passing them through a strong electric field. The mass of the particles would be much less than that of present fuels, but it is the momentum – the product of mass and velocity – of the propellant that gives a rocket its thrust. The net result could be a motor with a modest thrust, but able to work for a very long time – maybe for months on end. In this way the speed of the rocket could be built up gradually, over a year or so, to a formidable figure : not the few thousand miles per hour of existing rockets, but hundreds or even thousands of miles per second. But even then – if we assume all this has been done – the journey to Barnard's Star would take something like 100 years. Impossibly long for a human being; a long time, even, to wait for a piece of rather expensive hardware to travel – and it would be asking a great deal, too, of the radio engineers to fit it with a powerful enough radio transmitter to signal back news of its arrival – a signal that would itself, of course, take another 6 years to reach us.

Celestial Rip van Winkles

Altogether, the outlook for flight outside the solar system must be regarded as remote – certainly one must look much further than 25 years ahead. But that is not to say, of course, that it will never come about. A few years ago Werner von Braun gave his views on this in a BBC radio interview, and incidentally reminds us of a very interesting consequence of Einstein's relativity theory :

'Will men ever be able to go on interstellar travel? Well, I have learned to use the word "impossible" with utmost caution. Theoretically, if you travel at the speed of light – which of course you can never reach – time would stand still completely. If you could travel at say 99% of the speed of light, then only a week or so elapses in the spaceship while a year passes for the outside world ... so you could travel to a star 1000 light-years away, and return to the earth only a few years older ... the trouble is that when you return, several thousand years have elapsed on earth. So you are still a young person, but you meet your own great, great, great ... grandchildren – and for them you are almost a prehistoric animal.'

And if you need to come back?

Rip van Winkle brought up-to-date! Clearly the human and psychological problems, if such journeys ever do take place, would be appalling. But what *is* possible, or even likely, during the 15 to 25 years that we may reasonably try to predict? There is certainly the possibility of manned flight to another planet in our own solar system : to Mars, for instance, where conditions seem at least less totally hostile than on the scorching surfaces of Mercury or Venus. Some will disagree with me, but I find even this possibility rather remote – that is, within the 25 years we have taken as our arbitrary limit. This is not so much because the rocketry can't be built, but because the problems of enabling a human crew to survive a journey of 18–24 months in a very cramped capsule seem rather formidable. No : I believe that, for some long time to come, Man will confine his planetary exploration exclusively to instruments – though he may very well attempt within this century to bring some samples back to Earth for study, as he already has done from the Moon.

Let us leave the problems of manned flight with two thoughts. One : is it really necessary for men to go, if we can develop instruments as sophisticated as those sent already on the Mars missions to carry out analyses, and even return samples to Earth? And two : if manned long-distance flight *is* contemplated, might it not in fact be a woman – who weighs less and needs less food – who takes her place in the space-craft?

Machines, not men

Unmanned flights into space are quite a different matter. We will certainly see quite a number of these within the next two decades. The first in line will be an ambitious exploration of Venus, the launch of which is scheduled for summer 1978. The spacecraft will orbit Venus and send five probes down to descend slowly through the dense, hot atmosphere of the planet. Then, there will be a return to Mars : there will certainly be further landings with more sophisticated instrument packages, including an elaborate chemical laboratory controlled from Earth. Undoubtedly before very long this and other planetary missions will include the use of roving vehicles to gather samples – successors to the Russians' Luno-

khod, and the Moon Rover taken with them by the crews of
Apollo 16 and 17. Planning on some of these projects is well
advanced : Dr. James Martin, the Viking Mission Director,
told me recently, 'We already have a Moon Rover available – it will not be difficult to adapt this for Martian use –
and we would be able to go in 1981 if we get the go-ahead
soon.'

The more distant planets – Jupiter, Saturn and perhaps
Uranus – will also be explored at close quarters with fly-by
missions to follow up the very successful Pioneer missions
of 1975. The first two of these flights Voyagers One and
Two were launched successfully in 1977 and should, before
attempting to look at even more distant Uranus and
Neptune, give us our first good close-up look at Saturn
with its mysterious rings.

One ambitious project, the so-called Grand Tour of the
Solar System was, unfortunately for space enthusiasts, a vic-
tim of the budget cutbacks of the early 1970s. It had been
planned for launch in 1979, when a particularly favorable
alignment of the planets would enable it to use Jupiter's
gravitational field to help it reach the outermost planets –
Uranus and Neptune. The mission would have been active
for 8 to 10 years as it penetrated almost 3000 million miles
into space. A similar flight could quite possibly be revived
for the later years of this century and would give us our first
close-up report on these still little-known outer planets. But,
whether it is or not, one can say with confidence that future
unmanned flights will between them greatly extend our
knowledge of the solar system.

Is it all worth while?

When all is said and done, it must be admitted that most if
not all of the astronomical and space research described so
far has one purpose, and one purpose only – to satisfy Man's
curiosity about the universe he lives in. Not that this is in
any way to be scorned; without the spirit of exploration man
would be a poorer creature, both spiritually and materially –
for often in the end the adventure undertaken just for the

hell of it, 'just because it is there', as men said of Everest, has produced some practical and quite unexpected bonuses. Nor should it be forgotten that the great revival of astronomy 300 years ago had strictly practical aims – to help the seaman find his way in safety. 'With the stars, Man is not lost.'

Even so, we cannot expect that the actual exploration of space will advance mankind's happiness or well-being very greatly (except of course for the pleasure it undoubtedly gives those involved). And the vast resources, of both money and human skills, might well be put to better use in a thousand ways, down here on our polluted, overcrowded and resource-hungry planet. *That*, I suspect, was the real meaning of Sir Richard's curt remark, 'Space travel is bunk.'

There is, however, another side to the coin : a side that *will* develop, to the great good of mankind, even if changes in public opinion turn policy away from the more ambitious forms of space exploration, at least until we have solved our most pressing problems on Earth. Let us look for a while at the rapidly growing applications of satellite and near-space technology, and the varied and practical ends that they are already beginning to serve. Communications; weather prediction; crop and mineral surveys : the list is already long – and we are only at the beginning.

Putting space to work

It is little more than a dozen years since the first generation of satellites designed for specific, practical applications began to be placed in orbit. Many are given low or medium orbits, in which they circle the Earth at a height of a few hundred miles every 90 minutes or so. Typical of these were the TIROS and NIMBUS satellites, 17 of which were successfully launched, carrying increasingly refined sensing instruments to gather such data as temperature profiles of the upper atmosphere and complete global cloud cover pictures. But satellite meteorology took a great step forward in 1974 with SMS–1, the first of the global network of stationary meteorological satellites. Why stationary? All that this means is a satellite launched not into low orbit, but into one at a height of just over 22,000 miles. At this height the period of rotation becomes exactly 24 hours, what is called a synchronous orbit – for the satellite then rotates exactly in step with the Earth and stays always above the same point on the surface. From this vantage point it maintains a 24-hour watch on a quarter of the Earth's surface, scanning the ever-changing pattern of clouds and radioing complete images to its ground stations every half an hour. Comparing one image with the next gives the weathermen the vital information on wind-speeds that they need.

What does this mean for the art and practice of weather forecasting? Quite simply, the world's atmosphere is so vast – in depth as well as extent – that the 5000 ground stations and 600 weather ships and buoys in the Northern hemisphere have never been able to provide more than a fraction of the information needed to give a really accurate picture of the changing weather pattern – and thus make reliable forecasts possible. And even this existing network is extremely costly to maintain. Now, almost at a stroke, the available information is being multiplied many times over by the geostationary satellites. A second American one was launched a year after the first; a very powerful one, METEOSTAT, was launched in November 1977 as Europe's contribution; and there are plans for a Soviet and a Japanese satellite too. Between them these five satellites will be able to keep the whole world's weather under constant surveillance.

This is the basis of the World Weather Watch program – a vast network of ground stations, satellites, and data processing centers. And that is the second key to the program: to handle all the data will require the use of the biggest computers in the world. A formidable task; but with their help, the weathermen are hopeful that they will in a few years time be able to generate really accurate long and medium term forecasts for the first time. I share their optimism, provided that we are ready to wait a reasonably long time – ten or twenty years – for results of real value.

Something that will not have to wait twenty, or even ten

years to feel the impact of satellite technology, however, is telecommunications – for here, the revolution has already arrived. Starting with Telstar (remember Telstar?) back in 1962, we are now into the third, and vastly more powerful, series of communications satellites – the INTELSAT series. The story is told in detail in another chapter of this book: sufficient to point out here that they have already dramatically changed the whole pattern of communications, making possible the transmission of far more information of all kinds than would ever have been possible with the limited capacity of cable links. To give just one telling example, *over 800 million people* were able to watch the Montreal Olympic Games, live, as they happened. And nearly as many watched Man's first steps into space, in the Apollo missions.

All this may seem of more interest to a manufacturer in Detroit or a meteorologist in Bracknell than it is to the teeming millions in the cities and villages of Asia; the peasant farmers who don't know what a satellite is, who only know the endless struggle to scratch a bare living from the parched – or flooded – soil. But space is helping them, too, in ways they know nothing of, but which will do more than anything else over the next two decades to fight off the spectre of world famine. Two satellites are showing the way : originally called ERTS, for Earth Resources Technology Satellites, now simply known as LANDSAT 1 and 2. They went aloft in 1972 and early in 1975, and each now circles the Earth every 103 minutes, at a height of some 560 miles.

On each pass the television cameras and infra-red sensors on board cover a strip just over 100 miles wide, running almost but not quite North–South. So the strip slowly moves across the Earth and covers the whole globe every 18 days. With the two LANDSATS now in orbit, this means that every point on Earth can be photographed once every nine days. The images and data are sent down to three ground stations in the United States, one in Canada, and shortly to several others around the world – and from there the data is flashed to computer centers which can handle well over 1000 images a week. In fact, in its first $2\frac{1}{2}$ years in orbit,

LANDSAT 1 sent down over 100,000 images. Perhaps the most important point to make about all this information is that once the pictures have been processed, they are made freely available to anyone – any organization or government in the world – that wants to use them. And already there are over 100 users, in the United States and forty other countries, basing research and development programs on LAND-SAT data.

The uses of this information could quite easily fill the rest of this chapter : let me sum them up in a couple of sentences. By making maps of all kinds, plotting floods and forest fires, revealing new mineral deposits, surveying soil types and following the day-to-day progress of growing crops, the LAND-SATS are already proving of tremendous value – and especially to the developing countries, who need this help so badly. For these are just the regions, often poorly surveyed, for which orthodox surveys are both too slow and too costly. In Mali and Bangladesh, in Bolivia and the Sudan, in a score of other lands, the results are beginning to come in – results of real value in planning the food production and development of the world. And once more, we are only at the start of the road.

Technology for space

The actual development of the space program has also brought with it undoubted advances – the 'spin-off' sometimes rudely dismissed as the better non-stick saucepan. But it is in fact a great deal more than that. One corner-stone of modern electronics, for instance – the miniature integrated circuit used in pocket calculators – was first developed in the United States specifically to meet the demand for very compact and reliable circuits for spacecraft. No doubt it would have come, anyway; but not as quickly, had it not been for the stimulus of the space program. Then there have been a whole group of new materials, notably the new ceramics extremely resistant to heat shock. Developed first for the heat-shields that protect space capsules as they re-enter the atmosphere, they now have many uses, from the ceramic cooking ring through to key parts of modern aircraft.

The third field in which there has been significant advance is medicine. Not only have space flights taught doctors a great deal more about the human physiology, and endurance, in the curious conditions of weightless flight; they have led to the earlier development of a whole range of measuring equipment, notably the various telemetry devices which are worn by the astronauts to transmit reports on their body functions such as temperature, heart-rate and so on, to the controllers on the ground. This is technology that has already found its way into the hospital ward.

We have also taken the first steps towards a new technology of materials manufacture. For certain purposes, the space environment offers unique advantages. There is easy access to a near-perfect vacuum; but more important, certain things can be done only in the gravity-free environment of a spacecraft in free flight. Casting a perfectly spherical ball-bearing, for instance. However hard we try, it is just not possible to do this here on earth. Existing bearings are very good indeed – they *are* round, to within a few micro-inches – a few millionths of an inch; but even this is not quite good enough when those bearings are the essential part in an inertial navigation system, which has to be accurate to within a few hundred yards after a journey of perhaps millions of miles. This kind of manufacture – small-scale but important – may well be the first industry in space, and could easily come about within fifteen years, as the Space Shuttle program now to be described develops.

Working in space

Because Man is not likely yet to go to Mars – or live and work on the Moon – does not mean that he will remain entirely earthbound. Far from it; the brilliantly successful improvisation of the Skylab program is about to give way to much more sophisticated and ambitious plans. Skylab, as you will certainly recall, was a Saturn rocket stage ingeniously adapted to provide quite adequate living and working space for three or four astronauts for periods of several weeks. In it, the present endurance record for men in space – 84 days – was set up, and a whole range of observations and scientific experiments were carried out with great success. But access to Skylab was by the same crushingly expensive techniques of rocketry that have figured in all spaceflight up till now. The key to bringing down those costs – dramatically, by a factor of ten and more – is the new concept of re-usable space vehicles. The Space Shuttle, a cross between rocket and jet-liner, will make a rocket-assisted take-off and carry materials and men up to purpose-built space stations in orbit a few hundred miles up; and after transferring its cargo, it will glide back to earth and make a controlled land-ing on a suitable airstrip. With limited maintenance, it can then be refurbished for another flight, and another, and another – maybe up to 100 in all. The first Space Shuttle is already built – it was rolled out of its construction hangar for the first time in September 1976 and its flight testing has already begun. NASA plans to build five of these vehicles, and from the first manned orbital flight round about 1979, they will be the space workhorses of the 80s. Their great asset, apart from reducing dramatically the cost of taking men and materials into orbit, will be their versatility – they will be able to carry a 15-ton payload aloft in a huge cargo bay, some 60 feet long.[11]

What is this cargo to be? It could – and certainly will – take many forms, from smaller instrument packages to be launched into space to the components of quite large struc-tures to be assembled in orbit. (Already both Russian and American space crews have tested welding techniques in space very successfully.) But the Shuttle's principal object, in the early years, will be to carry aloft and service various models of orbiting laboratory. The first of these, Spacelab, is already being prepared for launch on an early Shuttle flight, probably in 1980. It represents Europe's most im-portant commitment so far to space, as it is entirely designed and financed by the European Space Agency, of which Ger-many, France, Italy and Britain are leading members. There will be room on board for a variety of science and technology experiments, which can be varied from flight to flight, and the crew of three or four scientists can expect to spend from 7 to 30 days aloft. And note that I said scientists; the flying will be done by the Shuttle crew, making it possible for the first time for the laboratory to be manned almost entirely by working scientists. Already, advertisements are appearing inviting the world's scientific community to bid for space and time in the orbital laboratory – in fact, the time is not so far off when it will be almost commonplace for a scientist to take a ticket for a month in space.

The Space Shuttle program and its successors, as they develop through the 1980's and 90's, will certainly in one sense be a further step towards the more distant goal of longer flights through space – for which an essential feature of course will be the ability to live and work for months on end in cramped and weightless conditions; and it will also provide a rehearsal for the future assembly of much larger space platforms, such as may be used for the launch of interplanetary flights. But in the near run, they will be far more important as a means of adding greatly to our knowledge of our own Earth and its near space environment – the magnetic and electric fields, for instance which surround us and play significant roles in our climate as well as our radio communications. And the Large Space Telescope with its 3-metre mirror which it is hoped to launch with the Shuttle around 1982, will provide astronomers with one of their most powerful tools yet.

Colonies in space

To end this short survey of the space science of the near future, let's just look for a moment to the end of the century, and perhaps a little further. For a long time Man seems to have had the slightly mysterious urge not merely to explore space, but to colonize it. What are the mainsprings of this wish? Some say population pressures here on Earth will force us to spread our wings. So, maybe, they will – in two or three centuries, or even more. Then there is talk of tapping vast quantities of solar energy, out in space where a 24-hour supply of sunshine can be guaranteed, and beaming it to earth by microwaves from the collector satellites. And yet again, there is the prospect of supplementing our dwindling resources by mining the Moon for the rarer minerals. Personally I find none of these arguments very convincing, if we are talking about the next century, at least. I suspect myself that the mainspring is psychological, the feeling that out there are new worlds, unlimited in extent and free of the pressures and restrictions that seem increasingly evident here. But whatever the motives, it seems that Man at last has, for the first time, the means – if not in his grasp, then close enough to conjure with. What are these possibilities?

It would certainly be feasible to set up a semi-permanent colony on the Moon; a giant dome, perhaps, inside which quite an acceptable environment could be created, and from which the space-suited occupants could sally to carry on their mining and exploring. The Moon would also, of course, provide a first-class platform for astronomical observation, with no atmosphere and no problems of stabilizing the platform in space. But the chief objection to using the Moon as a base is cost, which would be enormous, and this arises from the physical fact that it takes very much more energy to get a given load safely down on the Moon than to place it in stable Earth orbit. A simple comparison of the size of the Apollo rocket – as high as St. Paul's Cathedral! – with the Space Shuttle, comparable with a medium-sized airliner, makes the point.

For this reason, there has always been a strong school of thought which sees the future in terms of larger and larger space platforms – and eventually real colonies in space. One of the favored locations for such a giant platform would be far out in space, at a point on the Moon's orbit but 240,000 miles from both Earth and Moon; this position, called the Lagrangean point, has the advantage of complete dynamic stability – once established, no rocket power would be needed to maintain the station in place. There have been many proposals of this sort, but the most detailed, perhaps, resulted from a 10-week study session organized by NASA and the Stanford Research Institute in 1975. The plan is nothing if not ambitious : a wheel-shaped habitat, more than a mile across, housing in its rim 10,000 people, with shops, schools, industry and a special closed-cycle agriculture.[12] The whole colony would rotate, quite slowly – one revolution per minute would provide an artificial gravity much the same as that found on Earth. The authors of the study claim that such a station could be built out of material mined on the Moon and ferried into place by a kind of giant solar-powered catapult, and they add that the whole project could be based on known technology. However – whether or not one accepts this, and many of us have our reservations ! – there is no doubt the cost would be many times that of the Apollo project ... so I see it as a rather fascinating speculation, one that is more likely perhaps to come up for consideration sometime in the 21st century than in our own.

Looking still further ahead, do the planets offer any possibility for colonization? The problem with Mars, as with the Moon, is the almost complete lack of an atmosphere; as the Viking landers have shown. Yet all is not lost; it is at least theoretically possible to liberate sufficient gases from the planetary material by setting off underground nuclear explosions. However, Carl Sagan, a leading American space scientist and member of the Viking team, believes there are better chances of creating a useful environment on Venus. Here, the problem is not too little atmosphere, but too much, and most of it is carbon dioxide, the inert gas which does not sustain human life. But Sagan and his colleague Dr. von Eshelman believe this could be changed, using what they term planetary engineering. There are, for instance, certain micro-organisms which enjoy a diet of carbon dioxide and ultimately break it up into carbon and oxygen. Seed the Venus atmosphere with enough of these organisms, says Sagan, and you would eventually produce enough oxygen for human needs. And at the same time – since it is largely the carbon dioxide that makes Venus so hot, by keeping in the infra-red radiation that ought to cool the planet, one would end up with a more acceptable temperature.

But all these are speculations, for future generations far from our own. The next twenty-five years promise excitement and discovery in plenty, here on Earth – and in the space around us that we have already so firmly set out to master.

REFERENCES

1. F. Hoyle, *Nicolaus Copernicus* (Harper & Row, New York, 1973).
2. S. Mitton, *Exploring the Galaxies* (Faber & Faber, London, 1976).
3. L. John (ed.), *Cosmology Now* (Taplinger Pub. Co., New York, 1973).
4. Sir Bernard Lovell, *The Story of Jodrell Bank* (Oxford University Press, 1968).
5. J. S. Hey, *The Radio Universe* (Pergamon Press, Oxford, 2nd ed., 1976).
6. NASA Contemplates Radio Search, *Spaceflight* (London, October, 1976).
7. P. Moore, *The Sky at Night* (BBC Publications, London, 1976).
8. F. Hoyle, *Ten Faces of the Universe* (W. H. Freeman & Co., San Francisco, 1977).
9. J. Taylor, *Black Holes* (Fontana Books, London, 1976).
10. H. Rohr, *Radiant Universe* (F. Warne, London, 1972: English edition by A. Beer) – a beautifully illustrated summary.
    For the amateur stargazer, a useful book is: W. Widman, K. Shutte, *Stars and Planets* (Burke Books, London, 1969).
11. A very full account of the Space Shuttle is given in *Spaceflight* for September, 1976. This journal, published by the British Interplanetary Society, is the best source of news on space exploration.
12. G. K. O'Neill, *The High Frontier: Human Colonies in Space* (Jonathan Cape, London, 1977).

TRANSPORT AND COMMUNICATION

Arthur Garratt

Arthur Garratt took an Honors Degree in Physics at University College, London. He worked for a number of years in armament research and design.

He was Staff Physicist for the Festival of Britain Exhibitions and then went to the National Physical Laboratory where he was the liaison officer between the Laboratory and British industry.

Since 1958 he has worked as a consultant, broadcaster and author.

He has taken part in about 3500 radio broadcasts and 400 television programs, almost all of a technical or scientific nature.

Among his publications are *Energy from Oil*, *The Penguin Science Survey*, *The Penguin Technology Survey* and many scientific and technical papers in learned journals.

He is the Managing Director of Value Management Consultants Limited, Phillips and Garratt Ltd at Aldershot and of Cougar Films Limited.

He was Science Adviser to the British Government for the Moscow and Turin Trade Fairs and won the Wireless World Prize in 1961 for the best paper delivered before the Television Society (now the Royal Television Society). In 1969 he was adviser to Mullard Limited for the Faraday Lecture.

He is Past Master of the Worshipful Company of Scientific Instrument Makers.

'I'll put a girdle round the earth in forty minutes'

A Midsummer Night's Dream

Civilization has always been dependent on man's ability to move and communicate. The caveman whose radius of action was a few miles and whose only means of transmitting his thoughts and wants was by a succession of grunts stood little chance of improving his lot. As soon as he was able to tame an animal to carry him farther than he could walk, learned to make a simple boat to navigate the rivers and later the seas and, most important of all, learned first to speak and then to write, the way was open to limitless progress. In fact it would be fair to measure man's civilization by his mastery of distance and of time.

If we look back we can see that the greatest events in history were learning to speak, developing a written form of communication and then finding the way to duplicate words indefinitely by printing. Only marginally less significant are the landmarks in transport, the horse, the ship, the steam engine, the automobile and then the plane and rocket. These inventions have meant that we can now travel to anywhere in the world within the space of a day, can walk on the moon and can pick up a telephone and speak to almost any-one anywhere.

We might, looking at today's achievements, wonder if we are near the end of the line, whether there are new wonders ahead or whether the future may be no more than the con-solidation of today's knowledge and some of the ideas mooted for tomorrow are impossible. Wernher von Braun, the man who has been called Mr. Space, once said to me, 'I have learned to use the word impossible with the greatest caution !' What we must take care to do is to separate the truly im-possible which means the violation of established scientific principles, and the improbable. To give two examples, we have every reason to believe that perpetual motion is impossible. If it is not, then the entire foundation upon which we base our physics would crumble and this seems too unlikely to be worthy of serious consideration. On the other hand the chance of man visiting another solar system, as is discussed in another chapter, is highly improbable but could not be said to be impossible. This distinction between the impossible and improbable is a fine one but essential to any intelligent forecasting of the future. Even when the distinction is made, trying to predict the future is a thankless task because the technically possible may turn out to be impractical for political or other reasons.

Man walks

Let us start with the oldest and commonest method of transport – walking. What can technology do to help? Perhaps, surprisingly, quite a lot because it can remove the two major objections to walking as a means of transport as opposed to an exercise – these objections are exposure to the weathers and low speed, of the order of three miles an hour. Both these can be overcome by the Travelator or moving platform which is rather like a flattened out escalator. Travelling walkways are used in many airports to help passengers with their baggage to get from the check-in point to the aircraft and vice versa, as well as in transport systems such as the Travelator which connects Bank and Monument on the London Underground.

The travelling walkway is, however, only a slight improvement in terms of speed over ordinary walking because of the problems of getting on and off a rapidly moving belt. This has restricted the speeds of passenger carrying conveyors to the order of walking speed or a little less. Dunlop who built the Starglide conveyors at Montreal and Los Angeles Airports have solved this problem by an ingenious system now under development, it is called the Speedaway and combines the conventional moving platform with what is known as an Integrator. This is a device which enables a person to get on the system at slow speed and then speeds him or her up to a much higher speed on the conveyor proper. To exit one uses another integrator to slow one down to an easy speed for alighting.

At first sight the idea of a *continuous* moving belt which travels at 2 mph at one end and 7 mph at the other sounds impossible, but once more this is a dangerous word to use. All that was required was some creative thinking and the solution turned out to be, in principle, very simple. The trick is to turn the track and use a series of sliding platforms rather like logs in a stream. It is quite easy to imagine a succession of logs travelling down a river more or less in line with the current. If such a stream were diverted sideways into another stream leaving the river and this stream curves until it is almost at right angles to the original river, the logs could stay pointing in their original direction but now be side-by-side instead of end-to-end and be moving very slowly along the stream. This is the trick, simple enough when it's been thought out although clearly there are a number of mechanical problems to be solved.

The Speedaway could alter the whole concept of towns and cities. Added to the present-day trend of pedestrian precincts, it would provide protected travel at up to 10 mph, as fast as much city traffic, in air conditioned overhead walkways with main high-speed tracks fed at intervals by Integrators so that people could easily enter or leave the system. Prototypes are being built for use in Liverpool, Paris and London and it is reasonable to suppose that the next twenty years will see a proliferation of Speedaways carrying thousands of people an hour with all the advantages of walking but without its disadvantages.

Man rides – the ergonomic bicycle

The bicycle is not only a cheap, simple means of transport, it also provides useful exercise for the man or woman who needs to move about and keep fit at the same time. Already after three-quarters of a century of the conventional bicycle we have seen the introduction of the small-wheeled bike pioneered by the Moulton in the 1960s. Such vehicles are said to be easier to ride and safer than the large wheeled versions yet they still present certain problems. Because one's feet and legs are near the ground they are vulnerable to water and mud thrown up from other vehicles and, perhaps more serious, a bicycle is not really comfortable to sit on and ride.

But these problems are soluble, given good creative thinking. As is so often the case, one gets into a mental rut believing that because bicycles have been designed in a certain way for a century or more there is no other way of doing it. To give a lie to this philosophy a professor of production studies at Cranfield Institute of Technology has invented a completely new bike. It is based on sound ergonomic principles, ergonomics is a relatively new science which aims at matching man and machine. Prof. Cherry, the inventor, realized that one needs a comfortable seat with a back instead of a saddle which, after all is said and done, was designed to match the anatomy of a horse to that of a man. There is also the question of simple mechanics, to provide the force needed to rotate the pedals one should have something to push against, not merely use one's weight. The obvious thing to push against is the back of the seat, but this means that one must lift the pedals and put them in line with the rider and at right angles to the back of the seat. This logical thinking resulted in a completely new design with the rider perched fairly high up with his or her legs horizontal instead of vertical. The result is a remarkably comfortable bicycle eminently suitable for a city gent or housewife to ride round the town, on the way to the office or the shops.

'Puffing Billy' puffs no more

There is no doubt that one of the great inventions was the steam engine with its applications to both industry and transport. The railway changed the whole face of the world and it is interesting in retrospect to realize that it was the British invention of the railway train which enabled the young American Union to prosper, without the railroad they might have been hard put to make the great experiment in democracy work.

Yet for all its old glory, only a few years ago it appeared as though the Railway Age was over and serious suggestions were put forward that the old railway tracks could be made into low cost motorways. Countries that had developed slowly and had never built up a system of railways leap-frogged the older nations by setting up air communications which were fast and needed much less capital expenditure, while the railways everywhere languished.

Once more what was needed was logical creative thinking. Railways have one enormous advantage over all other forms of transport over land and this is the very low friction of a railway track compared with the friction of a tire on a road or the drag of air over the surface of an aircraft. In terms of cost per ton mile, a car is similar to a plane, while a train is about one tenth of this, ignoring of course the capital cost of the tracks.

This single fact makes the railway an attractive proposition if it can be modified to operate at much higher speeds, comparable to those of aircraft.

High speed is necessary not only to shorten travelling times but also, and this is even more important in many cases, to enable an existing track to carry more traffic. And it was here that the difficulties appeared : existing railway track, however good its construction, limits conventional trains to the order of 100 mph. At higher speeds what is called 'hunting' develops and the train sways from side to side causing discomfort and actual hazard. To overcome this problem it seemed as though new tracks would have to be laid, tracks specially designed for high speed travel. Such a solution is viable technically, but not economically except in places where the traffic density is very high and the capital costs can be absorbed. Such a place was part of Japan where in the 1960s there was only a small, largely narrow-gauge line joining Tokyo to Osaka and, with an undeveloped road system at the time, it was economic to lay a modern, high-speed track over the 320 miles separating the two cities. This was called the Tokaido Line and carried the first of the new generation of high speed trains travelling at speeds up to 130 mph.

A similar situation existed in France, a country in the forefront of new ideas in railways, where the electrified line joining Paris and Lyon was heavily overloaded and it was economic to lay a new track alongside the old to carry the French high speed train, the TGV or Train à Grande Vitesse. This train is powered by either a gas turbine or electricity and is designed to average 155 mph with speeds up to 185 mph on certain sections. In tests the train has travelled at 190 mph, very close to the world record held by two French electric locomotives which travelled at 205·6 mph over a $1\frac{1}{4}$ mile stretch. (This has been exceeded by l'Aerotrain, a French hovertrain.) Work started on the new line in December 1976 and 85 ten-car train sets have already been ordered.

In Britain, where the traffic density is lower, a different philosophy prevailed. Here it was decided that any high-speed train should be designed to run on existing track *without modification*. This meant a completely different kind of thinking. Instead of picking a speed and building a track to accommodate it, one had to design a train that would travel as fast as possible over existing track, all commensurate, of course, with comfort and safety.

The result of this thinking has been the development of the APT, Advanced Passenger Train, by British Rail at the Railway Technical Centre, Derby. Here a group of highly creative scientists and engineers, many of whom came from the aircraft industry, put their heads together to design what is certainly the most advanced conception in railway transport. They began from first principles by asking the question, 'Why does a railway train sway at high speed?' They started analytically by postulating a simple four-wheeled vehicle and writing down the equations which governed its motion. These equations, and they are very complex, were fed into a computer and the results which came out confounded railway practice which was over a century old. In the early days of railways, wheels had been designed with a special contour so that on a corner the outside wheels would ride up the track and the inside wheels ride down. In this way there was an automatic differential action so that there was no slip between wheel and track because the outer wheels, which had farther to go round the bend, had a larger effective radius. It was very ingenious and became standard railway practice all over the world. Unfortunately, as the computer showed, it was wrong and caused all the trouble. Interestingly enough railway wheels always began to wear to a different profile and after a certain time they were changed for new wheels with the engineer's profile. Here it is a case of the train knowing more than the engineer because the wheel was trying to get into the right shape and the engineer would not let it! A simple change in the profile of the wheel so that it can no longer ride up and down on the track overcame the sway and raised the maximum safe speed of a train *on existing track* by 50 mph or more at a stroke.

Armed with this information the engineers could proceed. They had a train that would stay on the rails, theoretically at least, now they had to make a practical train out of it. The first problem was to get the train round curves without throwing the contents, which included passengers, against the outside windows. This was a serious problem because roughly 50% of British Rail's major routes are made up of curves, some quite sharp with radii as small as $\frac{1}{3}$ mile. The simple answer was to bank the track but this broke the initial design criterion of leaving the track as it was. There was only one other possibility – bank the train instead.

British Rail have done just this with a system of jacks which tilt the carriages on the bogies automatically on cornering – this exactly counters the centrifugal effect and the passengers are unaware of the bend, just as passengers in a banking aircraft do not realize what is happening unless they look outside.

The next problem for a train designed to travel at 150 mph was reduction of drag. Wheel-to-track friction is very low for a train, this is what makes rail transport so attractive in terms of fuel used, but at such speeds the air drag can be very high. The answer was to make a train with a good aerodynamic know-how to such effect that the APT uses the same energy to travel at 125 mph as ordinary streamlined trains use at 100 mph, an effective energy reduction of 33%. The weight of the train has no effect at constant speed on the flat, but it is significant in terms of acceleration and hill-climbing. Here again the lessons learned from aircraft were put into practice and the APT has been built largely of aluminium alloys.

At such speeds as 150 mph or more, even a smooth railway track could produce enough bumps and vibration to make the ride uncomfortable, so the engineers devised a highly sophisticated suspension system with automatic self-levelling like some of the most advanced cars.

Another self-imposed restriction was that the APT should be capable of stopping from its top speed of about 150 mph in the same distance as an ordinary train could from 100 mph, with a $12\frac{1}{2}\%$ extra margin for extra safety.

The problem to be solved here was getting rid of the heat developed by the brakes while still keeping the unsprung weight low. This question of unsprung weight is critical in all sprung vehicles, because the behavior of a vehicle is dictated not by its absolute weight, as many people think, but by the ratio of the unsprung weight to the sprung weight. What this means is that for a well sprung vehicle, whether it is a car, bus or train, the wheels, axles and other parts that move up and down over bumps should be as light as possible compared with the weight of the parts which do not, or should not, follow the contours of the road or track. It is this question of the unsprung to sprung weight ratio that occupies the minds of car designers and has led to independent suspension systems to improve the ride. In terms of the APT this meant that the heat absorber or radiator could not be attached to the brakes themselves some of which are part of the unsprung weight. This led to the hydrokinetic brake. The principle is again simple. The rotation of the wheel being braked rotates blades not unlike those of a turbine which runs in fluid, actually water with glycol in it to prevent it from freezing. On the power bogies, where the wheels are driven through propeller shafts and universal joints, the brakes are mounted on the bogies themselves and are therefore sprung, on the trailer bogies the brakes are on the axles and the fluid carried to radiators on the carriage body. Needless to say the hydro-kinetic brake, efficient though it is at high speeds, does not work at all at zero speed, so it is supplemented by a relatively small hydraulic friction brake acting on the wheel treads.

The APT can be driven either by a gas turbine or electrically. For various reasons the electric solution is preferred and British Rail is now busy designing an overhead pick-up gear capable of operating at 125 mph or more. Here there is a special problem, the pantograph cannot be fitted to the carriage roof because of the tilting mechanism so there is a complex system of lazy tongs connecting it to the non-tilting carriage base.

The British Rail philosophy has been to build two high speed trains. The first, called simply the HST (High Speed Train) is really a conventional train stretched as far as possible, this is already in service and can operate up to 125 mph on existing track. It has limitations, however, and although the problems of hunting have been eliminated at such speeds, it cannot corner like the tilting APT. As a result it would take 4 hours 50 minutes to travel from London to Glasgow (401 miles) compared with 5 hours for the normal train with a top speed of 100 mph – the reason is that the route is a sinuous one. On the other hand, the APT with the same maximum speed as the HST (125 mph) brings the journey time down to 3 hours 58 minutes. The APT is designed for a maximum speed of 150 mph but will not exceed 125 mph in service.

High speed trains become serious competitors to aircraft for shorter hauls. Because of the time taken to and from airports a train can today carry passengers from London to Manchester City Centre faster than the airlines can, and do it independent of weather. The APT will extend this advantage from the present 200 miles to about 400 miles making the trip from London to Glasgow, or Boston to Washington, quicker than by air. Because of the lower drag it is cheaper too, so the railways, not so long ago the Cinderella of transport are now proudly wearing glass slippers. Efficient operation implies a speeding up of freight trains to the order of 100 mph, and British Rail has this in hand, coupled with a fully computerized signalling system so that the high speeds can be fully exploited.

Urban transport

Important though it is to travel hundreds of miles as fast as possible, more people travel daily on short hauls often within the bounds of a city or in its immediate neighborhood. Here the tendency is to dive underground or go overhead to avoid existing buildings and roads. Underground railways are now quite old, the first was part of the London system opened in 1863 using, believe it or not, steam locomotives – our great grandfathers were obviously immune to airborne pollution. Today the tendency is again to speed up systems and Paris has done this not by improving the existing Metro but by building a completely new system to serve the more distant suburbs of the city. This new system is called the Réseau Express Régional or RER (Regional Express Network) and serves areas 15 miles from City Centre. The distance between stations is large, on average about 1½ miles, so high speeds can be used and the trains run up to about 65 mph. By using rubber tyres and a good suspension the RER provides a quiet and comfortable ride. The stations and ancillary equipment are on a grand scale with shops, cinemas and other facilities at some of them. One of the stations on the RER has more escalators than the whole of the new Victoria Line in London.

The United States too is building high-speed underground railways of which the best known is perhaps the Bay Area Rapid Transport system or BART. This serves San Francisco and neighboring Oakland on the other side of the Bay, communication being under the sea. BART is completely computer controlled and is capable of operating up to 80 mph – so attracting the motorist who is used to travelling fast down the excellent freeways in the area.

Clearly, despite the very high capital cost, we will see many more underground railway systems in the next twenty years. They have the great advantage of safe high speed transport in areas otherwise cluttered by streets and buildings and the tendency is to go fully automatic. The Victoria Line in London has gone part of the way as the only job the 'driver' does is operate the doors, the actual operation of the train to run up to speed and slow down and stop, at the right place, is carried out by mechanical and electrical robots.

Going round the town

With the average motor car carrying only one or two persons, there is clearly a strong case in favor of public transport systems using buses or other vehicles which occupy perhaps twice the space of a car and carry ten or twenty times the load. The bus is put forward as a complete answer to the bulk of city transport by certain groups of people but, unfortunately, with rising labor costs it is becoming less and less attractive economically. At the present time bus drivers and conductors represent over half the running cost of a bus service, add ancillary labor and the cost rises to a staggering 85%. The result is that as labor costs rise, fares rise with them and passenger demand falls – currently on buses at a rate of about 10% per year. If this trend continues the problem will not be attracting motorists to buses, it will be having a public transport system at all.

The only answer appears to be unmanned public transport. Although this is expensive in terms of capital cost it can be economically viable on a long term basis. Various proposals have been put forward and some automatic systems already exist. Broadly speaking there are two distinct systems, shared transport and individual transport. The shared systems include *Minitram* with small tramcars holding about twenty passengers running on concrete tracks either on the level or above the street. The trams have no drivers, the entire system is automatic with each car carrying a mini-computer which receives signals from the side of the track and uses these to drive the tram. The entire system is under the control of one central computer. The speeds envisaged are about 45 mph with a five-second gap between vehicles. When one of the trams stops, signals are sent back to following vehicles to instruct them to slow down and stop, the whole system working exactly as though each vehicle were manned by a perfect driver. A variation of Minitram is a system by which one waits on a siding for a vehicle which comes off the main system and stops on receipt of a dialled message from a waiting passenger.

The individual automatic transport system, or self-routing taxi, is a variation of the Minitram with sidings, the difference being that instead of a car carrying twenty passengers, there are individual cars for each group of users. A taxi would come to a siding on demand and pick up the passenger or passengers who would instruct it to go to the desired destination, either by a punched card ticket collected from an automat or by direct instruction in the vehicle by a dial or push buttons – all these options are possible. The car would then re-enter the system and automatically route itself to the destination where it would disengage and pull up at the desired siding.

Such automatic systems sound something like science fiction, yet the technology is available and all the manufacturers need is the go ahead. Already automatic systems exist in the United States and one is under construction at Lille in France. Many cities have incorporated into their planning requirements the need for certain volumes to be left clear, usually above the streets, to accommodate Minitrams and the like in the future, often to be integrated with Speedaways. It is almost certainly true to say that if urban public transport is to exist in twenty-five years' time it will have to be completely automatic.

Fuel in the future

Generally speaking all the methods of transport we have considered so far can run without using fossil fuels. Electricity, the power used by railways, can be generated from water or by nuclear stations, not to mention other more esoteric methods like using the tides and the waves. But when we think of transport running along roads, crossing the seas or flying in the sky, then the power source has to be carried and, with a few exceptions, this implies fossil fuels. As is well known we are now beginning to worry about the future of oil and coal because, although we do not know for certain how much is still under the ground to be won, we do know that the amount is finite and that we are using it much faster than it is being laid down. So although the day when oil and coal becomes seriously scarce may or may not be a long way off, our consumption of fossil fuels is so immense that we have to look for other power sources for our children if not for us. Oil is a particularly convenient way of carrying energy and it is not feasible at this time to think of flying aircraft using any other fuel. Of course it could be that it may one day be economic to make oil from coal, as has been done for some years in South Africa, but in the foreseeable future the oil for our aircraft will come from conventional oilwells, either under the ground or beneath the seas. Much the same is true of ships and other floating or skimming craft.

Ships of tomorrow

Although today many more people travel by air than by ship, the bulk of the goods that cross the oceans go by boat, and there seems little chance that this pattern will change in the years to come. Air freight has special advantages, it is quick and often requires less protective packaging, but it is, and always will be, more costly than sea cargoes. The reasons behind this statement are simple, it costs much more fuel to provide lift than it does to provide motion and at relatively slow speeds ships are very cheap to operate in terms of the weight of cargo they carry.

Large ships have one fuel option that is not open to any other form of transport in a direct form, this is nuclear energy. We have had, for many years now, nuclear submarines and nuclear icebreakers and there have been experimental ships like the Savannah. But unless the cost of oil rises very significantly, such craft are not truly economic and rely on one special advantage, their range of operation without refuelling is enormous.

Soon after World War Two that great creative thinker, Sir Barnes Wallis, put forward a revolutionary idea, the use of *nuclear submarines* as freighters. His reasoning was theoretically sound. Most of the drag on a ship is caused by the formation of surface waves, such as the bow wave and wake, because these dissipate energy. A submarine at a considerable depth hardly affects the surface, so the drag on a submarine should be, and is, lower than an equivalent ship travelling at the same speed. The submarine has other advantages, operating well below the surface it is independent of the weather so avoiding the problems of cargo damage in heavy seas. Unfortunately the idea has not come to anything because of the necessity to build new and different dock facilities to load and unload submarines, a problem totally different from that of ships. But the idea could be resurrected and we may one day see a large fraction of our freight being transported under the sea instead of on top of it. A particular use of the submarine might be as an oil tanker where the special loading and unloading problems would not be particularly difficult to solve.

A ship, to be economic, must move slowly because of the large drag. If we wish to move fast we have to get the hull out of the water and there are two basic ways of doing this. One is the hovercraft which rides on a shallow cushion of air and has almost no drag due to the water. Hovercraft are in regular service in many parts of the world and have one unique advantage over any other form of transport – they can operate over any reasonably flat surface, wet or dry, so they can leave the sea and skim over a beach or operate in marshy territory where no other large carrier could be used. It is probable that it is in these places that we shall see the real worth of the hovercraft because on open seas there are serious difficulties in negotiating waves and the ride tends to be noisy and bumpy in most conditions.

More attractive in some ways is another craft which operates largely out of the water. This is the hydrofoil which lifts its hull clear of the sea and rides on wings in the water, hydrofoils instead of aerofoils. Hydrofoils are used in many places to carry passengers, for instance across the Adriatic between Yugoslavia and Italy, as well as providing an economic way of transporting men to and from offshore drilling rigs. Unfortunately hydrofoils, too, have certain limitations. One is that they are rather vulnerable to floating wreckage, another is their behavior in heavy seas. Here there are interesting possibilities, the idea of using a sensor in front of the craft to detect a wave before the craft itself reaches it. The signal can then be used to control the angle of the hydrofoils so that the craft in a sense 'steps over the wave'. We shall probably see hydrofoils used more and more during the next few decades.

Taking the air

There is no doubt that the aeroplane is the glamorous form of transport. Fast and comfortable, it is capable of flying over any part of the globe, be it desert, mountain or sea. Today it dominates passenger travel over a few hundred miles and seems likely to maintain this pre-eminent position.

Looking into the immediate future there seem to be three distinct developments. One is already with us, the supersonic transport plane exemplified by *Concorde* and the Soviet *TU 144*. Both these aircraft are interesting in the sense that they are the last generation of conventional aircraft. If planes fly faster then they will have to be built in a totally different manner using different materials. The reason is simple, supersonic aircraft get hot in flight, *Concorde* has a leading edge temperature above the boiling point of water, and materials such as aluminium alloys which have excellent strength-to-weight properties soften and fail at temperatures not greatly higher than those at which *Concorde* operates. So the next generation of SSTs will have to be constructed of exotic materials like titanium or stainless steel. This will be a completely new ball game and the development costs of the faster SST make the mind boggle remembering how much the relatively simple *Concorde* has cost.

There is little doubt that supersonic transport is here to stay – it is the logical development in aircraft technology. Just as the jets doubled the speed of piston planes and reduced the time between London and New York from thirteen hours to seven, so *Concorde* again halves the time to three and a half hours. For the first time it puts anywhere in the world within twelve hours of anywhere else. This in itself is a stupendous advantage on long runs like that from England to Australia but it is not the only pay-off, perhaps more important is the fact that supersonic planes can put in twice as many journeys in a year as subsonic aircraft – and this is where the eventual economic advantage lies. Two Boeing 707s carry as many people across the Atlantic in a year as the old ss *Queen Mary*, one *Concorde* will do the same – this is the true value of supersonic flight, not just knocking four hours off the time to fly to America.

The airbus

Another solution to low cost air travel is the airbus, such as the Boeing 747 Jumbo Jet. Here the economics are different. One accepts relatively slow speeds, a mere 660 mph, but one carries a very large payload, passengers numbered in hundreds. Clearly the Jumbo does a different job for a different market from the SST. *Concorde* is supreme for high speed travel, for the businessman who wants to go to a meeting in New York and return to London the same day with no jet lag problems, while the more sedate traveller is happy with the Jumbo. But the whole airbus concept depends on high density traffic, unless a plane is about two-thirds full it flies at a loss. So we can see a pattern emerging, supersonic transports on long haul, low density routes and taking the passengers prepared to pay a little more to get there fast on all routes, backed up by the Jumbo, a low-cost, slow carrier where the traffic is heavy. The two planes complement each other and are not really competitive.

Vertical take-off

The time taken to reach a city airport and the space needed by such airports in crowded countries like Britain both point to the advantages of an aircraft that can take off from anywhere and fly right to the passengers' destination. Most people think of helicopters as ideal in this respect, but unfortunately the convenience has to be paid for in very high operating costs and limited size. More likely is some form of vertical take-off plane which lifts by downward pointing jets and then flies normally. Such aircraft have been built for military purposes, planes like the *Harrier* which does just this. The civil version is likely to be somewhat tamer and to be a compromise, quick take-off from a very short runway rather than true vertical take-off. So far there is one serious problem – noise. We have many complaints about the noise at airports even though they are generally sited well away from cities, it does not seem feasible to suppose that the public would accept high level noise near a city center. Aircraft are almost necessarily noisy even though significant improvements are possible by clever engine design, the truly silent aeroplane is almost certainly a physical impossibility.

Lighter-than-air craft

One aircraft which can be quiet, can lift vertically and can carry heavy loads is the airship and there have been many serious attempts to popularize the lighter-than-air craft which lost favor overnight when the *Hindenburg* crashed in flames in New Jersey on May 6th, 1937, with the loss of 36 lives. The fire hazard no longer applies because modern airships would be filled with non-inflammable helium, albeit at some sacrifice of lift. But there is one feature of an airship which seems to put it right out of court, because of its size and poor aerodynamic shape, efficient operation implies low speeds, speeds of the order of 100 mph maximum. This means that airships are immensely vulnerable to head winds and in gale conditions might even fly backwards. So one feels that it is unlikely that we will see them as serious cargo carriers in the future.

The long view

What kind of aircraft will be flying in the distant future, say between twenty-five and fifty years from now? Forecasts like this run into fantasy, yet there are certain trends that can be observed as well as forward studies made by respectable organizations such as the Royal Aircraft Establishment and Rolls-Royce. These studies point towards two developments; one is the solving of the problem of unloading hundreds of passengers at a time, processing them through immigration, customs and health control and then getting them from airport to city centre. Here the thinking is in many cases the effective establishment of an interface between air travel and rail or road transport, including hovertrains and overhead railways such as the one connecting Tokyo Airport with the city. One interesting suggestion which has been mooted is to detach the wings and engines from a plane immediately on landing and then using the fuselage, with passengers and luggage still inside, as a rail car. In this way all formalities including baggage collection could take place on the journey between airport and city center thus eliminating at a stroke the problem of crowding at the airport and speeding up the journey for passengers.

The other trend is towards ever greater speeds, here one is thinking not of supersonics, i.e. speeds somewhat in excess of the speed of sound (about 760 mph) but *hypersonics* or speeds several times that of sound. Aircraft designers find it convenient to speak of *Mach Numbers*, named after a nineteenth-century Austrian mathematician, Ernst Mach. Mach 1 is the speed of sound, Mach 2 twice the speed of sound, and so on. *Concorde*, for example, flies at about Mach

2, the future HST (hypersonic transport) might well reach Mach 15.

As mentioned above one of the problems of flying at such speeds is the surface heating of the aircraft structure, particularly the leading edges of the wings and the nose of the fuselage. The *stagnation temperature*, i.e. the temperature of air brought to the speed of the leading edge, can be calculated for different Mach numbers and at Mach 5 it is over 900°C, hot enough to melt brass. Assuming for the moment that this problem of engineering can be solved, as it no doubt could be using a new breed of materials, there are still other massive problems associated with the aerodynamics and the power units to drive such a plane.

Taking first the engines, conventional jets can operate satisfactorily up to about Mach 4·5, above this the high stresses involved cannot be withstood by even the most advanced alloys. Fortunately there is a solution to this problem – use an engine without the rotating parts where the high stresses are developed. Such an engine is well-known: it is the *ram jet*, a primitive version of which was used in the German V–1 Flying Bomb. The ram jet is simple, it's sometimes described as a 'tube of wind' and consists of a pipe into which the air rushes at high speed due to the forward movement of the aircraft and in which burning takes place. The hot gases leave the rear of the pipe through a convergent-divergent nozzle – and that's it. Although the ram jet is simple and efficient it has one obvious snag, it will not work at zero speed and its efficiency is low below about Mach 3 when it is approximately the same as the conventional jet engine. At higher speeds the advantage moves to the ram jet but unfortunately other effects become prominent over Mach 5 making its top speed of the order of Mach 6. Of course Mach 6 is over 4500 mph at sea level so we are talking of a plane that could fly from London to New York, ignoring take-off and landing, in about three-quarters of an hour. But as we shall see in a moment, there is a possible way of providing thrust at even higher speeds.

Let us turn to the aerodynamic problems. Barnes Wallis

showed some years ago that there are positive advantages in changing the geometry of an aircraft in flight and he put forward the idea of a swing-wing plane. This was to provide suitable shapes for both subsonic and supersonic flight which ideally need quite different designs. The aircraft of the future will probably change its shape more than once in going from zero speed up to operating speed, while the engines too will have variable geometry. The change in the engines will be to turn a conventional jet into a ram jet by bypassing the compressor when the speed reaches about Mach 3 with a further change, this time to the inlet of the engine, at higher speeds.

The aircraft itself would change shape at Mach 1 by swinging its wings back and at about Mach 5 there would be another change, the plane turning itself into the shape of a wedge. This is to accommodate a revolutionary new idea in aircraft propulsion put forward some fifteen years ago by Dr. D. L. Dugger of Johns Hopkins University in the United States. He suggested and carried out experiments to demonstrate that a plane could be pushed forward by burning kerosene *outside the aircraft* on the underneath of a wedge-shaped wing. Wind tunnel tests at Mach 5 have shown that this system, known as Dugger's Wedge, works and it is at the moment the most likely propulsion system for really high speed aircraft of the future. It is taken quite seriously and Rolls-Royce have put forward a proposal for a plane to accommodate it. Such a plane would fly at Mach 15 at a height of about 200,000 ft, or, put another way, at 11,400 mph at an altitude of 38 miles. Such an aircraft would put Australia within an hour of Britain, travelling at about half the speed of a satellite in close orbit. Such proposals make *Concorde* look like a quiet run in the country and demonstrate that we are a long way yet from the end of the line in the development of transport systems.

The motor car
The motor car has reached a high state of development and future changes are likely to be dictated by external factors

such as air pollution and rising costs of fuel. Assuming a car runs on gasoline or diesel fuel, one is burning a hydrocarbon and ideally almost all the products of combustion should be pure water – which is, of course, harmless. Unfortunately there are small quantities of other materials produced, some of which are pollutants of an unpleasant if not serious nature. Perhaps the best known is soot and most of us have seen diesel trucks putting out large quantities of black smoke, almost always when the engines are out of adjustment. In addition there are oxides of carbon, carbon monoxide which is highly poisonous and carbon dioxide which is not poisonous but in large quantities can suffocate. A correctly adjusted internal combustion engine produces no carbon monoxide and almost all fatalities from car exhausts take place when the engine is cold and running on a choke. Nevertheless engines are not usually perfectly adjusted and some carbon monoxide is often emitted. In the open air the quantities are generally too small to matter although in traffic jams inside cities the levels can rise significantly. However, one authority has stated that a policeman on point duty under the worst conditions in a crowded city would inhale in twelve hours about the same amount of carbon monoxide as he would by smoking one cigarette.

More serious is the emission of oxides of nitrogen, the nitrogen comes not from the fuel but from the air. Such oxides when exposed to ultra-violet radiation from the sun can form particularly unpleasant acrid fumes, like the infamous Los Angeles smog. This only occurs in certain parts of the world, notably Los Angeles and Tokyo, where local meteorological conditions favor it and it has never represented a problem in less sunny Britain.

Almost all gasoline has lead added to it to allow the use of engines of high efficiency by raising what is known as the octane number of the fuel, meaning improving its anti-knock properties. Lead is known to be poisonous and although most of the lead falls onto the road surface, there may be some hazard alongside busy roads. Perhaps more serious is the fact that the lead in the fuel damages devices known as *catalysts* used to remove the oxides of carbon from exhausts, this is why all automobiles manufactured in the United States now run on non-leaded gasoline and such fuel has to be supplied in filling stations.

It is this whole question of pollution coupled with increasing costs of oil which may change the pattern of the motor car. High fuel costs point to smaller engines, yet other factors may prevent their general use. One such factor is the necessity in many climates to provide air conditioning which needs a fairly large engine to drive the cooling equipment. In fact it is probable that many countries which at present do not demand air conditioning in vehicles will want it in the future, just as car heaters which were rare before World War Two are now fitted almost universally.

There are a number of different options in power units for cars. Some of these, such as the steam engine, are not really attractive because not only would one have lower performance and efficiency but one still has to use fuel to heat the water and the most attractive fuel is oil with all its attendant problems. More interesting is the Stirling engine, devised by a Scottish clergyman a century ago and only recently developed to a point where it becomes a practical proposition. The Stirling engine can run on virtually any fuel and, because the fuel is burned at a lower temperature than in a conventional engine, air pollution can be greatly reduced.

What stands in the way of new forms of engine is the enormous capital which has been invested in the ordinary internal combustion engine which has been developed to a pitch of remarkable efficiency. To switch to, say, a Stirling engine would cost any manufacturer many millions and he would have to be certain of a return to make the proposition economically viable. The problem is not even totally economic, building the machinery required for the changeover might well consume more energy and produce more pollution than it would save. There is, however, an interesting solution which uses a normal engine running off hydrogen. At first this may sound horrifying remembering what happened to the *Hindenburg* but fortunately there are now ways of storing hydrogen which are quite safe. This follows the development of certain materials which soak up hydrogen rather as a sponge soaks up water. The driver of a hydrogen car would go into a special filling station and have hydrogen pumped into a tank containing the special material, such a tank would be perhaps thirty per cent larger than a gasoline tank for the same range. Hydrogen would be driven off by gentle heat, such as can be supplied by taking the exhaust pipe round the tank, and then fed through a special carburetor into an ordinary engine. Hydrogen is a high-energy, clean fuel which would produce rather less pollution than gasoline because it contains no carbon, although it would still produce the oxides of nitrogen which cause the Los Angeles smog.

The next question is, of course, where would the hydrogen come from? The answer is from ordinary water using perhaps nuclear power stations during the night when the load on the electrical supply is low. Hydrogen can be produced either by the familiar school experiment of passing a current through water or, and this seems better economically, by 'cracking' the water under heat. Experts say that, ignoring taxes, hydrogen would turn out cheaper than gasoline and that a hydrogen car, despite certain problems with the material to store the hydrogen, could probably be perfected in five years given the necessary support.

Another contender is the electric car. Here there are massive problems, which are probably insuperable, to make a flexible car in the sense that it could be fuelled in minutes and then drive several hundred miles on one filling. The electric vehicle will probably be used more and more for scheduled transport, such as delivery trucks and buses and even perhaps for commuters who would still need a conventional car for longer runs.

Why go?

So we can see that there are many exciting things to come in the transport of the future. But here we must pose a question which may sound trite, the question is, 'Why go at all?' In some cases the answer is obvious, one wants to visit friends, one wishes to see different parts of the world, but such travel is only a small part of the total passenger movement throughout the world which is made up of commuters going to work, businessmen travelling to meetings or going out to sell, experts getting together for discussions, a whole group of activities which might be carried out in a completely different way in the next twenty-five years. The way could well be the attack on space not by movement but by communication. Why should a man or woman travel to another city for a meeting when his effective presence could be there by the use of a television link? There is little reason why most meetings should not be through two, three or more TV links so saving time, fuel and money, because a TV link is cheaper than moving people about and accommodating them.

Even the person who now commutes to work could perhaps not go at all. A teletype keyboard at home or in a nearby office can often do as much as the actual presence of the individual, so lifting the bulk of the load off the transport system near large cities as well as shortening the effective day and adding to the comfort of the individual. Thinking this way we can see that there is a close link between transport and communications quite apart from the obvious links associated with navigation and traffic control.

Communications

The most popular and simplest form of communication at a distance has been the letter ever since the penny post was introduced in 1840. The delivery of mail has been automated to some extent but failing a means of reading written names and addresses, it still demands human sorters and men or women to deliver. This makes it economically unattractive in a world of rising labor costs, the problem is similar to the operation of transport systems, and mail is becoming so expensive that other methods of satisfying the same functions must be investigated. During World War Two, to save scarce shipping space, air letters were photographed onto microfilm, sent in this miniature form and enlarged at the destination. We may use an electronic version of this system in the future, it is quite practical to send what are called facsimile signals down a line or by radio – this is how photographs are transmitted rapidly across the world today – and it will almost certainly become economic to transmit ordinary correspondence in this way from place to place and reproduce it by electrostatic prints – as the Xerox system – for final delivery. People who would object, quite rightly, to unauthorized persons seeing their correspondence need have no fears – to be economic this system would have to be fully automatic.

The telephone has already reached the stage when most people in the world can dial a subscriber in another city, country or continent and this trend will continue until every telephone will be capable of being dialled from any other. Electronic exchanges already developed can store calls in case lines are busy and put them through automatically when the line is disengaged and we are just seeing the introduction of the Videophone. Not only do television telephones like this enable caller to see caller, they also enable photographs, drawings and sketches to be studied at a distance, making the telephone call much more like a real visit and often eliminating the need to travel. Unfortunately the old Chinese proverb that a picture is worth a thousand words has its analogy in communication systems, a good quality television picture occupies as much communication space as a thousand telephone calls. At the present time communication channels are expensive but new developments such as the use of waveguides which as their name implies, guide radio waves along tubes instead of using conventional cables. Waveguides are capable of carrying literally millions of communication channels down one pipe at the same time and will eventually increase communication capacity a thousandfold at a stroke. This is one area of man's activity where the future is fully assured. Waveguides are waiting for the time when communication demands rise to such a level that their expense is justified. Meanwhile there is another development, the use of light transmitted down minute glass fibres as a carrier for speech and pictures. The light is generated by a laser and both Bell Telephones in USA and the British Post Office are experimenting with optical links rather than wire links in the future. Both these developments will increase the capacity of the communications network sufficiently to make instruments like the Videophone practical and economic.

Telephone traffic

One tends to think of a telephone network carrying nothing but speech, yet even today the world's systems are involved with many forms of transmission. In addition to speech there are pictures – stills by facsimile transmission – television pictures, five unit code to operate teleprinters and, recently, a great deal of data transmission much of which is used in association with computers. Not all though, in some cases biological signals, like heart beats, are sent into a central point for analysis and diagnosis and we will expect more of this kind of traffic as the years pass.

The electronic revolution

Electronics is still a relatively young science, the electron itself was only discovered in 1897 and the vacuum tube or valve a few years later. In 1948 the transistor was announced and with it a revolution in electronics, no longer were we dependent on fragile and power hungry tubes working at high voltages. Early in the 1960s came another revolution, the full implications of which have not yet been realized by most people, the introduction of the *integrated circuit*. Now on a 'chip' little bigger than a pin's head, not just a transistor, but a complete circuit of an amplifier or radio set can be formed. Such integrated circuits are very small, cheap to make and, perhaps most important of all, almost 100% reliable. It is the integrated circuit which has made it possible to build a pocket calculator as cheaply as a slide rule with more complex calculators doing as much as a roomful of electronics could do in the early days of computers – just 25 years ago!

Tomorrow's radio and television

Already a radio set can be built inside the tuning knob and we can expect this trend to continue. The actual size of the integrated circuit is no longer significant, it is already much smaller than the rest of the equipment like the loudspeaker or headset. It is easy to forecast that in the near future people will listen to stereo radio using headsets into which the receivers are built. Such apparatus will be cheap and we can expect a family in the future with grandfather listening to Elvis Presley or the Beatles on one headset, mother with Bob Dylan or the Who on her set and the children listening to, well if I knew the answer to that I would be a millionaire!

The idea of everyone having a wristwatch-sized transmitter and receiver to communicate with other people anywhere in the world is still in the realm of science fiction and is unlikely ever to move out of it. For short-range working wristwatch-sized transmitters and receivers are technically totally feasible – we can see this from the minute radio bugs used for espionage purposes – but for long range transmission, i.e. over thousands of miles, one still needs aerials of reasonable size and sophistication. Yet Citizens' Band radio, already well developed in the United States, will certainly spread so that with relatively simple equipment the size of a car radio anyone can talk to anyone over a limited distance – say a few miles. Such equipment supplements the ordinary walkie-talkie which can now be built on a really small scale, on the lines of a radio-microphone used for TV.

Equipment now exists where walkie-talkie sets can use the ring main electrical wiring of a family home to act as an aerial.* The idea of a watch-sized television is not completely impractical and the 2-in. screen TV is being introduced in 1977 by Sinclair Ltd, although one wonders whether a minute viewing screen is really worthwhile except as something of a gimmick.

What we can anticipate is a great increase in the use of

*A group of houses in North London uses this technique for continuous conversations between the housewives during the day!

closed-circuit links, the kind of link we now see for security in supermarkets. The certain fall in real costs of such equipment as techniques improve and production runs increase, will make such links feasible in the home for baby-watching or at the front door so that one can look at a caller before answering the bell. Already in industry we see many applications of TV links to transmit information, either monitoring machinery or even a bank of dials, from one place to another and these applications will grow with lower costs and greater familiarity with the possibilities.

For some years now there has been promise of a large, flat TV screen which would hang on the wall like a picture and provide a picture in full color. The hopes have not materialized yet and it may be that we shall never see such a system in operation. Yet one development, at present in its infancy, known as the coupled-charge device (CCD) may lead to just such a display. If it becomes possible to do away with the large and cumbersome cathode-ray tube, not only could a set be made smaller with the same screen size but it would also eliminate the need for high voltages, so simplifying the circuitry and significantly reducing the cost.

Apart from this possible improvement we can expect the home TV in the next twenty-five years to change in detail rather than principle. What we can anticipate is the use of satellites to bring us TV from different countries always on tap. Experiments in India and Canada with satellites have shown the feasibility of people picking up signals on home sets direct from satellite using a simple dish in the garden as an aerial. The satellite would be in the now familiar 'stationary orbit' meaning that it really orbits once every 24 hours and so remains stationary above one spot on earth. It can easily transmit perhaps a hundred channels of TV plus an almost limitless number of speech channels so it will be possible to provide many entertainment programs as well as using it for education and for the transmission of vital information, e.g. from a doctor to a hospital who could send details of his patients, or even couple the patient directly to a transmitter so that his or her heartbeats, for example, could be sent to specialists – who might well be in the form of a computer – for expert diagnosis. The possibilities are nearly limitless and, as we shall see in a moment, the system can be coupled to data displays to provide printed material at the same time as another program is being watched.

Data displays and the encyclopedia at the end of your phone!

Within the last few years much work has been carried out in several countries to transmit data by television or telephone which can be 'called up' and read off a conventional TV screen, a system known as Teletext. In Britain both the BBC and IBA have pilot transmissions operating under the names of CEEFAX and ORACLE – the two systems are compatible.

The great advantage of both these systems is that information is transmitted down a TV link *at the same time as the normal picture*. This is done by using part of the time when no picture is being transmitted, the so-called 'fly-back' time. The principle of this is quite easy to understand. A television picture is scanned so that each part of a scene is looked at in turn, rather in the way we read a book. After each line is completed the spot on the cathode-ray tube has to fly back to the left-hand side and start again while after each frame, i.e. the last line at the bottom, has been scanned, the spot flies back from the bottom right-hand corner to the top left-hand corner of the screen. During this time – this is the fly-back time – no picture is being transmitted although there are pulses which synchronize the scanning at the set with that at the transmitting studio. The TELETEXT information is ingeniously fitted into this 'dead' period, so no picture information is lost. (I have ignored subtleties like interlacing, a trick by which each picture is scanned twice, the second time on alternate lines, but this does not affect the validity of the explanation.)

TELETEXT displays the information as ordinary print, with capitals, lower-case letters and numerals. Each character can be displayed in one of seven colors. A page of TELETEXT consists of 24 rows each containing 40 characters, a

total of 960 'bits' of information, or can be in the form of a
line drawing. Each television channel can accommodate 800
separate pages giving a total storage of over three-quarters
of a million bits, equivalent to a 24-page newspaper. The
viewer has a switching device which he can first of all select
TELETEXT in place of the normal TV picture and then
choose the page he or she wishes to read. This is done by
flicking through the pages electronically at high speed until
the required one is located. If this turns out to be the last
of the 800 pages, it can be selected within three minutes.

TELETEXT began operating experimentally on both
BBC and IBA networks in 1975. An attachment to any television set enables viewers to use the service, but such attach-
ments are not yet available commercially, when they are it is
hoped that they will cost less than $90. Information can
be fed into the system as it becomes available so that one
should be able to have in one's home a completely up-to-date
newspaper, weather information, items of general interest
such as cooking recipes as well as the exact time and, perhaps,
selected advertisements of general interest such as what's on
at the local cinema. An additional feature is that a channel
can be selected to 'inlay' information into the normal TV
picture so that emergency messages like hurricane warnings
will appear over the picture automatically. It is possible to
record the displays permanently by electrostatic printing
similar to the Xerox system.

TELETEXT necessarily carries information for the general public, but another system recently announced by the
British Post Office can provide material tailored to the requirements of the viewer. Like TELETEXT the information
is displayed on an ordinary TV screen, but instead of being
transmitted with the TV picture it comes down the viewer's
telephone line. This system is called VIDEOFILE and it can
be used to supply information from a central computer. One
possible application is as a reference book. Using a keyboard
like a typewriter anyone could then 'request information',
e.g. the height of the Niagara Falls or a short biography of
Cleopatra, the information would then appear via VIDEO-

FILE on the screen. VIDEOFILE started pilot trials in 1975, a market trial was started in February, 1978, with 80,000 pages on tap. Later this will, if the system is a success, be extended to several millions. Pages are selected by 'asking for' a given page using a selector, after which the information starts to appear after only about five seconds. When reading the information the viewer presses a key to 'turn over' and the next page appears.

VIDEOFILE differs from TELETEXT in an important way, in that the viewer is able to answer back. For example a store might demonstrate its goods and a viewer could select what was required and send an immediate order. He will also be able to send information to a selected subscriber – perhaps play games by telephone. The possibilities seem endless; there is no doubt that TELETEXT and VIDEO-FILE are going to make life easier and more entertaining for everyone before long, indeed such interactive video terminals are now spreading fast into all walks of life.

As mentioned, a system like VIDEOFILE will make many books redundant. Reference books, like encyclopedias, are out of date before they are delivered, so fast does information accumulate, and it seems reasonable to suppose that in the not too distant future such books will all be stored, and continually updated, in a central computer. Then anyone wanting information will ask for it, probably using a keyboard, and it will appear on the screen within seconds. It may not be only data, a subscriber might ask for something to be worked out, such as the solution to an equation or, at a more mundane level, the thickness of wood for a shelf needed to carry a heavy load. The computer could do such a calculation at high speed, select from the answer the nearest available thickness of wood on sale and even give a list of shops holding it in stock with the prices and finishes available – all quicker than it takes to read these words.

Recording television

We are now quite familiar with the idea of recording sound programs off the radio, whatever may be the legal problems involved. Some people already have videorecorders capable of recording TV programs in full color or, on the other hand, recording the output from home TV cameras – a kind of electronic home movie.

Picture recording is at present somewhat expensive, a videorecorder costs about as much as a color TV set, but there are indications that the prices will fall. In which case we may see the end of home movies on film. Television is simple to use, enables monitoring of the picture during shooting, immediate playback and the re-use of the recording material. As everyone has a TV set, part of the cost has already been paid and even today a videorecording system compares favorably with an expensive home movie outfit including a projector. There are snags of course, portable TV cameras are bulkier than movie cameras and editing is, at the moment, limited. But technical improvements will doubtless eliminate these really minor problems and we can confidently predict that most people will have the equivalent of a movie studio with pictures in color with full sound available for showing without even having to set up a projector.

This is the home market of do-it-yourself recording, but there is another important field, the field of hiring or buying pre-recorded programs as we now hire or buy films. This is an area where an immense amount of work is going on behind the scenes by many of the large electronics firms to develop a system which will become a standard. There are several possible methods of recording for playback through a normal domestic set. Videotape is obvious but is expensive both in terms of the tape itself and in the cost of copying which has to be carried out by recording from a master. Film is possible; this is expensive in terms of material but much cheaper to copy.

Several manufacturers are experimenting with disc which

is cheap but poses enormous problems of actually recording high quality television signals with their large bandwidth onto a fine groove. Certainly one of these systems will eventually triumph and the day is not far distant when we shall be able to buy videorecords of one kind or another and watch our favourite artist perform as well as hear him.

Commercial communications

Although we are naturally especially interested in communications as they affect us directly, by far the greatest use of the media is for scientific and commercial purposes. Already as we have seen we can substitute electronics for actual objects in many areas of medicine, science, commerce and government and the limitations are generally lack of information channels. Until 1956 our only method of speaking across oceans, say from Britain to America, was by radio. Then came the first transatlantic speech cable followed very soon by the first active communications satellite, TELSTAR, which could send TV pictures as well as speech, still pictures and telex. Today we have many cables and satellites spanning the world, but demand for communication space is continually rising as, in particular, computer data channels increase. So there is a constant race on the long distance circuits between demand and supply. On the shorter haul, as we have seen, the technology is available when the need arrives so there are no signs of the communication explosion abating.

Communications outside the earth

Lunar landings and space probes which have travelled round our solar system have demonstrated what can be done in terms of long range radio using small, light equipment on the spacecraft even though the ground equipment is massive and complex. Perhaps the most exciting aspect of space communications is the possibility that one day we might communicate with intelligent life somewhere in the universe. Although it seems as though there is no such life in our own solar system, we have no information that it does not exist on a planet orbiting some distant star. We have almost no evidence that such planets exist (see though Chapter 8) but it would seem highly improbable that there are not other habitable worlds and improbable that some of them are not inhabited. This is why there are space listening posts already in operation trying to receive intelligence from somewhere else.

Such posts do not expect to receive speech, instead they are using computers to analyse the crackles and hisses which come in to us all the time on radio waves from space into something organized rather than random. If a distant civilization were trying to communicate with us it is interesting to consider what might be transmitted. Many people believe that messages would be deliberately simple and easy to understand probably in the form of two sets of dots, perhaps consisting of three and five dots, followed by their sum, eight dots. Such a message would demonstrate that it was sent by some intelligence and could form the beginning of a conversation. The conversation would necessarily be protracted as our nearest star, neglecting the sun of course, is $4\frac{1}{2}$ light years away, meaning that it would take nine years to get an answer back to any question.

Impossible, you say? Remember those words of von Braun about *impossible* being a word that should be treated with the utmost caution. One thing we can say, there is no single endeavor of man more important than communication and we shall all take to heart the words of Thomas Erskine:

> When men can freely communicate their thoughts and their sufferings, real or imaginary, their passions spend themselves in air, like gunpowder scattered upon the surface – but pent up by terrors, they work unseen, burst forth in a moment, and destroy everything in their course. Let reason be opposed to reason, and argument to argument, and every good government will be safe.

Words as true today as when they were spoken at the trial of Thomas Paine in 1792.

Publisher's postscript – extra-terrestrial communication – linked to space research

Man's first serious attempt to reach out to another civilization began on March 3rd, 1972, when Pioneer 10 was launched.

Pioneer 10 is the fastest space ship yet launched from Earth – yet it will not enter the planetary system of any other star for over 9 billion years. Only a civilization sophisticated enough to locate and recover the space ship will know of our presence. The visual message aboard was devised by Carl Sagan, one of the leading US astronomers and a leading thinker on the implications of the universe of which we are such an infinitesimal speck.

His design has been widely reproduced and it is symbolic not just for possible receivers but for us. It is written in the only possible shared language – science.

Top left is a schematic representation of the hyperfine transition between parallel and antiparallel proton and electron spins of the neutral hydrogen atom – the most abundant atom in the galaxy.

The radial pattern fixes the position in time from where the space ship left in relation to 14 pulsars (neutron stars).

An advanced extra-terrestrial civilization would be able to plot the launch position to an area comprising 'no more' than 1000 stars in the Milky Way. Of these 1000 only one star has the number and range of planets as represented at the *base* of the diagram. That star is our sun.

The approximate sizes of our sun's planets are shown and there is a representation of Pioneer 10 launching from Earth – and passing Jupiter.

Any extra-terrestrial beings capable of intercepting this ship must inevitably be *much more* advanced than we are and would find decoding the schematics no problem – it perhaps puts *us* in context to think it is the drawing of the man and woman which is likely to puzzle them! – for the chances of life evolving on other planets similar to ours in form or even biological system are remote.

The reverse idea – trying to monitor – listen for possible

broadcasts from advanced extra-terrestrial civilizations is beginning to be explored seriously. A special USA/USSR joint conference in 1971 concluded that the time was ripe and the practical impact and possible knowledge reaped would be immense.

At the same time, the probability of success is not high, although there are 250 billion suns in *our* Milky Way galaxy alone and billions of other galaxies. There must be millions of other earths. Our evolution as a planet must have been repeated millions of times but it is doubtful whether the life evolution has been similar to our animals and humans – so much of our evolution has been through accidents and chances.

A chance eavesdropping on extra-terrestrial signals is almost nil – for example 100 years ago we had no radio signals leaving Earth – 100 years from now it is possible that cable TV developments and optic beam and light beam transmissions from satellites will mean little stray radio signals will escape.

Quite possibly an advanced society has developed communication methods (i.e. not radio) quite beyond our comprehension. An excellent parallel is that if a native who used drums as communication was asked to imagine a sophisticated method he would probably answer 'a loud drum' – while all the time radio waves were passing through him. In just such a way we tend to imagine sophisticated communication as more powerful radio. (Another example of closed thinking.)

Conversation there will not be. The vast distances between stars means that if life were found on planets round our *nearest* star – a radio signal out takes 5 years – the answer back 5 years. In fact it is far more probable we would find communicative life 100 light years away – in which case a quite different generation would pose the questions to that receiving them!

More and more as we consider space and astronomy it is its value in giving us *perspective* of man's place in the universe that becomes clearer. It is no accident that the surge of interest in ecology – and quasi-religions – appears to date from the first satellite pictures of Earth – a small, vulnerable and beautiful blue sphere hanging defencelessly in space. Carl Sagan – in his thought-provoking book *The Cosmic Connection* explores a number of points that repay repeating.

Firstly, as the title implies, we *are* creatures of the stars in a literal sense. Our galaxy, sun and planets were formed billions of years ago from material thrown violently into space at the 'death' of a giant sun – a supernova explosion. A star is born out of gas and dust, evolves by thermonuclear fusion in its interior (converting 4 atoms of hydrogen into 1 of helium) and in its later death stages temperatures rise further and heavier elements are formed. These are precisely the same elements most common on earth. The rock we stand on, the iron in our blood, and carbon in our genes were produced billions of years ago in the interior of a giant red star – and were thrown out to form planets and new second generation suns. A further direct link between supernova explosions and humanity is cosmic dust – minute particles thrown out from the dying star; some hit Earth and cause mutations – variations in our gene blueprints, DNA. Mutations are generally adverse but now and again a freak results in a major advance. Mutations are responsible for evolution – just as the rare leap of imagination is responsible for the truly heroic invention.

Our own sun is growing brighter and 4 billion years hence the Earth's surface will become too hot for habitation – the greenhouse effect. Interestingly the *same* solar heat increase will change Mars from a $-100\,°F$ average to a similar temperature to that of the world. Our ancestors should note!

Secondly, the study of planets can provide lessons of what to do and to avoid on Earth. Venus, for example, despite being similar in size to us and not so near the sun as to preclude life, has a totally inhospitable environment – the main cause is a massive concentration of CO_2 and water in the atmosphere preventing the escape of infra-red heat emissions which has ultimately produced a climate of $900\,°C$. This is not just of academic interest as we may learn from the love planet how to avoid the same runaway greenhouse effect on Earth being caused by the burning of fossil fuels. Oil and coal increase CO_2 in the air which blocks the infra-red emissions from the Earth's surface – thus trapping heat and raising temperature – heat pollution!

Thirdly the need for fully automated planet vehicles is already forcing the pace of developing robots to help take over drudgery jobs.

Fourthly Sagan makes the point that the huge expenditure on defence is partly self-perpetuating. 'The military establishments in the US and USSR owe their jobs to each other, and there is a very real sense in which they form a national alliance against us.' Similarly powerful interests are vested in the armaments industries. Sagan suggests we should direct this collection of talents and organization towards the exploration of space – the technical requirements of both defence and space being similar.

Fifthly we are overcrowded, psychologically as well as physically. October 12th, 1992 is the 500th Anniversary of Columbus's discovery of the New World – it could be that one mental escape for us will be the search for new planetary worlds. (Note, not *stellar* worlds, for with the current technology the Pioneer 10 space ship would still take 80,000 years to travel to the *nearest* star.) The infinity of space and some of its more grandiose possibilities will keep *some* men from boredom for ever.

Possibly one of the most dramatic areas of study currently is the death of stars (the study of supernova). When a large star runs out of hydrogen to burn, it explodes, becoming brighter. At the same time an alchemist's dream come true. We have already seen how the upward conversion of elements results in oxygen changing to magnesium, etc., at a later stage elements like gold and uranium are created from iron! The explosions rip away most of the elements (star stuff destined to become second generation stars and planets) and what remains are white drawfs comprised just of the nuclei of atoms with the electrons stripped off. The nuclei therefore no longer repel each other with negative electricity and move together to become *incredibly* dense material – so dense that a thimbleful would weigh a ton. Some white dwarfs are pure carbon – a star made of diamond!

For larger white dwarfs the cycle does not end there and the crushing pressure does not stop until the inner nuclear force equals the outer density. Such a star can condense to about a mile wide and matter is so dense a speck would pass through our world like hot butter. These are neutron stars or pulsars.

The most incredible type of star discovered so far, however, must be the Black Hole. It seems to happen that when a really massive star collapses not even the counter-balancing inner nuclear force can stop its continuing contraction. The pull of its gravitational field becomes so strong that not only matter cannot escape but nor can light – and because light cannot escape the star becomes invisible. Hence Black Hole – an area of intense gravitational activity.

Black Holes are certainly provable theoretically – and one may now have been actually identified as Cygnus X–1. The incredible properties of Black Holes do not end with their invisibility – at least in theory. Because there is no way out of a Black Hole, any object sucked into its gravitational influence would disappear – maybe (runs the speculation) into another universe – or to reappear some*where* else some*time* else. For this reason, Black Holes have been propounded as the ultimate answer to near instantaneous stellar travel.

Whilst Black Holes are irrelevant to the real objectives of Science Fact – we could not pass them by without some comment.

LIVING SEA

Carolyn Roberts

Carolyn Roberts studied Zoology, Botany and Geology, specializing in Marine Studies which culminated in a research project on sea-birds of the Farne Islands. For the past 7 years she has been editor of *Aquatic Sciences and Fisheries Abstracts*, which is a module of the international information system of the same name.

'And God said, Let there be a firmament in the midst of the waters, and let it divide the waters from the waters. And God said, Let the waters under the heaven be gathered together in one place, and let the dry land appear, and it was so. And God called the dry land Earth; and the gathering together of the waters called he seas: and God saw that it was good. And God said, Let the waters bring forth abundantly the moving creatures that hath life ... And God created great whales, and every living creature that moveth, which the waters brought forth abundantly ... and God blessed them, saying, Be fruitful, and multiply, and fill the waters of the seas.'

– From the Book of Genesis.

Introduction: the world ocean

About 71 per cent of the Earth's surface is covered by water, at an average depth of 2 miles. If you imagine that the Earth is the size of a grapefruit, then the film of water would be the same thickness as this sheet of paper. The vast volume of water is called the world ocean, and in it live the vast majority of the Earth's inhabitants. The ocean affects all our lives, via the weather. The immense stretches of open water are a source of water vapour in the atmosphere that produces rain. The seas redistribute heat, transporting it from the tropics in great currents like the Gulf Stream and the Pacific Kuroshio into higher, cooler latitudes. It was in the ocean that the first living creatures were born : primeval ancestors, tiny blobs of life which, after millions of years, eventually evolved into man himself. Now, ironically, man has turned on his watery birthplace, and has done much to alter its character – sometimes deliberately, like in the building of the Suez Canal; sometimes unscrupulously, by overfishing, and sometimes thoughtlessly, by dumping waste matter in it.

All of us learn in school to identify the names and positions of the continents and oceans. This shows us that all the land masses are surrounded by ocean, but it is slightly misleading, because it suggests that the different seas and oceans are separated geographically. We should instead try to think of a world ocean that is continuous and completely intercommunicating. This body of water extends from the Arctic to the Antarctic, and although it is forced to twist its way around the continental land masses and forms distinct basins which we name, all of the basins are connected. Of course, the different parts of the world ocean are not identical physically and chemically : some waters are warm, some cold; some saltier, some fresher; some dirty, others relatively unpolluted. However, the differences in chemical content are comparatively slight. Tests have revealed that in the relative proportions, ocean basins at all depths revealed striking similarities. This implies that the oceans have been well mixed, and that the relative proportions of various chemicals flowing in from the rivers has been removed by a stirring process, the winds and currents acting like giant egg beaters, to blend in all the ingredients. The complete mixing time is thought to be about 1000 years. It is important to remember this unity in time and space when we consider the sea as a resource, because any change that we cause will influence all areas.

Marine study has changed its emphasis and scope throughout history. The ancient mariners traversed the sea in search of new lands, or to transport goods from one part to another. Their knowledge of the sea was concerned with winds, currents, sailing conditions and navigation. In the late 19th century, however, scientists began to study the world ocean for its own sake. The early scientists accomplished the first ocean-wide survey of the marine environment. Probably the most famous expedition, which opened the era of ocean exploration, was made between 1872 and 1876 by HMS *Challenger*. The *Challenger*, with its crew and 7 scientists, crossed the Atlantic, Pacific and Antarctic Oceans, travelling over 125,000 km (68,900 nautical miles). The expedition observed weather, currents, water chemistry at all depths, temperature, bottom topography, sediments and marine life on a global scale. These measurements provided the factual information which was the foundation for the science of oceanography.

Why should we wish to study the ocean at all, except for purely scientific fascination? Well, the ocean represents an incredibly large natural resource for all mankind. The sea is a treasure trove, providing us with chemicals, minerals, food and energy. Much of the bounty is as yet relatively unused, but as population pressure increases, so will the demand for resources. The ocean basins contain approximately 1,370,100 cubic km (350 million cubic miles) of seawater, which is in itself an important resource as a possible supply of drinking water. Every cubic kilometre of seawater contains about 39 million metric tons of dissolved solids, of which only common salt, magnesium and bromine are being extracted in large quantities. Other valuable ocean resources are found on or under the sea floor. Deposits on the bottom surface include minerals precipitated from seawater, sand, gravel and oyster shells. Offshore gas and oil wells account for 17 per cent and 6 per cent respectively, of total production in the western world. By 1980, one third of the world's oil and gas will come from offshore wells.

Food resources of the sea are as considerable as physical resources. The annual income from the world marine fisheries is 8 billion dollars, far in excess of the monetary return from all other types of exploitation. Geographically, the Pacific Ocean yields 53 per cent of the marine harvest; the Atlantic Ocean yields 42 per cent; and the Indian Ocean 5 per cent. The low figure indicates a lack of exploitation rather than a poor resource.

How can we summarize the properties of the sea? The most obvious property is the ocean's vast area, great depth and huge volume. Seawater is the most abundant substance accessible to man, and all continents are in fact islands in a world-surrounding sea. The shoreline of the sea gives many people access to it and enables us to enjoy its benefits. More important, the shoreline allows relatively easy access to vast areas of land.

The fluid character of the water on our planet is the miracle that makes life possible, but it also means that the oceans fill all the low places on Earth. Because of this geographical fact, the oceans are the ultimate receptacle of the wastes of the land, including the wastes that are produced in ever-increasing amounts by human beings and their industries.

The relatively high density and low viscosity of seawater are essential qualities that make the sea surface a broad and travelled highway. Large ships and heavy cargoes can be moved fairly rapidly across the ocean with comparatively little motor power. Maritime commerce would be impossible if seawater were as viscous as syrup, and the fuel required to carry the huge cargoes of modern ships would be positively expensive if water were as light as air! At the same time, the combination of low viscosity and high density gives rise to the principal hazard of the sea, the giant wind-waves caused by storms that crush small ships and attack coastal structures. If the water were more viscous, the wind could not build up high, steep waves, and if it were much lighter, the wave force would be insignificant.

The high surface tension of seawater, although not of fundamental importance, is a matter of considerable convenience to human beings, because it means that water does not stick to surfaces with which it comes into contact, but runs off easily, and leaves them comparatively dry. What a messy business seafaring would be if the oceans were the consistency of oil!

The high density and fluidity of the water create a serious difficulty when men attempt to lower themselves or their equipment below the surface; the enormous hydrostatic pressure at great depths crushes all but the strongest vessels, forms air pockets in submarine cables, and produces high stresses in equipment made from materials with different compressibilities.

The saltiness of seawater is its least desirable quality; man can neither drink it, nor use it to water his crops. Radio-waves penetrate only a short distance, because the ionized salts of seawater make it a good conductor of electricity, and the salts plus dissolved oxygen cause the seawater to be highly corrosive for most man-made objects.

Seawater is much less transparent to visible light than air is, but more transparent than other substances. This intermediate transparency, combined with a high heat capacity and a high heat of vapourization, makes the ocean a giant thermostat for the Earth. Most of the energy of sunlight passes virtually unimpeded through the atmosphere into the ocean, where it is absorbed and transformed. Nearly a third of all solar energy reaching the Earth's surface goes to evaporate seawater, which eventually falls as rain.

People who study the sea come from all branches of

science. In an oceanographic institution you can find bio-logists, chemists, physicists, geologists, meteorologists and engineers. Their common bond is the sea, even though their education and techniques may be completely different. This is because it is difficult to separate the study of the ocean as a subject on its own – oceanography is an overlapping study of many subjects. For instance, a scientist studying the effect of the marine environment must understand the character-istics of the sea bottom (geological oceanography), the nature of the ocean currents (physical oceanography), the chemistry of the water (chemical oceanography) and of course, the biological community itself (marine biology). Hence, we refer to oceanography as a multidisciplinary science.

The seas and oceans have always held a fascination for mankind. The ocean has an impact on all our senses : the unique sea smell, the crashing sound of the breakers, the gleam of the waves under sun and moon, the feel of the spray, the salty taste of the water. Part of the spell is, as the poets have said, one of mystery and timeless motion. The English painter J. M. W. Turner was fascinated by the sea, and in order to make observations for his famous paintings, he found a house by the sea in east Kent. He allowed the locals to think he was an eccentric sea captain named Puggy Booth, who even in retirement, could not stop looking at the sea. In the following sections, we attempt to explore some of the mysteries of the sea, and to discuss some of man's most recent studies. The rate of increase of oceanographic facilities and studies is accelerating, yet, even with this total effort, our knowledge of the world ocean is still inadequate to solve many of the practical problems that exist. As we shall see, there is still much to be learnt.

Communication in dolphins, porpoises and whales (Cetacea)

'In the world of mammals there are two mountain peaks, one is Mount Homo sapiens, and the other Mount Cetacea.' This quotation is how Prof. Teizo Ogawa of the University of Tokyo describes the intelligence of the whales and dol-phins (Cetacea). The world of the dolphin is mainly one of sound. A great part of its large brain (larger than that of man) is concerned with the transmission and reception of sound. Part of this brain is concerned with the dolphin's sonar, the narrow beams of high frequency sound which it emits through its forehead. Some of these can be heard by man as clicks, but most are ultrasonic. These sounds are bounced back by solids in their path and are received by the dolphin through the lower jaw and ear. In this way, a dolphin finds its way around and hunts for food, and man has copied this method in his own version of sonar fishing. In addition to the sonar, dolphins communicate with each other by whistling sounds, and they can even make shrieking noises in air to communicate with men.

Dr. Lilly, a neurologist who has done much work on the brain of dolphins, claims that the dolphin mimics human speech 'in a very high-pitched Donald Duck quacking-like way'. Dr. Lilly claims that 'within the next decade or two the human species will establish communication with another species: non-human, alien, possibly extra-terrestrial, more probably marine'. The press has nicknamed him 'the man who will make fishes talk'. Should he prove to be right, all sorts of ethical problems could arise! Could our whaling fleets continue to kill animals who could answer back?

Dolphins are amazingly willing and co-operative during experiments. On one occasion, Dr. Lilly found that the dolphin was experimenting on him rather than the other way round. Dr. Lilly was making the dolphin produce a whistle of given pitch, duration and intensity, in order to gain a reward. Every time it whistled, it was possible to see the blow-hole twitch. After a few minutes, the dolphin became bored with this game, and added a new rule. It raised the pitch of each successive whistle. Soon it was out of Dr. Lilly's acoustic range, but although he could hear the notes, he could still see the twitching of the blow-hole. The rewards were stopped, and the dolphin returned to whistles Dr. Lilly could hear.

This same dolphin was kept in a small tank for several days during cold weather, before it was released into a larger tank with two other dolphins. The cold and the restraint had caused its back to stiffen into a painful S-shape, and it could not swim properly. It immediately started to give the dolphin SOS call, and the two other dolphins swam to it and lifted its head out of the water. A great deal of twittering and whistling took place between the three dolphins, and then the two healthy ones started to swim under their injured companion, raking the base of its tail with their dorsal fins. This caused the injured animal to bring its flippers down in a reflex jerk, thus forcing itself up in the water. They kept this up for several hours, thus saving the life of the injured one, which would have drowned if it had sunk.

The inter-dolphin language seems to be very complicated. There are a large number of calls which humans are beginning to identify, for example, the babies' call on separation from their mothers, the SOS call given by sick or injured dolphins, the 'I am bored' signal whistled by dolphins separated from dolphin or human companionship, the 'keep together' whistles of the school, the mating call, and the 'hands off' warning from one male to another. In addition, there seem to be far more sophisticated communications, and recordings show that they do not whistle at the same time, but take it in turns, as though having a conversation. There is evidence that dolphins and whales may communicate to each other descriptions of strange objects, their position, and whether or not they are harmless. Recordings of dolphin sounds have proved difficult to interpret, and much more research is needed to explain how messages can be passed on without complex descriptive language.

Dolphins trained for warfare

It is a very sad fact that man is now considering using dolphins and small whales in warfare. They have already been used for peaceful purposes under the sea, for tool-carrying and the guiding of divers in several underwater experiments. However, the United States Navy is at the moment using dolphins for purposes which are undisclosed, and a group of them were brought back from Vietnam! The Russians are also running a research program on the military use of dolphins, and it seems likely that they are being trained for submarine detection and for the carrying and placing of mines. How ironical that such a gentle and peaceful animal should be used as a tool in man's urge to destroy himself!

Life of the sea: the floating pastures

All living organisms need energy to live, and marine animals derive their energy by eating organic matter. They do this by feeding on other animals or plants (live or dead), or on organic debris brought into the sea by rivers. Marine plants, however, like land plants, are able to manufacture their own energy supply. The fuel for this process is sunlight, which is intercepted by the green pigment (chlorophyll) of plants and used to bring about the reduction of dissolved carbon dioxide to carbon, and thence to carbohydrate. This reaction is probably the most important in the world, because it ultimately fuels all life processes. It is called photosynthesis, and can be stated chemically as follows:

$$6CO_2 + 6H_2O = 6C_6H_{12}O_6 + CO_2$$

This reduction is made possible by the sun's energy; it will not take place in the dark, nor in the absence of chlorophyll. Because energy has been put into the reaction, the product, carbohydrate, contains energy which can be recovered. The plant uses this energy to grow and reproduce, and itself provides energy for the animals which feed upon it.

Sea plants

In the sea there are two kinds of plant: the familiar seaweeds of the seashore, and tiny one-celled plants which are part of the plankton; namely, the community of minute animals and plants which form an important source of food for larger animals. Although the seaweeds of the shore do contribute to the energy supply of the sea by photosynthesis, more than 90 per cent of organic material is synthesized within the lighted surface layers of the open water by the plant (or phyto-) plankton. We can regard this thick green soup of plant cells as a marine pasture which is grazed upon by animal (or zoo-) plankton and by some small fishes. These in turn are prey to various carnivorous creatures, large and small, who have their predators also. The debris from activities in the surface layers of the sea settles in the dimly lit and unlit layers, and provides a source of food for the inhabitants there. The line of organisms one feeding upon the other is sometimes referred to as the 'food chain' of the sea, but when one studies the infinite variety of organisms and the ingenious ways they have of obtaining food, one can see numerous interacting chains forming a complex network rather than a straight line. Nowadays we tend therefore to refer to the 'food web' of the sea.

It is not immediately obvious why the most important production of organic matter in the sea is carried out by single-celled plants. After all, on land, photosynthesis is carried out by trees as well as grassland. Why are there no 'trees' in the ocean? The answer is that in order to develop, disperse and reproduce in the sea, there are many advantages in being small. Currents cannot damage these small plants, and their mobility ensures that they are continuously swept into nutrient-rich areas. Marine plants are also small enough to be dispersed in storm-borne spray, in bird feathers, and by birds and fish in undigested food.

The zooplankton mainly consists of arthropods, namely, animals with external skeletons that belong to the same broad group as insects, crabs and shrimps. The planktonic arthropods include the abundant copepods, which are in a sense the marine equivalent of insects, and are represented by some 10,000 species. Other important arthropods are the shrimp-like euphausiids which occur in vast shoals and are known as krill, constituting one of the principal foods of baleen whales and some plankton-eating sharks. Zooplankton provides food for organisms in the pelagic zone, or the middle depths. This region contains some of the largest and most superbly designed creatures of the Earth, and includes jelly-fish, squid, tuna fish, multi-colored dolphinfish, the conversational porpoises, swordfish and toothed whales, seals and sealions, and sharks awaiting injured prey. Marine birds also feed on pelagic life, diving, plunging and skimming after their prey.

The ocean's mammals are one of its most finely adapted groups. The order Cetacea, to which the whales, dolphins and porpoises belong, is one of the most fascinating in the animal kingdom. Among them is the largest species of animal that has ever lived, the large blue whale which can grow to a length of 100 feet. Cetaceans are so much adapted to aquatic life that stranding results in their death. There is much for us to learn from marine mammals. Their ability to dive to great depth and to surface quickly without suffering the cruel 'bends' is due to a physiological adaptation which we imitate by giving divers mixtures of gases to breathe while diving. Their built-in echo-location system we imitate as sonar. A lesson we still have to learn from the whale is how to effectively harvest the krill. The baleen whale gathers and absorbs it by pushing its vast mouth through the 'swarms'. It is well adapted to take on very small creatures in quantities sufficient enough to nourish its huge bulk. If the whale had had to rely on a feeding mechanism resembling trawls towed at great expense from high seas trawlers, it would have starved towards extinction instead of being blasted by the harpoon gun! Fishing gear resembling the open-mouthed scoop method of the whale must therefore be developed in order to harvest this valuable resource, which is now much more plentiful than it was, due to the decrease in whale stocks. How we would use krill is not yet certain, but it would almost certainly provide a good source of animal feed, or even, if made palatable, 'plankton biscuits' for human consumption!

The floor of the sea provides a site for attachment for invertebrates which filter detritus from the water. Among these are some of the most primitive creatures reminiscent of the earliest forms of sea life: glass sponges, sea lilies (once thought to be long extinct) and lamp shells. At one time it was thought that the abyssal floor was sparsely inhabited, the populations being supplied with food from the slow fine rain of detritus from the upper layers. The latest explorations have shown that this is not so. Until recently, the deep fauna was studied mostly by trawl or grab samples that were extremely time consuming, technically difficult and expensive. Two new techniques have made a more extensive study possible. The simplest is drop-lining. Fifty baited hooks or traps are set on the bottom and brought up after a few hours. The second system is a pop-up camera developed at the Scripps Institution in California. A baited, free-fall camera, which is dropped to the bottom, takes a programmed sequence of still photographs or cine sequences and can then be acoustically commanded to drop its ballast and pop up to the surface for recovery. The speed at which bottom fishes and sharks accumulate around the bait suggests a more active community than first thought, and that their main source of food may be the carcasses of large whales or tuna fish which sink rapidly to the bottom on death. Further exploration of the abyssal realm will undoubtedly reveal undescribed creatures including members of groups thought long extinct, as well as commercially valuable populations.

Man is the ocean's most voracious predator, at the very top of the food web. The present ocean harvest is about 55 million metric tons a year, bringing an annual income to the world's fishermen of 8 billion dollars, more than twice the revenue from the sea's oil and gas! More than 90 per cent of the harvest is finfish; the rest consists of whales, crustaceans and molluscs, and some other invertebrates. In the century 1850–1950, the world catch increased tenfold – an average rate of about 25 per cent per decade. In the next two decades, it more than trebled, and this rapid growth is still continuing. There has been a recent trend for less of the catch to be used directly as human food, and for more to be reduced as meal for animal feed. The proportion of the catch preserved by drying or smoking has declined in recent years, as has the proportion sold as fresh fish. The consumption of canned fish has not changed substantially, but that of frozen fish has grown to approximately 12 per cent of the world catch.

Although, with certain exceptions, the traditional fisheries in the colder waters of the Northern hemisphere still dominate the statistics, the emergence of some of the less developed countries as modern fishing nations and the introduction of long-range fleets mean that tropical and subtropical waters are beginning to contribute significantly to world production. Two countries have emerged in recent years as new fishing powers : firstly, Peru has become the leading country in terms of sheer magnitude of catch through the development of the world's greatest one-species fishery, 10 million tons of anchovies a year, almost all of which is reduced to meal. Secondly, the USSR has emerged as a fishing power of global dimension, fishing for a large variety of products throughout the oceans of the world, particularly with large factory ships and freezer trawlers.

What effect has the enormous increase in the world's fishery operations had on the fish stocks of the ocean? Increased fishing tends to reduce the level of abundance of stock progressively if the catch is greater than the natural increase in population. This is called 'overfishing'. Unfortunately, each fishery tends to expand beyond its optimum point unless something such as inadequate demand hinders its expansion. This is because it will usually still be profitable for a fisherman or ship to continue fishing after the catch is no longer increasing, or even when it is declining. By the same token, it may continue to be profitable for the fisherman to use a small-meshed net and thereby catch young as well as old fish, but in so doing he will reduce the catch in future years. There is, then, a limit to the amount that can be taken year after year from each natural stock of fish. Is it possible that the ocean could one day run out of fish? The drastic decline in baleen whales in recent years provides a dramatic example of the effects of overfishing, and there is no reason to suppose that similar declines could not occur in certain of our fish species. The extent to which we can expect to increase our fish catches in the future will depend on three considerations : first, how many as yet unfished stocks await exploitation, and how big are they in terms of potential yield?; secondly, how many of the stocks on which the existing fisheries are based are already reaching or have passed their limit of yield; thirdly, how successful will we be in managing our fisheries to ensure maximum sustainable yield from the stock?

Several major international conferences have attempted to examine the state of marine stocks on a global basis, and it has been found that fully fished or overfished stocks include some tunas in most ocean areas, the herring, cod and ocean perch in the North Atlantic, and the anchovy in the southeastern Pacific. The classical process of fleets moving from an overfished area to a fully populated one cannot continue indefinitely. It is true that new stocks are being discovered, mostly in the Indian Ocean and the eastern Pacific, but in another 15 years, very few accessible stocks of commercial interest will remain underexploited.

The Food and Agriculture Organization of the United Nations is attempting to predict the production of fisheries in 1985 and beyond. With the cooperation of a large number of scientists and organizations, estimates are being prepared in great detail on an area basis. One fact has become clear: our knowledge in quantitative terms of the living resources of the ocean is still very scrappy. There are several ways of estimating present and future fish populations: directly, through knowledge of present stock and mortality rates; by counting fish individually using echo-sounders; by age studies based on the growth rings in fish scales; by counting fish eggs in the plankton; and by making comparisons between areas with similar conditions. It has been concluded from all these studies that the potential catch is about three times the present one. Given the rate of fishery development in the past, it would be reasonable to suppose that providing the stocks were managed properly, the present maximum sustainable world catch of 55 million tons could reach 200 million tons by 1985, or at least by the end of the century. But will it be worth it, economically? We cannot assume that past growth will be maintained – fishing will have to be even more efficient in the face of depleting stocks, and nations will have to agree on regulations to prevent overfishing.

There is still much to be done in increasing the efficiency of fishing gear and ships. We are only just beginning to understand exactly how trawls, traps, lines and seines really work. The trawl is not a simple bag collecting more or less passive fish. Fish are lively, sensitive creatures, and are not predictably active. Much research needs to be done on fishing gear relating hydrodynamics to the physical and biological environment. The gear of the future will be able not only to fish more deeply, but will also be able to home in automatically on deep-dwelling concentrations of fish and squid, using acoustic devices. If such methods were used in the under-exploited Indian Ocean, what a life-saving protein source would be available to the hungry people on its shores! The two principal developments are :

Sonar fishing

The development of sonar fishing has provided man with an important underwater probe with many applications. It is particularly useful in fish hunting and fish counting. Sonar works by giving out beams of pulsed acoustic signals, and where the beams intersect an object in the sea, the energy is then back-scattered. These echoes are recorded by the vessel, and in this way, shoals of fish can be located. This information can be fed into a computer, which estimates the density of the shoal. The advantages of this method is that it cuts out a lot of fruitless searching, because the vessel does not have to pass over the fish to find them – the shoal can be detected before the boat reaches them. Some vessel owners estimate that sonar cuts their operating costs in half.

Submarine trawlers

Most fisheries at present are centred mainly in the upper layers of the open ocean and in the shallower waters over the continental shelves. However, there are also large aggregations of pelagic animals that live farther down and are associated particularly with the 'deep scattering layer', the sound reflecting stratum observed in all oceans. The more widespread use of submersible research vessels will reveal more about the layer's biological nature, but the exploitation of deep pelagic resources awaits the development of suitable fishing apparatus for this purpose. The most logical development would seem to be an adaptation of the usual trawling gear, towed not by a vessel on the surface, but by a specially fitted 'fishing submarine'. The discovery of an abundance of fish in the abyssal zone, previously thought to be sparsely populated, indicates a new area for fisheries exploration, and there is no doubt that we shall see in the near future the development of submarines equipped with sonar and nets which enable this new area to be explored.

We can also learn a lot from traditional primitive methods of trapping still used in Asia and South America, methods which are based on generations of intimate knowledge of the habits of the fish. Successful fishing depends on using knowledge about where the fish have concentrated : where they have gathered to feed or reproduce. In future we will attempt to cause the fish to congregate artificially. Lights and sounds are already used to attract fish, and these methods will become more sophisticated. Concentrations of fish occur naturally in areas of 'upwelling' where seasonable winds and currents combine to cause a periodic enrichment of the surface waters, and a subsequent phytoplankton bloom. There would be a considerable advantage in being able to predict these occurrences of high biological production, which makes a study of the weather regime very important in fisheries science.

The ocean fishermen of different countries all share a common resource, and they are at the mercy of each other's governments to ensure controlled exploitation. The private fish farmer, however, is a law unto himself, and if he is efficient and thrifty in his husbandry, he can benefit directly, because his farm is his private property. Marine fish farming (a branch of aquaculture) is done on a substantial scale in the brackish waters of estuaries and salt marshes. About 1 million acres are in production, and, at an average rate of production of 400 pounds per acre per annum, contribute not less than 200,000 tons of fish and prawns annually. The stock must be easy to feed and easy to breed, with naturally spawned young which are cheap and easy to obtain, and fast growing. The most commonly cultivated are eels and grey mullet in temperate waters, and milk-fish, grey mullet, tilapia and prawns in tropical waters. Shellfish have been cultivated for many years, because their sedentary habit makes them very suitable.

Much higher rates of production are obtained in fish farming than in wild fisheries. This is because (1) only herbivorous fish and prawns are cultivated, not carnivorous ones. These are higher up the food chain (i.e. nearer the level of plant or primary production) and are therefore more productive. (2) The stock is manipulated to exactly the right level, avoiding wasteful understocking or stressful overstocking. (3) Unwanted predators, competitors, pests and parasites can be controlled or eliminated. (4) The natural fertility of the culture area is increased by manuring with organic fertilizers.

Mussels feed directly on phytoplankton right at the level of primary production, and are therefore very productive. They are grown on ropes suspended from rafts, raised clear of the usual predators. The enormous rate of 300,000 kilograms per hectare per annum is feasible. The fish grown in marine fishponds feed on the algal pasture growing on the bed of these shallow ponds. The rate of production of this algal pasture may be as high as 28,250 kilograms per hectare. There is great scope for expansion of this well-established kind of fish farming. Nearly 2 million acres are available for development as fish farms in the Niger Delta of West Africa, 6 million acres in Indonesia, half a million in the Philippines, and incalculable acres in the deltas of South America. It may also be possible to give the natural fertilization process of the sea a helping hand by pumping up the richer waters from the ocean depths up to the lighted surface where it could support greater crops of phytoplankton which in turn provides food for the fisheries. The idea is being tested of pumping up these rich waters into the confined ring-shaped coral atolls, thereby conveniently creating a natural fish farm. It has been calculated that an atoll lagoon, continuously replenished with this deep water, and stocked with edible seafood able to make full use of the increased phytoplankton, could produce enough protein per square kilometre to supply the needs of 4,600 people continuously! It is even possible that additional lagoons can be created by making artificial reefs from discarded motor tires. Off the coast of Florida, tire reefs create a haven for fish breeding and support the state's sport fishing industry. As tires have an estimated life of 2000 years, we can assume that such reefs will afford shelter for some time to come.

What is life for the fish farmer like? Like a land farmer, he must understand the breeding habits of his stock, be willing to work long hours, and a seven-day week. His 'fields' are long vertical rows of pools, and his tasks involve ditch digging, drain cleaning, feeding and treating and cleaning fish, all very often in uncomfortable weather conditions. He may have to be prepared to dive to check underwater equipment such as rearing cages floating from rafts. Akin to fish farming, is fish ranching which is improving on natural stocks by implanting and transplanting young fish or shellfish in suitable areas. The young are cultured in the laboratory and then released in the sea. Much work has been done to increase the world salmon stocks in this way, and similar projects are being done with scallops in Tampa Bay, Florida.

In spite of the extensive use of fish for animal foodstuffs, and in the increase of frozen products, we are still over-conservative, unimaginative and wasteful in our use of the sea's living resources. Most fishing methods produce a proportion of the favorable inland areas in the world have already been explored for oil, but a much smaller proportion sponges, soft corals and molluscs. Until recently, these were discarded, but the advent of the factory trawler is remedying this. In these ships, the 'trash' material is used for fish meal and oil. Attempts to market high-grade fish meal for human consumption have so far failed, but there is some hope of making it acceptable by incorporating it into high-protein biscuits and cereals. Scrap fish and shrimp are also being processed to make fish sauces and pastes. We could use so many more fishery products if we were not so set in our ways as regards our tastes. In Britain, for example, the main preference is for cod, and frozen food manufacturers have found it difficult to introduce less familiar species such as blue whiting. In Germany, coley or saithe is popular, but virtually unknown in Britain. If we are to continue to buy reasonably priced fish, we must learn to overcome entrenched attitudes, and to develop new tastes. New products will undoubtedly emerge in future; experiments are being made now with shrimp 'sausages', and with successful marketing campaigns, we shall no doubt see a greater variety of fish in the shops in the coming years.

Riches of the sea – physical resources

Men have caught fish and extracted salt from the sea for thousands of years, but only in the past 15 years have they begun to appreciate the full potential of its non-living resources. This awakening has been due to three factors: firstly the development of oceanography as a science, and secondly, new techniques have made it possible to extract resources once inaccessible, and thirdly, with the growth of industrialization, there are new demands for every kind of raw material. The ocean is a treasure trove, whose valuables include not only the oil and minerals, but also the seawater itself, and the adjacent shoreline.

Let us first consider seawater as a resource, together with the minerals dissolved in it. As we have seen in the section on living resources, the sea acts as a processing plant whereby the sun's energy is converted into protein by marine plants. Seawater is also a storehouse of dissolved minerals and fresh water, a receptacle for wastes, a source of tidal energy, and a medium for new kinds of transportation, such as the hovercraft. The 350 million cubic miles of ocean water constitute the Earth's largest ore body – if you like, a vast, floating 'mine'. Each cubic mile (4·7 billion tons of water) contains about 165 million tons of solids. Although most chemical elements have been detected and (probably all are present) in seawater, only common salt (sodium chloride), magnesium and bromine are now extracted in significant amounts. The economic recovery of other chemicals is questionable, because even with modern technology, extraction costs are too high. We know that valuable metals such as iron, cobalt, copper, gold, silver, uranium and zinc are dissolved in the sea, but in very dilute amounts. We also know that marine organisms are capable of concentrating these metals many thousands of times within their bodies. Unfortunately, we do not know exactly how they do this. *The man who discovers a marine creature which can concentrate gold for our benefit would certainly make a fortune!*

The potential resource of seawater that has proved most difficult to extract economically is water itself, that is, fresh water. However, as requirements for water for domestic use, agriculture and industry rise, and as modern technology improves, the process becomes more feasible. There are two practical methods of desalination, namely, distillation and freezing. Both require a fairly large energy input, which makes desalinated water expensive, but it is competitive in water-deficient areas or where a local water supply is unfit for consumption. Considerably lower costs will be attained within the next decade where large-scale desalting operations are combined with nuclear-fuelled power plants, to take advantage of their output in heat. Nearly 700 desalination plants with a capacity of more than 30,000 gallons of fresh water a day are now in operation throughout the world, and many more are planned for the next decade.

Now to consider the resources of the sea bed. Unlike those of the essentially uniform overlying waters, these occur in scattered, highly localized deposits on top of and beneath the sea floor. They include, firstly, unconsolidated deposits on the surface, which can be dredged, such as heavy metals, oyster shells, sand and gravel, diamonds and manganese nodules formed from the precipitation of seawater. Economic exploitation has so far been confined to the continental shelf, i.e. in waters less than 350 feet deep, and within 70 miles of the coast. Secondly, there are consolidated deposits beneath the surface, such as coal, iron ore and other metals, which so far are only mined from tunnels originating on land. Thirdly, there are fluids and soluble materials such as oil, gas, sulphur and potash, which can be extracted through boreholes.

Oil and gas represent more than 90 per cent by value of all minerals obtained from the oceans and have the greatest potential for the near future. Offshore sources are responsible for 17 per cent of the oil and 6 per cent of the natural gas produced by non-Communist countries. By 1980, a third of the oil production (four times the present output of 6·5 million barrels a day) will come from the ocean; the increase in gas production is expected to be comparable. A large proportion of the favorable inland areas in the world have already been explored for oil, but a much smaller proportion of the continental shelf has been surveyed. Prospects are therefore encouraging for additional large oil finds, such as the recent one presently being extracted in the North Sea.

Sulphur, one of the world's most important chemicals, is found in salt domes buried within the sea-floor sediments. It is recovered inexpensively by melting it with superheated water piped down from the surface and then forcing it up with compressed air. A critical shortage of sulphur and a recent discovery of sulphur-bearing domes in the Gulf of Mexico have stimulated an intensive search for off-shore sulphur deposits.

Undersea mining can be traced as far back as 1620, when coal was extracted in Scotland through shafts that were driven out to sea from an offshore island. To date, over 100 subsea mines with shaft entries on land have recovered coal, iron ore, nickel-copper ores, tin and limestone off a number of countries in all parts of the world. Coal extracted from beneath the sea level accounts for almost 30 per cent of Japan's total production and more than 10 per cent of Britain's. With present technology, subsea mining can be conducted economically as far as 15 miles offshore, and this distance should increase to 30 miles by 1980 with the development of new methods for rapid underground excavation, especially laser boring. Eventually shafts may be driven directly from the seabed if ore deposits are located in ocean-floor rock far from land.

For gathering unconsolidated sediments, dredging is an attractive mining technique, because of low capital investment, quick returns and operational mobility. So far it has been limited to nearshore waters, less than 235 feet deep, but as knowledge of resources in deeper water increases, dredging technology will undoubtedly advance also. Of the surface deposits, sand and gravel are the most important in money terms; oyster shells come next as a source of lime. As metropolitan areas spread out on land, they cover, and therefore render inaccessible, dry-land deposits of the very construction materials being used to build them; hence the importance of marine deposits.

The only known minerals on the floor of the ocean that appear to be of potential economic importance are man-ganese nodules. These are widely distributed, with concentrations of 31,000 tons per square mile on the floor of the Pacific Ocean. The nodules average about 24 per cent manganese, 14 per cent iron, 1 per cent nickel, 5 per cent copper and just under 5 per cent cobalt. It may be the minor constituents that prove attractive economically, and the key to profitable exploitation is successful metallurgical separation, which will no doubt be solved in the near future.

Few discoveries have created more excitement in recent years than the location, in the 1960s, of three undersea pools of hot, high density brines in the middle of the Red Sea. The brines contain minerals in concentrations as high as 300,000 parts per million – nearly 10 times as much as is commonly dissolved in seawater, and overlie sediments rich in such heavy metals as zinc, copper, lead, silver and gold.

Resource exploration is advancing on many fronts. Chromite has been found in the Indian Ocean; zirconium, titanium and other heavy minerals have been detected in sediments off the Texas coast. Methane deposits have been confirmed in the Adriatic Sea, and new oil fields of value have been found off Mexico, Trinidad, Brazil, Dahomey, Australia and the British Isles. Surveys of the East China Sea and the Yellow Sea indicate possibly one of the richest oil reserves in the world.

So much for offshore resources, but what of the shoreline itself? The coastal margin has only just recently become to be recognized and treated as a valuable and perishable resource. Our coastal zones are narrow and fragile, under constant demands from increasing populations, and shrinking under the pressure of natural forces such as hurricanes and wave action. Pollution can have disastrous effects on the coastal zone; for example, more than a tenth of the 10·7 million square miles of shellfish-producing waters bordering the United States is now unusable. Dredging and draining projects and chemical mosquito control have a devastating effect on coastal aquatic life. The trouble is, there is such great competition for the zone's resources. We need the coasts for industrial and housing development, for ports, shipbuilding, for recreation, for commercial fisheries and for waste disposal. All this affects our estuaries and marshes, beaches and cliffs, bays and harbors, islands and peninsulas. Tidal wetlands in the United States are disappearing rapidly; 45,000 acres were lost along the Atlantic coast between 1955 and 1964, and of these, 34 per cent was dried up by being used as a dumping ground for dredging operations, 27 per cent was filled for housing developments; 15 per cent went into recreational developments (parks, beaches and marinas); 10 per cent to bridges, roads, car parks and airports; 7 per cent to industrial sites, and 6 per cent to rubbish dumps. Unfortunately, many of the demands on the coastal zone are conflicting. Private beach development restricts public access; dredging and filling affect commercial fishing; offshore drilling limits freedom of navigation, and what about swimming, boating and skin diving? All of these seem to have been found off Mexico, Trinidad, Brazil, Benin, Australia and the British Isles. Surveys of the East China Sea at the same time public and private interests, and taking a long-term as well as a short-term view.

Vast potential of untapped resources

Man has in the sea, a storehouse of under-developed physical resources, and what is more, new tools of science and technology are at his disposal. This combination of new ground, new knowledge and new technology may be a unique experience in human progress. With this new understanding, it may be possible to harvest mineral wealth, maintain water quality, stop beach erosion and create modern ports and harbours simultaneously. However, choices must be made about priorities and courses of action, and the right legislation passed to effect them.

Is the sea the sewer of the world?

Although water pollution has been much discussed in recent years, pollution itself is not a new phenomenon. The waters of the Earth have been contaminated in one way or another since life began. Rotting vegetation, dead bodies of plants and animals, and excreta from man's communities have long fallen or been thrown into the waterways of the world. However, this is all natural, organic pollution. In early times, if the amount of pollution became too great and caused odor, decay, and disease, a community would usually move away and establish a new home. Gradually, the water left behind was able to cleanse itself through the action of waste-consuming bacteria until it became fresh again. As the population of the world grew, and as communities gave up their nomadic habits, water came under assault from sewage, waste and garbage to such an extent that there was too much material for the natural waste-consuming bacteria to digest. So the water in heavily populated areas became polluted and remains polluted. The growth of industry and technology has added to the problems. Waterways and the sea are contaminated not only with organic human waste, but also wastes of industrial processes. Pollution therefore comes from several sources : human sewage, agricultural chemicals and radioactive waste.

Oil tanker disaster causes havoc

A particularly persistent human attitude has been that the ocean is the ultimate answer to waste disposal. The idea of the ocean as a dump seemed sensible initially, because of its vastness and relative depth. Anything toxic or obnoxious, from atomic wastes and old war gases or ammunition, down to scrap cars, sewage and general urban debris has been placed in the ocean by man confident that he has seen the last of it. Added to this, all the outpourings of pollutants from streams and rivers reach the oceans. However, the ocean is not a limitless dump, and in the 1960s and 1970s people first began to realize this. Perhaps more than anything else, the wreck of the *Torrey Canyon* off the coast of England in 1967 brought the situation to the public eye. The *Torrey Canyon* was a giant oil tanker used to transport oil from the fields of the Middle East to refineries in Europe or the United States. In March 1967, it ran aground off the coast of Cornwall and spilled its cargo of 36 million gallons of petroleum into the seas. The beaches were soaked, the holiday season was ruined, and thousands of seabirds became entrapped in oil and died, despite efforts to save them. In spite of a campaign to clean up the oil using detergents and napalm (themselves potential pollutants) the damage was done. The oil spread along the coasts of England and reached the beaches of France. Investigations were launched to determine the extent of the damage to fisheries, intertidal life, marine mammals and birds and to recreational facilities. The *Torrey Canyon* episode was to be the pollution 'test case' of the decade, and dozens of scientific papers have been written on its effects.

Not long after the storm had died down, the scientific world was provided with another example. In the Santa Barbara Channel off the coast of California, an oil well used to pump oil from the fields that lay deep beneath the channel waters began to leak. Oil poured upward to the surface of the water, formed an enormous slick, and began to move towards the beaches of California. Again, the *Torrey Canyon* pattern was repeated, and despite all efforts, great damage was done. The news from around the world continues to record the wrecks of tankers, the flushing of oil bunkers by ships at sea, the leakage from undersea oil wells, and other continuing sources of pollution from oil. In 1970, Thor Heyerdahl made one of his exploratory voyages in a papyrus ship from Africa to the West Indies, and he reported encountering floating petroleum or tar almost throughout his Atlantic voyage. At one point, the ocean water was so filthy that he and his companions could not use it for washing.

An even more dramatic example is the Cuyahoga River in the USA which was so polluted by tire and steel factory effluent that when a lighted stick from a fire fell into it in 1969, the *river* caught fire and actually burnt.

Contaminated fish can kill

Yet oil is only one of many marine pollutants. Organochlorine pesticide residues (of which DDT is the most widespread) are now being recorded from all parts of the ocean. In 1970, it was discovered that the eggs of many seabirds were sterile, due to accumulated DDT. Plankton in coastal waters was contaminated with DDT which is used as a pesticide to protect land crops. The DDT is washed into rivers in the run-off from agricultural lands, and hence enters the sea. It enters the food chain via the plankton, and then goes on to affect fish and sea-birds. There is therefore a danger to man, not only through the destruction of our food species, but also to ourselves if we eat contaminated fish. Catches of mackerel off the California coast have shown that they contain 20 times more DDT than is the allowable 'safe' level for human consumption. Deaths and reproductive losses of marine animals have also been attributed to DDT. It has been estimated that as much as one billion pounds of DDT alone might be circulating in the biosphere. It has been found that even at a level of 100 parts per billion in seawater, DDT could dractically reduce photosynthesis in plant plankton, the basis of all oceanic food chains.

Studies carried out at the University of Miami have revealed an unholy alliance between two major pollutants: oil slicks in the ocean appeared to concentrate oil-soluble pesticides such as DDT and dieldrin. One oil slick investigated off Miami revealed 10,000 times more dieldrin in the thin surface film of oil than in the water immediately below. This could bring great danger to surface feeding marine life, including animal plankton. Petroleum has also been found in the form of 'tar balls' throughout the ocean. Again, these were found to contain dissolved organochlorine pesticides, as well as equally dangerous polychlorinated biphenyls (PCB's) in high concentrations. Since these tar balls attract marine organisms, probably for shelter, the likelihood of their poisoning is increased.

In controlling pesticide pollution, we are in a dilemma. The World Health Organization, faced with controlling malaria and other insect-borne diseases does not want to abandon DDT, which is relatively cheap and effective. Likewise, agriculturists who use DDT to protect their crops do not want to sacrifice an increased yield. In some countries (e.g. Sweden and Switzerland) DDT is banned, and in others (e.g. Great Britain and the United States) its use is controlled. However, it is still widely used in the developing countries. The answer is not simple, but experience has shown that with high crop diversity, the need for pest control is reduced, and *biological control* (using predatory insects to attack pests) has proved as effective as chemical poisons.

105 people die of mercury poisoning

Mercury pollution is usually the result of industrial waste, particularly from pulp mills, which produce paper. In the late 1960s and early 1970s mercury was found to be widespread in North American waters, and vigorous action was taken to close down obvious polluters. However, in 1971 the world was shocked by the discovery of high levels of contamination in marine tuna and swordfish. These fish feed across open oceans on other fish which feed on oceanic plankton. For these fish to have been contaminated meant that mercury, like DDT, had become a global pollutant and would have to be tackled at an international level. Some industrial plants with highly toxic wastes have devised recirculating systems whereby waste water is processed, the chemicals reclaimed for further use, and the water rendered pure enough for re-use. However, these systems are expensive to install and slow to take effect, and in the case of mercury, came too late. The horrifying consequences of mercury poisoning were seen in Japan in 1953 when 105 people, many of whom died, suffered a dreadful disease which affected the nervous system. The culprit was a chemical plant, the mercury having been dumped in the water and concentrated in the fish which were eaten by the local people. It is to be hoped that the recent measures taken against chemical dumping will ensure that such a catastrophe will never happen again, but the consequences of the more widespread lower concentrations may be still to come.

Atomic power plants – the next catastrophe?

Atomic power plants are the newest sources of electric power, and during the recent energy crisis and increase in world oil prices were discussed as being the main power source of the future. However, atomic power plants use tremendous quantities of water for cooling, and they discharge heated water back to the river. This artificially heated water can have a harmful effect on the aquatic community and is known as 'thermal pollution'. However, even more sinister is the possibility of radioactive pollution from nuclear power plants. Tritium is one potential pollutant, a serious one because it enters the sea via the waterways, and accumulates in plants and animals along the way. It is one of the few substances in the world that is not purified out by evaporation in the hydrological cycle. Its primary threat to man is that it may, via fish and seafood in our diet, damage the DNA molecule, which is at the root of man's genetic inheritance. In most industrialized countries, there are now laws to regulate the amount of pollution by radioactive waste, but surely it is wrong to declare certain amounts of radioactivity 'safe'? Any amount is potentially dangerous, because it *accumulates* in the body, and *because its effects may be passed on from one generation to another*. There is also well-documented evidence that radioactive pollution can cause cancer, in particular leukaemia. The sad thing about pollution from nuclear power plants is that the technology exists that makes it possible to avoid serious damage. Water may be recycled and the wastes containerized. However, such methods are expensive, and even when nuclear wastes have been containerized, further expense is required to dispose of it safely. Governments all over the world dispose of concrete canisters containing radioactive wastes by dropping them in the ocean. On occasion, the pressure at great depths causes these canisters to explode, and their highly dangerous wastes are scattered with the tides. Off the coast of California, canisters dumped by the Atomic Energy Commission exploded, and fish caught 2 hours later were showing abnormal levels of strontium 90, a radioactive isotope highly dangerous to man. Much research is obviously needed to develop container substances that can be dumped at sea or preferably buried underground without risk of leakage in the future at all.

Not all pollutants are chemical ones : the 'red menace' off the coast of Florida is a biological one. Every year the ocean in this region turns red. Fish leap about in a frenzy, and thousands are eventually washed up dead on Florida's beaches. The occurrence is known as the red tide, and has been observed with increasing frequency in recent years. The killer is an organism named *Gymnodinium breve* – a minute creature which is a cross between a plant and an animal. When the red tide occurs there is a fantastic population explosion of these organisms which excrete a toxic waste which immobilizes the nervous system of fish, causing convulsions and death. No one knows exactly what causes these blooms, but outbreaks coincide with periods of heavy rainfall, when the rivers discharge nearly twice their normal amount of water into the sea. Undoubtedly, the red tide is triggered by a pollutant in the river, but which is not yet known.

Dead fish and oiled birds in their thousands are dramatic visible effects of pollution, and are alarming events. Unfortunately, they are soon forgotten after each crisis has blown over. But supposing these effects became widespread and continuous? Is it feasible that the entire fish supply of the world could die out? With population growing and technology advancing, this is not merely an alarmist scare. If the waters of the Earth are polluted until they can no longer support life, then the Earth will lose the supply of oxygen (via plants) that keeps all creatures alive.

You may wonder what as individuals, we can do about pollution. To take effective action, it is necessary to be fully acquainted with the facts, and here, newspapers and TV are invaluable, but having had the facts, we must learn how to make use of them. We must learn to develop a healthy suspicion of new actions that are likely to affect the environment, and also to make our views known to the politicians. Governments must learn that if they fail to respond to these views they will forfeit the goodwill, and therefore the votes, of the electorate. Some difficult decisions as regards priorities may be involved, and higher taxes and production costs may result. However, these costs are small compared to the destructive costs of uncontrolled pollution. Emotional outbursts are not really much good – they might shock people into paying attention for a minute, but people get immune to shock tactics and conservationists can not afford to be dismissed as indulging in yet another fad. We must try to create a new climate of thinking that regards all kinds of pollution as antisocial. Only an educated public can push government and industry into action. We can all start now, in our day-to-day lives. Remember, everything that goes down your sink may one day end up in the sea. Be stingy with detergent, and choose brands that contain little or no phosphates. Avoid chemical fertilizers and weedkillers in your garden. Find out where the sewage goes from your home. Does it cause pollution there? Join a local conservation society. Ask what can be done to fight pollution in your locality. If there is no society near you, why not start one? When you see pollution, make a fuss. Write to newspapers, TV, local radio stations, to your elected officials.

It is difficult to believe that man who has the knowledge and technological know-how to walk on the moon is at the same time wantonly destroying the ocean whose tides are controlled by the waxing and waning of that same moon. Scientists have knowledge to perform miracles of technology, but sometimes seemingly lack the wisdom to direct their talents for the long-term good of the Earth. Perhaps the responsibility for that rests with you and me?

Ocean technology and the future

During the past 20 years, marine technology has improved enormously, and in many ways. As we have seen in the preceding sections, these improvements have brought new mineral provinces within reach, and made food from the sea more readily available. Let us review the most important of these developments. Firstly, the development of the 'super-ship'. A 'supertanker' used to carry 35,000 tons. Now a fleet of ships with nearly 10 times that capacity has been developed. Second is the deep-diving submarine. Men can now descend to the deep ocean bottom in an underwater balloon submersible. Several techniques have been employed to solve the problem of how to make a submarine hull that is strong enough to resist great pressure, yet light enough to return to the surface. Thirdly, there is now the ability to drill in deep water. Virtually all floating drilling equipment, including semisubmersible platforms and self-propelled vessels, has been designed and built in the last 15 years. Fourth is the ability to navigate precisely. A ship 1000 miles from land can now fix its position within 1 mile, 500 miles from land within 0·01 miles, and 10 miles from land within 10 feet. The techniques that make this possible include orbiting satellites, inertial guidance systems and electronic devices. The last important development is the ability to examine the ocean bottom by means of television and sonar, in the same way that land is surveyed by aerial photography.

There has been much research in the past decade into marine materials. A good marine material should be light, strong, easy to connect, rigid or flexible as required and inexpensive. Ocean engineers have laughingly named this mythical substance 'nonobtainium', but none the less, research has produced materials which have some of these qualities. For example, the steel available for marine purposes has improved as a result of demands for submarines that can withstand pressure of great depth, drill pipe that must survive high bending stresses, and great lengths of oceanographic cable that must not twist. A new kind of steel with a high nickel content is tougher, more resistant to notching, and less subject to corrosion fatigue than steel formerly available. Also available for marine purposes are new high-strength aluminium alloys, titanium, glass, fibre-glass and plastics. Glass is less fragile than people think, and is being used widely for small submarines that must withstand compression, but also require a large window. Fibre glass is used to build hulls, which are light and strong, and need little maintenance – a boon to small boat owners. Among the plastics, polyvinyl chloride is used in marine pipelines subject to severe internal corrosion. Nylon and polypropylene for ropes and fishing nets and Dacron for sails are favoured by fishermen and sailors, because the materials are light and elastic, and do not rot. Ship-bottom paints, designed to reduce fouling by marine organisms such as barnacles, have been greatly improved.

A remarkable variety of marine vessels has been developed in the last decade. Ships now exist that go up, down, sideways, and stand on end ! They can skim, fly or dive. Some are amphibious and some go through ice, over ice, or under it. The ships that go down are of course submarines, some fitted as underwater laboratories. Those that go up are hovercrafts and hydrofoils. The hovercraft can run up a beach and cross mud flats, ice and smooth land surfaces. The ship that can stand on end is suitably named FLIP, operated by the Scripps

Institution of Oceanography in the United States. It has two positions of stability. While it is on tow, it lies on the surface, like a barge. Once stationary, it ballasts itself and floats on end, and is not affected by the motion of the sea. Thus positioned, it is used as a platform for making underwater sound measurements.

Jacques-Yves Cousteau perfected the aqualung and scuba (self-contained underwater breathing apparatus). The freedom of action of the scuba diver and his ability to operate at considerable depth – using helium or other gases such as hydrogen – led people to think seriously about exploration of the Continental Shelf areas (average depth 35 fathoms). The idea of numbers of men working for prolonged periods under water led in turn to plans for some sort of dwelling on the sea-floor, maintained at sea-floor pressure and into which divers could come and go without need for constant decompression. The first experiment to achieve this was made under Cousteau's direction in the Mediterranean. In September, 1962, two men lived for a week in a steel cylinder 33 feet. In June, 1963, five men spent a month at 36 feet, while the surface. They made several dives daily as far down as 85 feet. In June 1963, five men spent a month at 36 feet, while two more spent a week in a smaller 'house' at 90 feet. Both crews lived normally, worked, made dives down to 165 feet, and even received visitors.

A year later, four United States Navy men spent 11 days in Sealab I at 192 feet, despite constant pressure of nearly 100 pounds per square inch. Sealab II – a 57-foot long, 12-foot diameter, 200-ton cylinder with laboratory space and accommodation for ten men – was lowered to a depth of 205 feet on August 26th, 1965. Aquanauts of the first team stayed down for 15 days, diving and performing experiments. They were lifted to the surface in a sealed capsule, which was then locked to a decompression chamber. Decompression took over 30 hours. No ill effects were suffered, except for one scorpion fish sting, and sealions, attracted by the numbers of fish surrounding Sealab II, peered in the portholes at the inhabitants. Another of Cousteau's underwater habitats,

Conshelf II, had an underwater hangar into which his diving saucer could be manoeuvred and then winched up for servicing.

In 1969, participating in a project named Tektite, four American aquanauts spent 60 days in, and working from, an underwater chamber at a depth of 42 feet. In 1970, two British scientists made a simulated dive – in a chamber whose air pressure equalled the water pressure found at corresponding depths – to depths ranging from 1000 to 1500 feet. The divers spent $5\frac{1}{2}$ days below 1000 feet, of which $3\frac{1}{2}$ days were below 1200 and 10 hours at 1500 feet. There seems to be no limit to the size or equipment of these undersea homes or laboratories. If man can build 8000-ton submarines, a sizeable complex of semi-mobile structures of this size should pose no great problem. Perhaps in the future man will descend to even greater depths, for although a can of peaches collapses at the simulated depth of 300 feet, a mouse can survive a pressure equivalent to 3000 feet with no ill effects. The underwater vessel *Deep Diver* (able to go down to 1300 feet) was the first vessel to include an 'artificial gill' which admitted a mixture of helium–oxygen into an internal diving chamber. When the pressure inside is equal to the deep-water pressure outside, the hatch is opened, and the divers slip out. When they return, the pressure is gradually reduced and decompression begun. This obviates the need for a long decompression process at the surface.

Sonar, the sound-ranging device, is not new, but still figures prominently. It has been refined considerably, and is used in navigation and in detecting fish shoals. Hydrophone arrays detect low frequency sound created by a series of gas explosions and enable rocks under the sea bottom to be examined in great detail. Such continuous-reflection seismic profiles have revealed features in sub-bottom rocks to depths of as much as 15,000 feet, and have found many new undersea oil deposits. Satellites are used to determine a ship's position accurately and for weather forecasting. They transmit photographs of cloud cover and the state of the sea, and enable charts to be drawn up with reliable up-to-date information.

Buoys moored in the ocean hold instruments for measuring, recording and transmitting sea and weather conditions at minimal expense. Shipboard computing methods have become accepted, in order to plot the ship's position continuously, and to relate it to other accumulating data about the sea below.

To consider some trends in fishery technology : one of the most revolutionary of these has been the Puretic power block, which handles fishing nets. It is a simple, wide-mouthed, rubber-lined pulley, driven by a small hydraulic motor. It has been adopted by many fishing fleets, and now accounts for over 40 per cent of the world's catch. With the block, it is possible to handle much larger nets with fewer men. As a result, the tuna industry has changed almost entirely from line fishing to net fishing. Fish-processing equipment on boats has become common in recent years, and fish can now be filleted automatically, quick-frozen and packaged, so that a finished frozen product can be delivered at the dockside. Scallops can be shucked and eviscerated on board ship, so that the scalloper can be at sea for a week at a time, and return with a cargo of ready-to-eat scallops.

Deep diving is receiving increasing attention by ocean technology. Men go into deeper and deeper waters by means of methods that grow more and more complicated, involving a variety of chambers, hoists, gases and instruments. The major difficulty in man's free submergence to great depth arises from breathing the nitrogen in air at high pressure. Because nitrogen diffuses slowly through the lungs, it is retained for a long time in the body. When the pressure is reduced during the diver's return to the surface, the nitrogen is released as bubbles in the bloodstream resulting in decompression sickness known as the 'bends'. The bubbles block the blood circulation and so deprive the nervous system of oxygen, producing severe pain causing paralysis and even death. By allowing a diver to rise slowly to permit diffusion to occur, without inducing bubble formation, it is possible to avoid the bends, but the time involved is considerable. For instance, if a man made a 150-metre dive (500 ft), it would take him 3 days to come up! The ideal situation is that a man should ascend as quickly as he descends, and underwater life support system technology is chiefly concerned in bringing about this ideal. There are two main methods at present : the bounce dive and the saturation dive. In a bounce dive, the diver goes from atmospheric pressure to the required depth in a chamber, breathing a mixture of gases that changes with depth and physiological requirements. He works for a few minutes, then returns to the surface in a pressurized chamber for slow decompression on the mother ship. In the saturation dive, the diver's body is saturated with inert gases while he lives in a pressure chamber while on board ship. He then moves into a similarly pressurized capsule that is lowered to the bottom. He can work much longer than the bounce diver, and since he lives on the surface but at the pressure of the bottom, he can keep going back, and the process can be repeated for many days. The diver then takes a slow decompression to atmospheric pressure. Divers using each of these systems have reached 1000 feet.

International co-operation on the way?

One subject that has not been mentioned in the preceding sections is that of international co-operation. Most of the world is ocean, and most of the ocean lies beyond the limits of national jurisdiction – in other words, it belongs to nobody. How then can the resources of the sea be apportioned, and who will regulate the activities of the sea so that it brings maximum benefits rather than harmful pollution? Two factors affect this problem : there are firstly the vastness of the ocean, and secondly its unity, which we discussed in the first section. Investigation of the world ocean is inherently an international affair, requiring co-operation ranging from the simplest exchange of information to the most complex integration of research programs. To this end, the Food and Agriculture Organization have set up the Aquatic Sciences and Fisheries Information System. This consists of an accumulation of information on the marine environment, gleaned from publications from all over the world. This information is being put into a computerized data base and is also published in a monthly abstracts journal. Hopefully, information on all aspects of marine science will be immediately accessible to the research worker, via the information system. This is essential, because information about the sea is accumulating far too rapidly for one person to keep abreast of it. Until about 20 years ago, the community of oceanographers was so small that most of the members knew one another, and could exchange information on personal basis. Now the number of oceanographers is much larger so that personal evaluation of reported work is often impossible.

Even with a free exchange of information, political and legal problems still remain. Freedom of research is essential, and investigations must be made on the basis of phenomena, rather than political boundaries. However, as coastal states contemplate the resource potential of their shelves, they become more reluctant to permit foreign scientists to work there. It would obviously be best for permission to be granted on condition that all findings are made freely available and that scientists of the coastal state can participate fully in the research. As regards the oceans as a whole, the idea has been put forward that the whole sea be divided up and put under UN control. The idea was that the ocean would be divided into 30,000 blocks measuring up to 20,800 square kilometres (8000 square miles) each. If a particular government did not wish to use its allocation, it would be able to sublet it to others. Unfortunately, with this scheme, nations with long coastlines would gain the lion's share, and those without shorelines nothing at all. It is not within the scope of this chapter to enter into political controversy, but one has seen recently with the North Sea fishing disputes that subdivision of the sea between bordering nations can lead to argument. Whatever scheme is finally adopted, either free-for-all or allocation, it will not come about without much debate and conflict.

It is interesting to speculate on how far we can sustain life under the sea. Will the ocean depths become a day-to-day environment for the future? We can draw a useful parallel here between problems of interplanetary (outer) space and space beneath the sea (hydrospace). In both environments, humans are exposed to hostile conditions, and fail-safe life support systems need to be evolved for man to survive them. Thus we have research modules and 'Skylabs' in space, but the environment has as yet proved too hostile for permanent human settlement. However, perhaps hydrospace development is several steps ahead of outer space. After all, there are today hundreds of submersible vessels in the ocean, some built for oceanographic research, but many are designed to operate commercial mining and drilling operations on the seabed. In future, we will no doubt have fish farms operated from submersibles, too.

In an early experiment, two French divers remained one week below the surface in the Mediterranean in Cousteau's first underwater house, Conshelf I. In the United States, Sealab experiments have proved that man can survive long periods on the sea bottom at considerable depths if the breathing mixture is stringently maintained at correct levels. So – will man ever live on the bottom of the sea? My answer is 'yes', and in the near future, too, but only for a limited time, and for the specific purposes we have mentioned – not on a day-to-day basis. The cost would be too great to provide people with watery homes just because they liked the idea of fish swimming past the bedroom window! Underwater package holidays, however, are certainly feasible in the future, as are cruises to the ocean depths. The underwater city is not so easy to contemplate – imagine, your house would be a submersible, or maybe just a unit in a submerged 'block'. Your drinking water would enter your house via your own desalinator, and your hot water would be a spin-off from a nearby nuclear reactor. Every so often you would have to leave your house (don't forget your gaseous mixture!) to harpoon a few squid to make a change from the eternal diet of fish-meal biscuits delivered to your door from the local submarine factory. And for a holiday, you would go through a decompression process and visit that strange land above water! Technically, all this is possible, but we have forgotten one thing, and that is the psychological factor. Divers under the sea and astronauts in space have said the same thing – their main difficulties were psychological rather than physiological or technical ones, and the feeling of isolation and disorientation are difficult to cope with over a long period. Man's remote ancestors may have once come from the sea, but there is no doubt that he is really a landlubber at heart!

FURTHER READING

The Ocean (a Scientific American book). (W. H. Freeman Company, San Francisco, 1969.)

R. F. Dasmann, *Environmental Conservation* (John Wiley & Sons, Inc.; 4th ed., 1976).

C. F. Hickling and P. L. Brown, *The Seas and Oceans in Colour* (Blandford Press, London, 1973).

R. Burton, *The Life and Death of Whales* (André Deutsch, 1973).

A. C. Hardy, *The Open Sea: The World of Plankton* (Houghton Mifflin Co., 1957).

A. C. Hardy, *The Open Sea: Fish and Fisheries* (Houghton Mifflin Co., 1959).

J. L. Mero, *The Mineral Resources of the Sea* (Elsevier Publishing Co., 1965).

R. Carson, *The Sea Around Us* (Oxford, New York, 1951).

T. A. Olson and F. J. Burgess, ed., *Pollution and Marine Ecology* (Wiley, New York, 1967).

Aquatic Sciences and Fisheries Abstracts (published monthly by Information Retrieval Ltd, London).

ENERGY

Arthur Conway

Arthur Conway is a chartered engineer and writer. His literary output, mainly on technical subjects, includes articles for the press, talks and features for radio, and books. He took his degree at Imperial College, for some years did design and research work in the iron and steel, aircraft and nuclear power industries, and has been the editor of three technical journals, *Control, Instrument Review* and *The Engineer*.

ENERGY FOR LIVING

'Energy', said the great theoretical physicist, James Clerk Maxwell, 'is the go of things.' That is, of course, *all* things, including the living ones and their habitat. Habitat means house and car (among other things), and even houses have become machines for living. A house today is typically a compartmented box, rather inadequately insulated thermally, filled with appliances demanding electrical energy and possibly gaseous, liquid or solid fuels, and flanked by a garage for a gasoline-driven (but not gasoline-saving) car and a shed for powered gardening and other tools. The house of the future could be more mechanized than this, could even be automated to a high degree. It could be a dream-house, at least in house-agents' terms.

House-agents are not the only ones promoting dream-houses today. Dream- (or nightmare-) houses abound in both technical and polemical literature. At one end of the spectrum is the autonomous house, virtually independent of centrally organized services, perhaps drawing all or most of its necessities from a new-style farm which produces everything but waste, all its operating cycles merging so that what comes out of one always goes into another. At the other end of the spectrum is an accommodation unit in a wholly integrated city, built as high as it stretches horizontally, perhaps on an artificial island, or underwater on the sea bed, with all its services automated and everything on tap.

Fiction into fact

Between these extremes (dreams or nightmares according to your taste) come such fancies as solar houses, with all heat, ventilation and power – even refrigeration – run off the sun's radiation. Or all-electric houses, with their own generating facilities to drive heat pumps and other modern conveniences – perhaps robots to fetch and carry, make the beds and so on. Or all-gas houses, supplied with hydrogen instead of mains-electricity, for hydrogen can bring in all the necessary energy to provide heat, cold, light, ventilation, power – *and* the electricity for entertainment and telecommunication. Approaches to such dream-houses are already in being, as working models that purport to show at least their technical feasibility.

In a wholly integrated city there might be dream-cars too, finding their way to the occupant's destination at his touch of a button. Moving belts might convey all foot-traffic. Magnetically levitated trains might whisk people and goods through evacuated tunnels between cities at speeds that only aircraft can achieve today. But the art of visiting people in their homes, autonomous or integrated, might be lost, for telecommunication could become more efficient and more comfortable, as well as more rapid, than even the swiftest transport.

This book is about science fact, not science fiction. Yet some science fiction does in time acquire an uncanny resemblance to emerging fact. Only in retrospect does history assume that convincing look of inevitability. At any instant many futures may seem about equally possible, or about equally fictitious. Take the future of that machine for living, the house. How it will turn out is anybody's (or anybody's computer's) guess. One future that most of us prefer not to contemplate is a return to mud-huts, igloos or caves, or a return to hard labour with no powered devices to make life pleasant for us. But this is also a possible future. Predicting *the* future, as distinct from *possible* futures, is a hazardous occupation, as nearly everyone has found who has tried to predict the endurance of earth's fossil fuel reserves. Most estimates have proved wrong, and Table 1 shows how clouded the crystal ball can get.*

It is important to distinguish between *resources* and *reserves*. Resources are total quantities, some part of which (and possibly the greater part) may be inaccessible. An ex-

*Table I is from *Energy for the Future*, a report by a working party of the Institute of Fuel (FINA).

World's total energy consumption The present total annual figure is about 7500 million kilowatt-years, equivalent to roughly 17,500 million tons of coal. In fact oil accounts for about 3500 and gas for about 2000 out of the 7500 figure.

Table 1. WORLD ENERGY RESERVES: VARIATION IN ESTIMATES

		Present known			Potential future[2]	
	Reserves	Life at 1971[1] consumption rates	Life at future consumption rates	Reserves	Life at 1971[1] consumption rates	Life at future consumption rates
Oil						
Lowest	80 b tons	32	16	250 b tons	100	30
Highest	90 b tons	36	18	360 b tons	140	40
Coal						
Lowest	130 b tons	60	30	1100 b tons	500	150
Highest	2200 b tons	1000	190	4800 b tons[3]	2200	250
Natural gas						
Lowest	34 t.cu.m.	33	15	90 t.cu.m.	90	25
Highest	48 t.cu.m.	45	19	340 t.cu.m.	330	40
Uranium						
Lowest } Highest	0·9 m tons[4]		16/(50–100)[5]	1·3 m tons 3·2 m tons[6]		20/(500–100)[5] 37/(50–100)[5]
Shale/Tar Sand						
Lowest	97 b tons	39	Extend oil by 9	280 b tons	110	Extend oil by 10
Highest	120 b tons	48	Extend oil by 11	500 b tons	200	Extend oil by 17

Table 1. WORLD ENERGY RESERVES: VARIATION IN ESTIMATES

This table was compiled from a variety of sources including the following:

Boxer, L. W.   *Combustion*, December 1971.
Boxer, L. W.   Paper 678 to Fourth U.N. International Conference on Peaceful Uses of Atomic Energy, 1971.
British Petroleum statistical review of the world oil industry, 1971.
Hubbert, M. K.   *Resources and man*, Chapter 8, W. H. Freeman & Co., 1969.
World Oil, 173 p. 52, 1971.
World Power Conference Survey of Energy Resources, 1968.

Notes:

Units: Oil, and oil shale and tar sands, are expressed in billion (10^9) tons of oil.
Coal is expressed in billion tons of coal (multiply by 0·7 to give billion tons of oil equivalent).
Natural gas is expressed in trillion (10^{12}) cubic metres (multiply by 0·86 to give billion tons of oil equivalent).
Uranium is expressed in million (10^6) tons of uranium.
Life is in years.

1. 1971 consumption rates assumed are: *Oil:* 2500 MTOE; *Gas:* 900 MTOE; *Coal:* 1500 MTOE.

2. Potential future reserves are usually estimates of recoverable reserves, allowing for improvements in technology of extraction and price rises, unless otherwise stated.

3. Several sources quote 7,600 billion tons of coal as potential total reserves, which is much higher than recoverable reserves.

4. Known reserves recoverable at less than \$20/kg.

5. The life of uranium resources is considerably increased when fastbreeder reactors are considered, rather than existing light water or similar generation reactors (hence two figures for life in the table). Due to uncertainties in the future development of nuclear power the lifetimes quoted are particularly speculative.

6. Reserves of 3·2 million tons of uranium assume a recovery cost of not more than \$30/kg. However, it has been estimated that 60 million tons of uranium are available at costs up to \$200/kg. In addition there are some 4 billion tons of uranium which are inaccessible with present technology and economic conditions.

7. This figure represents total resources rather than economically recoverable reserves.

ample is uranium. Its concentration in the earth's crust has been estimated to average nearly three grams in every million. Digging no deeper than a kilometre into the continents we could, theoretically, find ten million million tonnes. The sea is also well salted with uranium, perhaps 3000 million tonnes of it, some fraction of which may one day be economically extracted. But a needle in a haystack may cost too much effort to recover. Which brings practical folk down to discussion, not of *resources* but of *reserves*. Table 1 shows their state to the best of our present shaky knowledge. Shaky though it is, it does indicate that the Great Oil Binge, at least, is nearing its end.

Power politics and energy policies

So there appear to be two energy crises. One is definite and immediate and everybody knows about it. It is a crisis of cartel pricing and power politics. The other is less definite and may not feel at all like a crisis to some people. Nevertheless, Table 1 suggests that, historically, it is a real one. The prospect is a drastic decline in oil and natural gas supplies within a few decades – 'around the turn of the century' in the fashionable phrase. What do we do then? We could of course go back to burning coal or coal derivatives, squandering that precious chemical raw material as we have squandered oil, but thereby sustaining our power supplies for perhaps centuries longer. We could also, or alternatively, make what pessimists call a Faustian pact with the Devil and go all out for nuclear power. This could postpone the inevitable by many centuries, during which we could have almost any dream-houses we liked, and dream-cars too, if there were any countryside left to go driving in. Or could we? Would the Devil's price for an energy cornucopia be intolerably high? The easy response is to shrug these off as long-term questions to be dealt with as they arise : *if* they arise.

It is true that prophets who pronounce in centuries and in millennia do not need to worry much about what happens to their reputations if they are proved wrong, as well they may be. But the energy prophecies that sound most threatening are the ones being made in terms of decades. If the prophets are right, you and I will be feeling the pinch increasingly as the years pass. Not just because the people who sit on the oil wells can demand any price they like, but because the wells themselves will be becoming scarcer and less productive. Must our dreams fade, then, and must 'developed' countries eventually go back to muscular power, with limited help from wood, wind and water, while 'developing' countries never truly take off?

These are general but vital questions motivating much of the work that is now being pressed forward in the world's research laboratories. However, the questions are not simply technical ones, with straightforward and unique technical answers. There may be a choice of answers, and there are certainly non-technical criteria to guide selection. All the same, to exercise the power of choice sensibly one must understand enough of what science and technology are offering us, or hoping to offer us. This chapter is an introduction to some of the goods. It cannot do more than lead you to this fascinating subject. Fortunately there are plenty of books if you want to dig deeper.[1]

MORE POWER UNDERFOOT

The energy stored underfoot is not all in oil, natural gas and coal. There are other minerals from which it is possible to derive energy. There is also the earth's own internal heat, already being tapped in some places and (if current researches are successful) promising a great deal more.

So-called tar sands or oil sands are plentiful but mostly in the Americas. These are deposits of sandstone impregnated with a tarry oil or bitumen. Oil sand is already being mined on a commercial scale in Alberta, Canada. The deposits there have been described as one of the world's largest single energy sources. There are about 600,000 million barrels (a barrel is about 160 litres) of oil in place and about half of all this may be extractable once all the processes have been fully developed. The highest initial hope is for the shallower deposits, which can be mined by open-cast methods and the bitumen subsequently separated from the sandstone by hot water and steam to form synthetic crude oil. The sand is pumped back into the mined site to help reclaim it. But environmental damage is inevitable, and the more it is reduced the higher is the cost of the synthetic crude. Nevertheless, as oil prices jerk upwards the commercial success of the venture looks more likely.

The same cannot yet be said of another fossil hydrocarbon resource, also plentiful, and indeed more plentiful world-wide than petroleum. The biggest reserves of this mineral, oil shale, appear to be in the Americas, China and the Soviet Union, but Britain also has some. In spite of its name it is not truly a shale and it contains no oil. Geologists would call it a marlstone with a bituminous content. The latter is a solid substance, kerogen, which decomposes when heated to about 370°C to give an oil as distillate. Oil shale has to be mined like coal, and although abundant it is of very variable quality. Good shale yields 400–500 litres per ton of rock. Bad shale yields hardly anything. Both the mining and the waste disposal problems are tough – the waste product is more voluminous than the mined rock – so once again there are environmental snags. Dr. Armand Hammer of Occidental Petroleum claims to have a potentially exploitable technique of underground blasting followed by gas-heating, which produces oil that can be drawn off. Proponents of fluidized combustion (details later) suggest that oil shale, and even oil sand, might be best exploited by burning them in power stations at or near the mines and transmitting electricity, rather than producing and transporting costly synthetic crude oil.

Evidently these alternative fossil fuels have not been quite mastered by the technologists yet, but they are waiting in the wings. Meanwhile other technologists are looking to underground energy of quite a different sort. They want to mine *heat*. The origins of geothermal energy are not entirely certain : some of the heat has presumably been left over from earth's creation, but the greater part is ascribed to the radioactive decay of such elements as uranium and thorium. In some areas this heat is sufficiently concentrated to produce hot water or steam. More than half of Iceland's domestic space heating is provided by such natural hot water, and in Italy, Japan, New Zealand and the United States hundreds of thousands of kilowatts of electricity are generated by power stations driven by natural steam.

Heat wells

Altogether about 32,000 million kilowatts of thermal power are being conducted to the earth's surface from the interior, and less than 1% of this emerges in hot springs and volcanoes. The hot springs are already being put to use, and countries favored with them have ambitious expansion plans, but what about volcanoes? According to Dr. G. P. L. Walker of Imperial College the volcanoes are dissipating every year more energy than man generates in his electrical power stations, and it would prove no more difficult to turn volcanic energy to account than it was either to explode the first atomic bomb or to fly to the moon. If the incentive ever becomes as great as it was in those two instances, Dr. Walker will no doubt be proved right. But there are perhaps easier ways to be tried first.

One that is being experimented with is the drilling of heat-wells in dry rock. An oil-man's technique called hydro-fracturing is being used for this. Hydrostatic pressure is applied several kilometres down to make a cleft in the rock. Oil-men employ the technique to ease the flow of oil, but the heat-miners make a large-area cleft to transfer heat. The cleft is held open by sand pumped into it. Two pipes connect the cleft to the surface, one reaching to the top of the cleft and the other to the bottom. Water is pumped down to the bottom, gets heated as it rises through the cleft, and comes up through the other pipe to the surface. The engineers developing this technique at Los Alamos, New Mexico, have already proved that they can make the holes and the clefts. If they can make the system produce economic power over a long period of time they will open up geothermal possibilities for countries not blessed with hot springs or volcanoes.

There are at least ten times as many dry-rock thermal deposits as wet ones. However, seismological evidence from mining areas has been adduced to suggest that circulating water through those deep clefts could start tremors, accumulating into earthquakes.

As already mentioned, much of the earth's heat is probably due to the radioactive decay of uranium and thorium. These elements, too, are part of the energy bounty underfoot. They can provide fuel for nuclear power stations.

NUCLEAR POWER [2]

Fission power

Nuclear energy is the most controversial alternative to fossil fuel energy. Its civil use is at present restricted mainly to electricity generation, although other industrial applications (e.g. to metallurgical reduction, coal liquefaction and coal gasification) have been proposed. The difference between nuclear fuel and fossil fuel is striking. Coal, oil and natural gas undergo a certain amount of processing before they are fed to furnaces, but they remain crude materials compared with the manufactured assemblies of fuel 'elements' that go into a nuclear reactor. There is also a great difference in bulk. A tonne of uranium in a contemporary nuclear power station will generate as much electricity as 50,000 tonnes of coal.

But these stations consume uranium wastefully by comparison with another sort that nuclear engineers have spent decades of work to develop. This is built around 'fast' reactors, so called not because they react quickly but because the neutrons that fissure the atomic nuclei in their fuel (as explained in the next section) are energetic and fast-moving. These neutrons can transmute natural uranium that is *not* fuel (it is not fissile) into another element that *is* fuel. The infissile uranium is called 'fertile' because it gives birth to the new and fissile element, which is plutonium. So a tonne of uranium, instead of equalling a mere 50,000 tonnes of coal, can equal two million tonnes. Uranium is not the only fertile material in the world. Thorium is fertile too, and it is three times as abundant.

These energy riches sound enticing, but they take a great deal of engineering to be enjoyed. The evolution of low-power experimental fast reactors has been continuing since the world began to use nuclear power stations, but only now do the engineers feel ready to build on a commercial scale. The various technical problems have taken time to solve, partly because they have demanded novel solutions. For example, a tremendous quantity of heat is generated in a small space, and to carry it away fast enough the station has to pump molten sodium-potassium metal through the reactor. (Helium gas is being considered as an alternative heat-transporter but collaborative multinational research would be needed to introduce it now.) Such unconventional methods take time to perfect. Moreover, what constitutes perfection is differently perceived by different observers. While nuclear protagonists see plutonium as a highly desirable fount of energy, created anyway in both civil and military plants and stockpiled by the military for weapons, and better 'burnt' in power reactors than kept, there are antagonists who see the metal as one of the most toxic substances in existence, a danger to keep anywhere, to handle, to transport, and to tempt unscrupulous thieves or terrorists who might want to threaten violence with it.

A fateful dilemma

Nuclear power stations have another output besides valuable power – radioactive material. A little of this is valuable too. The rest is merely nasty. If it got accidentally into the environment it could in serious cases damage the health of entire communities and generations of their successors. The chance of accidental release from a reactor has been reduced to a minimum by the designers, and they estimate that the probability of a failure that would release radioactivity in fatal amounts is so low that the risk of death to anyone living nearby is less than one in a million per year's residence.

But besides the reactor failure risk (which human ingenuity reduces but human fallibility keeps above zero) there is the problem of waste disposal. The pro-nuclear faction says that the compactibility of nuclear waste is an advantage. Orthodox fuel burners disperse their pollutants in a creeping and perhaps irreversible attack on the environment. Processes of vitrification are being perfected that could lock up the waste from the nuclear generation of 340 million kilowatt-hours of electricity in a solid glass cylinder 750 millimetres high and 500 millimetres in diameter, so in Britain six or seven hundred such glass repositories could store a year's waste-product out of harm's way. The anti-nuclear faction is unappeased, denying our right to foist upon our descendants the guardianship of materials the worst of which can keep their fatal potency for hundreds of thousands of years. The pros on the other hand say that *their* wastes *do* ultimately decay, and some of them quite quickly, while other industrial poisons stay deadly for ever. And so the debate continues.

Society is thus faced with a fateful dilemma. Should the great fission-power experiment be carried through to its technically logical commercial conclusion, and should the nuclear engineers be allowed to tie up what they regard as the remaining loose ends of their technology? Then man's electricity and heat supplies might be secured (albeit at rising prices for ever-harder-to-get raw materials) for thousands of years. Or should nuclear risks be banned from the list of hazards, some of them statistically more threatening, that industrialized society already accepts? There are opponents of nuclear power who warn of encroachment on civil liberty by the elaborate security measures needed to safeguard nuclear plant and materials. They foresee violent resistance to the 'imposition' of large-scale nuclear programs'. Some environmentalists are turning these into self-fulfilling prophecies. In Australia they have reportedly threatened to detonate home-made fission bombs in cities if uranium mining and export are not banned, adding that if their bombs prove damp squibs, they will at least have contaminated city water supplies !

So, one way and another, the future of fission power seems rather obscure at the moment. Is there no form of nuclear power that is unquestionably safe? Some people place their highest hopes on power from nuclear *fusion*.

Fission and fusion

Fusion is the direct opposite of fission, yet the result is still a profuse outrush of energy.

In fission the nucleus of a very heavy atom is struck by a neutron and split into two other, lighter nuclei. Two or three neutrons fly from the collision scene and, if they hit other fissile nuclei, cause repeat performances. Thus a chain reaction can be brought about. Somewhat as the energy of stretched elastic binds projectile and catapult together before the fingers are opened, but then whangs the projectile into its trajectory, so dozens of protons and neutrons are bound into a heavy nucleus by energy that, after fission of that nucleus, expresses itself in flying fragments and radiation.

Now it may sound paradoxical but it is an experimental fact that binding energy can also be released if the nuclei of certain very light atoms collide so hard that their major parts stick together and send some of their other parts flying. There is more than enough binding energy in each of the original nuclei to fuse those major parts into one heavier nucleus, so the surplus energy expresses itself again in flying fragments and radiation. Nuclei that can be forced to fuse generatively in this way include those of hydrogen (hence the hydrogen bomb) and of helium.

It is believed that fusion reactions power the sun.

Fusion power

Fusion power appeals for a number of reasons. One is that the source of it seems virtually unlimited, for there are vast amounts of hydrogen in water. However, things are not quite as straightforward as that. The most suitable hydrogen nuclei, because giving the most energy while being the least difficult to fuse, are not the simplest hydrogen nuclei. The simplest ones consist in fact of just one proton each. But there is a 'heavy' hydrogen nucleus (the kind in heavy water) that is a partnership between a proton and a neutron. When electrically neutralized by an orbiting electron, such a nucleus becomes an atom. It is an atom of the sort or 'isotope' of hydrogen called 'deuterium'. There is a great deal of deuterium in the world's water. Yet another isotope of hydrogen, called 'tritium', carries two neutrons in its nucleus. Tritium is unnatural. To obtain it the scientist does what alchemists would have given their souls to do, though today it is a commonplace : he transmutes the elements. As in fission reactors, the agent is the neutron. But to produce tritium the neutrons are made to irradiate a fairly plentiful element, lithium, which yields helium as a by-product. (One day that could be very interesting to airship engineers and the designers of closed-cycle gas turbines.)

So one can imagine a reactor in which deuterium and tritium are made to fuse, emitting neutrons into a lithium blanket to produce more tritium. Which brings us to another appealing feature of fusion. The helium that is the principal material product of the reaction is not radioactive, and there is therefore no radioactive waste-product problem. The structure of the reactor must become radioactive, however, as must that of a fission reactor, so *all* outworn reactors become monumental sepulchres, haunted by physical rather than psychic emanations.

Imagining a fusion reactor is much easier than designing one and getting it to work. To make the nuclei of deuterium and tritium hit each other is not easy. In the natural way of things they repel each other, both being positively charged. To get fusions going on a large scale it is necessary to make a mixture of deuterium and tritium gases very hot : and by 'very hot' is here meant something over a hundred million degrees C. The atoms' electrons are torn from them well before this, so there is a chaos of electrically charged atomic fragments. This is called a 'plasma'. Plasma at a hundred million degrees costs a lot of energy to create. If fusions in the plasma are to give an energy profit the plasma has to be held together for a short time. It *is* only a short time – about a second – but even that is intensely difficult to reach. The atomic fragments dart about at millions of kilometres an hour. To bottle them up for even a second has been a task beating scientists for quarter of a century. Can it be done?

One method that is being investigated is *magnetic* bottling. Electrically charged particles are affected by magnetic fields, so a suitably shaped magnetic field should in theory be able to contain plasma. A popular type of magnetic bottle turns out not to be bottle-shaped at all. It is more like a tire. Only this tire-like tube, known technically as a 'torus', is not inflated – it is evacuated. Running round inside the torus, well clear of the wall, is a thread of plasma, following this path because held to it by powerful magnetic fields.

Bold European bid to harness fusion power

The most promising form of toroidal magnetic field being experimented with in the world's fusion laboratories today is a Russian idea and called a 'tokamak', a name constructed from parts of the Russian words for toroidal magnetic container. One of the leading tokamak projects is a European one, popularly known as Jet (an acronym from Joint European Torus). Jet was designed at the UK Atomic Energy Authority's Culham laboratory by a multinational team. It is not intended to be a power reactor. All that is hoped for is the knowledge that could lead to a power reactor later. But if and when built, Jet will be a formidable plant. The toroidal vacuum tube is specified to be made of eight stainless steel sections, welded together into an eighty-tonne whole. The various magnetic fields are to be generated by water-cooled coils weighing 460 tonnes. The iron core of the transformer is designed to weigh 1500 tonnes. Those figures are enough to convey the idea. The plant should take about five years to build, and after that it would be taken step by step up the temperature ladder, rising from the present laboratory levels of less than ten million degrees towards the hundred million threshold.

Hydrogen microbombs to provide power of the future?

The energy abundance that fusioneers are working for is so alluring that the USA, USSR and Japan are all putting money into tokamak development. But tokamaks and other magnetic bottles are not the only devices that fusioneers are working on. It has been suggested that the kind of accelerator employed in fundamental nuclear physics research could fire protons hard enough at each other to set fusions going. This is only the germ of an idea at present. Another possibility is exciting some visionary scientists more than any 'electrical machine'. They hope to fuse plasma nuclei without relatively lengthy magnetic bottling. Their scheme is to aim a battery of lasers onto minute pellets of deuterium, frozen or encapsulated. (For enhanced output they might mix tritium with the deuterium.) The laser beams, converging on these pellets, could if powerful enough vaporize the pellet skins so suddenly and with such force and heat that nuclear fusions would occur. The pellets would in effect be miniature hydrogen bombs. Nobody knows yet whether so powerful an array of lasers is feasible, but early experiments have not proved the concept too far-fetched to work on.

Ambition vaults higher than laser-induced fusion. Some scientists have looked beyond that to the possibility of triggering-off *fission* reactions with laser-induced fusions. Given a temperature approaching 3000 million degrees C, protons become energetic enough to penetrate the nuclei of boron atoms and split them. The products of fission in this case are not radioactively 'dirty' – they are clean helium nuclei. So we could have fission power without the radioactive waste problem. All that is necessary is boron pellets filled with deuterium and tritium, and a powerful-enough battery of lasers. Intense laser pulses will raise the temperature in the pellets to the hundred-million-degree level and the fusions will follow. Hydrogen in the mixture will provide the necessary protons.

Taking the various considerations into account, optimists look forward to proving fusion power feasible by the mid-eighties. More cautious prophets think in terms of next century, and point out that controlling the hydrogen-bomb process may be the most difficult technical feat that man has ever attempted – much more difficult than getting men to the moon or robots to the remoter planets. No political planner dare think of a fusion power program this century, at any rate. Planners of research programs, on the other hand, are likely to have fusion very much on their minds.

Now I did mention that fusion reactions are believed to power the sun, so there would seem to be one sizeable fusion reactor ready-made for us, a mere 150 million kilometres away. Why not use that?

SOLAR POWER

A popular observation is that all our energy originated in the sun. Our planet supposedly gestated there, and Earth's hot interior is a molten memento of the birth pangs. The energy stored in fossil fuels came from the sun and was fixed for us by the vegetable and animal life of millions of years ago. The energy that sustains life today is also from the sun. So is the energy that sets the winds blowing, the sea-waves running, the rain falling and the rivers flowing. The tides created by the sun's planetary system might seem exceptional, but if the planets are also the sun's progeny then one can trace that energy back to the sun too. So the only energy that we do not owe ultimately to the sun is from the stars and their company in outer space.

The sun is a prodigious provider. It is believed to be a nuclear fusion 'furnace', though there are doubts about what exactly is going on. Whatever the mechanism of generation may be, the amount generated is huge. The total radiant power of the sun is 38,000 million million million kilowatts, uninterrupted. This goes out in all directions, of course, and our relatively tiny planet (150 million kilometres away) intercepts only a slender beam of it. But that slender beam's average power is about 170 million million kilowatts. A space vehicle, orbiting above the atmosphere, can turn a square metre of flat surface at right angles to the rays and catch energy at a rate of 1·3 kilowatts on just that area. Evidently there is more than enough power there for every-body – thousands of times more than the world is consuming at present – if only we knew how to catch it and turn it all to account.[4]

By no means all that sunshine gets down to the ground or the sea. Part is reflected back into space by clouds and so forth. Part warms the atmosphere, creating winds and other disturbances. What does nourish the earth's surface varies with the weather, the locality, the season and the time of day. In mid-Britain the noonday sun on a midsummer's day can beam 0·9 kilowatts onto a horizontal square metre. In midwinter, on a clear day, this solar power can drop to less than 0·2 kilowatts. If we add up the receipts of solar energy on the earth's surface we see that, in theory at any rate, suffi-cient solar power for our *total* energy needs reaches us. Even Britain's annual direct insolation (or intensity of radiant energy on a horizontal surface) is well over half that at the Equator. If we were to add up all the solar energy reaching the earth's surface we see that we receive free potential energy equal to about 800,000 million million kilowatt-hours a year, which is slightly over half what was received at the top of the atmosphere, so a lot has been lost on the way. But what has got through has an energy esti-mated to be equal, in a mere *three days*, to the sum total of the earth's proved reserves of all other fuels.

Solar energy caught by satellites in space

The ideal way of catching solar energy must be to go out into space, collect it every hour of the day, and deliver it just as continuously down to earth. That may sound ideal but impossible, yet it has attracted a great deal of thought. An American, Peter E. Glaser, of Arthur D. Little Inc., suggested in 1970 that a fifty-square-kilometre island could be assembled 40,000 kilometres away in space. The objective would be to turn solar energy into electricity with silicon semiconductor 'solar' cells (like those used in space vehicles), or with closed-cycle turbines driving electrical generators.

However generated, the electrical power would then have to be beamed down to earth. For this purpose the space island would orbit synchronously with the earth, so that it stayed always above the same spot, but it would orient itself always to receive sunlight full on. Electricity from the converted solar energy would be used on the island to generate microwaves which would be transmitted from a giant aerial – perhaps a two-kilometre-diameter dish – in a focused beam to a receiving station on the ground. The receiving station would be an aerial 'farm', turning the microwave radiation into an output of electrical power. Enthusiasts for such satellites have claimed that their islands in the sun, or at least pilot-scale islets, could be out there and working before the end of the century.

The concept is audacious but fascinating. If successful it might even muffle the critics who see non-stick saucepans as the only worthwhile spin-off from expensive space research. Just think of what satellite suntraps could do. They would not be subject to alternations of day and night, so they would generate all the time. Microwaves of about ten centimetres wavelength could be generated on the island very efficiently, and they would penetrate the atmosphere with little loss. All round the clock (a Boeing Aerospace Corporation study tells us) nearly sixty square kilometres of satellite-borne solar collector would generate electricity to beam microwaves at 13·5 million kilowatts into an aerial farm a few kilometres in diameter. Not all of this power would get into the terrestrial electricity grid because there would be unavoidable losses from the system. But a splendid ten million kilowatts of uninterrupted electrical power could be expected to emerge. Pipe dreams? The US government is spending almost a million dollars to find out whether they can be turned into reality.

Boeing talk of an eventual thirty such stations, but they do not pretend to know how many millions of dollars each would cost. Construction would be a long step forward from the kind of space link-up that has been achieved so far. Many arts would have to be developed well ahead of their present state. Failures in parts of the system could be disastrous. Microwave radiation damages living tissue. What if the beam flicked sideways in a death-dealing arc across the earth? Engineers say that any spill-over, flicker or wander could be detected at once by instruments and the satellite would be signalled as soon as these monitors sensed danger. Failing automatic redirection of the rays the beam could be defocused and spread out harmlessly, or just switched off. Hence the need for plenty of satellites, for standby supplies, and hence also more columns of airspace forbidden to traffic (but no less fatal to wild life). There do remain questions to be answered. For instance, would large-scale ionization of the atmosphere be injurious? Would a steerable beam be a potential weapon for war or blackmail? One lesson we have all learnt is that strategically sited minorities (or cliques or countries) can, by threatening to stop the flow of energy, hold everybody else to ransom.

Giving up the twenty-four-hour service of a satellite and putting up with such other vagaries as the weather, one can bring down to earth the solar cells that are carried by satellites to turn sunshine into electricity. The original silicon solar cells were expensive and, although much has been done to bring their price down, they remain so. Also, they are not very efficient, managing only about 12–14%. The more recent aluminium gallium arsenide cell has exceeded 20%, however. Gallium is not as exotic as perhaps it sounds, for there is nearly as much of it to be had as of lead. Silicon is of course abundant – sand is full of it. The cost soars because the materials have to be refined by complicated processes and the cells manufactured with the aid of high technology. If developed for mass production, solar cells might be made less expensive. Sanguine forecasts suggest that they could be brought down to thousands of pounds per kilowatt, compared with hundreds per fossil-fired or nuclear kilowatt. Solar cells are in fact powering counter-corrosion gear for desert pipelines, and isolated electronic equipment such as radio and television relay stations, air traffic beacons, offshore platforms and marine buoys. Wider applications are still being sought.

Solar cells of quite another sort are attracting research attention of a more fundamental kind. These are cells relying, not on inorganic semiconductor crystals, but on organic materials with semiconductor behavior. While the theoretical conversion efficiency of inorganic crystalline semiconductors is limited to 24%, very much higher figures might be possible with organics. The mechanism of electric charge generation and transfer is not the same in organics as in inorganics, but the much higher conversion efficiency is manifest in nerve and other biological systems.

The 'solar eyeball'

One of the complications of capturing solar energy is that the direction of the rays is always changing. Ideally the capturing surface should always be at right angles to the rays. Technically this is quite feasible, but at a cost. And if motors are used for continual reorientation, some of that cost is in partial diversion of precious energy output from the solar device. An imaginative idea for dealing with this problem is due to Derek Mash, of Standard Telephones and Cables. His device is called a 'solar eyeball'. This gargantuan organ keeps its gaze fixed, however, as no living eyes should be, on the sun. Instead of a retina it has a gallium arsenide solar cell. It is a soccer-ball-sized plastic sphere, hermetically sealed, with a Fresnel lens to concentrate the sun's rays on the solar cell when the 'eye' is looking into the sun. When the sun's relative position has moved far enough for the focused sunspot to drift off the cell onto one or another of four neighboring air-vessels, the air is heated and moves pistons along a pair of arc-shaped tubes. The pistons are magnets and they interact with magnets fixed outside the eyeball. The net effect is to roll the eyeball into a sun-stare angle again. You can imagine a huge tank of water, bubbling with eyeballs, all floating and rolling virtually without friction or wear. Stray solar heat need not be wasted : it can warm the water to some purpose. One day plastic eyeballs might be mass-produced for solar farms in the world's deserts.

The oldest way of turning solar heat into electricity is also attracting research. The method is to concentrate the rays onto a boiler for an engine-driven generator. Research is leading to new kinds of boiler, and vapors other than steam are offering possible advantages. So those solar farms in the world's deserts might equally sprout vast plantations of mirrors, perhaps on the lines proposed by Prof. Giovanni Francia of Genoa, Italy. In Francia's demonstrations the arrays of mirrors are made to follow the sun by clockwork and they reflect the rays onto a honeycomb arrangement cunningly designed to trap them, not reflect them. Another dramatic-looking solar furnace is at work in the Pyrenees, part of the solar power research program in France.

All radiation-into-electricity converters have one obvious failing in common. However efficient their transactions, they spend as fast as they earn. In other words, they do not store energy. What some ambitious scientists hope for in the long term is emulation of biological solar-energy conversion, which does make storage possible. Plants catch the rays while they can, convert the radiant energy into chemical energy, and thus cope with the variability of sunshine.

Certain plant cells have parts called chloroplasts. These contain the green pigment, chlorophyll and other pigments. The chloroplasts supervise construction of carbohydrates from raw water and carbon dioxide, taking the necessary energy from sunlight. The 'exhaust gas' from this process of photosynthesis is oxygen. Plants draw on the energy store by reversing the process, oxidizing the carbohydrates and 'exhausting' water and carbon dioxide. But oxidation in this case is a subtle, stepwise process controlled by enzymes, not the crude combustion process we use to extract heat from fuel. All life – animal, vegetable, bacterial – relies on the biological oxidation process, respiration, to get its energy. Biologists tell us that the primary energy-storing capacity was first acquired about 3000 million years ago by single-cell plants (blue–green algae), and it is still monopolized by the vegetable kingdom. Worth contemplating is the fact that this process continues to be the origin of most of our fixed carbon and oxygen.

In biological terms the process is highly successful. In engineering terms, on the other hand, it is not highly efficient. At best, photosynthesis can fix about 27% of received solar energy in carbon compounds. Plants generally fix less than 1%. Man, growing crops of plants or algae under artificial conditions, may boost the conversion efficiency to 6–8%. It is easy to see why hundreds of millions of years have been taken to accumulate the buried energy of fossil fuels, and why their natural rate of replacement cannot keep pace with man's rising demands.

Vegetation acquires only about 0·2% of earth's solar radiation income for safe deposit, but even that small proportion is in absolute terms a great deal of energy. All life depends on it, and activities such as agriculture and silviculture are designed to channel more of the benefit manwards. Renewable sources of fuel from agriculture are bagasse, cotton stick, dung and hay. Agricultural and other biological waste is a source of fuel gas consisting mainly of methane. Silviculture gives wood and derivative fuels – charcoal, methanol and producer gas. Research is being done to make the utilization of these fuels more efficient. Longerterm investigation aims to make the various biological processes more efficient from man's point of view. In the future algae may be farmed to produce gaseous or liquid fuels. Such fuels may be needed to drive motor vehicles, whatever other means may become available for driving large electrical generators. An attractive possibility for motor vehicles is methanol, which can be made from biological gas or wood. Methanol from the United States' forest lands, it has been suggested, could meet the entire current energy needs of Americans.

It is difficult to foresee just how large-scale intensive 'energy farming' would work out in practice. For high yields an energy subsidy is necessary – fertilizers, mechanical harvesting and so forth. Pessimists predict that on balance more might be consumed than produced. Conversion of large tracts of land to energy farming could have unexpected results. Violent changes in the ecosystem can be harmful, as inheritors of deserts know, so prevailing equilibria must be changed with due caution. This calls for research beyond the technologies of energy capture and utilization to the consequences of those technologies. Here the best and nearest approximation we can make to prescience is by science.

Man may not have to cling so closely to Nature's apron-strings, however. Photoelectrochemists may make new departures possible. A practical battery may be developed that can be charged by exposing it to sunlight. Or a photolytic solar-stimulated cell may at last be found that economically splits water into oxygen and hydrogen. And hydrogen, some say, is the fuel of the future.

Before we go on to the larger question of alternative fuels, however derived, there are some more ways of using solar energy to consider. Sterling work is being done on ways of turning solar heat to account in buildings and their services. Solar storage has been brought nearer by the researches of such scientists as Prof. H. Tabor in Israel, who capture the heat in salt water ponds. Dissolved salt makes the hotter water go to the bottom. Solar ponds are probably the cheapest means known yet for collecting solar energy in a big way. In some circumstances the water temperature can approach $100°C$. Work is also being done to improve specially radiation-absorptive surfaces. Such surfaces are employed in solar water-heating panels, which are common on Mediterranean roofs and are now beginning to appear under our own grey skies as boosters or economizers for domestic and industrial systems.

An imaginative use of absorptive surfaces is being experimented with by Graham Stevens. He has devised a plastic balloon in the shape of an inflatable raft. The absorptive surface is applied inside the raft, on the bottom skin. The top skin is transparent. In sunshine the air inside gets hot and the whole thing blows up into a hot-air balloon. It lifts off the ground and would float away if not tied down. Stevens, a visionary, sees the device as an almost infinitely adaptable sun-supported surface, providing shelter in tropical desert, a condensing surface for desalting water there, or a magic carpet actually lifting water into the sky and irrigating the desert. . . . No less dramatic, perhaps, but certainly less fanciful, is a sun-powered irrigation pump invented by Dr. Colin West of Harwell. This is a new variation on an old theme – an engine invented over 150 years ago by the Scottish clergyman after whom its thermodynamic cycle is named. Dr. West's Stirling-cycle pump can be demonstrated in principle with a few bits of plastic tubing, making a model which rocks in the sun's rays (or other heat) as if of its own volition. This pump is being developed at Harwell by Dr. West, with collaboration from an Indian company which might well be the first to exploit the idea.

But technology is not all. A solar-powered pump that was technically a sucess was developed by French engineers and installed in a Saharan village. The project was carefully planned, with such nice touches as the dovetailing of the children's schooling with their duty as water-carriers : they brought well-water to a solar panel that cooled their school as well as heating water. The water's heat drove the engine that pumped the water from underground, and the increased supply of water should have benefited the village. But the extra water was monopolized by the two rich men, and they got richer while everyone else got poorer. As a final irony, the same sort of solar pump *has* proved beneficial, but in another and wealthier part of the world.

Now for yet further uses of solar energy – but not as we have been looking at it so far. Solar energy comes in other guises, and perhaps more alluring ones, at least in some countries.[5]

Wind power[6]

Quite a slice of the sun's supply of energy goes into stirring up the atmosphere. The amount of energy blowing about our planet's surface in wind has been estimated at 13 million million kilowatt-hours a year. The fact that the wind is as capricious in some places as the sunshine has not deterred man in the past – windmills and sailing ships are ancient inventions. The steady power generated by fuel-burning engines ousted both the windmill and the sailing ship, but now that the power of those engines is liable to interruption by fuel monopolists of one sort or another, the variability of the wind seems less unfriendly. Also, the wind will go on blowing long after the wells and the mines have been exhausted.

The classical windmill, with its sails on a shaft revolving in bearings at the top of a tall tower, is expensive and inefficient by today's standards. Attempts are being made to bring the concept up to date by employing modern materials and design techniques. American enthusiasts thought likewise quarter of a century ago, and they built the biggest-ever wind-driven electrical generator at a place called Grandpa's Knob, in Vermont. The tower was over thirty metres high and the two stainless steel blades spanned over fifty metres, tip to tip. All this, on a windy hill, could generate power approaching 1500 kilowatts. But the 'modern materials and design techniques' of those years were not up to it, and one of the blades snapped off in a storm. Today there are such materials as light alloys and glass-reinforced plastics and other composites, which aeronautical engineering demands have summoned into existence. The US Energy Research and Development Administration is following up a hundred-kilowatt experiment at Sandusky with two two-bladers rated at 200 kilowatts in low winds, and then two high-wind machines designed to generate 1500 kilowatts apiece. Costs go down as generator size goes up, so the bigger plants are more attractive economically. A 45-metre blade (i.e. for a ninety-metre-diameter wind turbine) is being built experimentally.

Some wind-power researchers prefer to experiment with a principle that, in one respect, goes back to the earliest windmills. These rotated about a *vertical* axis, not a horizontal one. Such machines do not need a high tower and they do not need to be steered round to face the wind. But the modern vertical-axis machine is of a distinctly twentieth-century configuration. One such configuration is named after its

French inventor, G. Darrieus, the other after its Finnish inventor, S. J. Savonius. Both brought their ideas out about fifty years ago.

The present-day development of the Darrieus principle is due to Raj Rangi and Peter South of the National Research Council of Canada. Their embodiment of the principle can be likened to a pair of helicopter rotor blades, bowed into vertical arcs like stretched longbows and joined at the middle by a vertical strut. The actual shape of the arc is not exactly that of a longbow but is designed to distribute the stresses evenly in the whirling rotor. It is a shape known as a 'troposkien' and would be assumed naturally by a rope whirling in place of the rotor blade. The blade has an aerofoil cross-section, and the machine as a whole is lighter and cheaper than any classical wind turbine. Engineers in North America and Europe are working on variants of the Canadian Darrieus design. A British one, recently developed at Reading University, departs from the fixed bow shape and has straight blades on hinges. In low winds the blades are vertical, giving the highest efficiency when it is most needed. In high winds the blades can lean outwards, against the pull of a spring, to reduce blade stresses. This construction could turn out to be an economical one for Darrieus turbines.

The Savonius machine, or S rotor as it is sometimes called, looks S-shaped on a horizontal cross-section through the blades. The rotor spins in a cross-wind because the concavities of the S resist the airflow more than the convexities. Small S rotors can be seen at work in ventilators. By comparison, Darrieus turbines are fast runners. As Darrieus turbines do not have much urge when starting, it has been suggested that the best answer is a combination of Darrieus and Savonius. Another suggestion is to place Savonius machines alongside a bluff vertical cylinder, diametrically opposite one another at the points where the airflow is most speeded up by the cylinder's presence. Thus the rotors are blown round more vigorously.

Quite another idea is what its American inventors call

the tracked vehicle aerofoil concept. Blades of aerofoil section are fixed upright on an endless train, travelling round a track shaped like the one every schoolboy must have played with – two semicircular portions joined by two straights. As the carriages enter the bends their blades' angle to the wind is changed automatically. By the time the carriages have got round to the other straight they are being pushed the other way by the wind's thrust on their blades. So the train is being pushed in opposite directions along each straight, and the train as a whole is pushed round and round the track. Computer analysis is said to have proved the superiority of this wind train over conventional wind machines – an eight-kilometre track with blades twelve metres high and three metres wide could, on these calculations, supply the electricity needed by about 15,000 people.

Another American fairground-style idea, reminiscent this time of a helter-skelter instead of a railway, comes from the Grumman Aerospace Corporation. This too is an attempt to get away from large rotating structures and the stresses they suffer. The Grumman idea is called, graphically, a tornado system. Basically it is a round tower with louvered vertical slots in its side. The louvers are adjustable, and admit the wind so that it is guided into a vortex flow (but upwards, instead of the helter-skelter's downwards). The vortex eventually spirals away out of the top of the tower, but not before it has sucked air through a vertical-axis turbine driving a generator to put out a million kilowatts or more.

Dr. J. Swithenbank of Sheffield University looks ahead to times when North Sea gas has been drawn off and exhausted wells sigh for use. Why not anchor a floating chemical works over the wells to extract hydrogen from the sea and keep the wells topped up with this gas for ever and ever? The power for the floating plant would be fed down to it from an airship, tethered to the vessel. Generating the power on board the airship would be a huge horizontal-axis wind-driven turbine. Less daring spirits have suggested wind-driven generators on North Sea oil and gas platforms, sending electricity ashore through sea-bed cables, but, as Dr.

Swithenbank points out, the wind blows harder at altitude.

Coming down to earth we can observe what might seem relatively plodding progress with a wind-driven greenhouse-heater. This is an NRDC-backed design being developed by the Wind Energy Supply Co. and it has a fully automatic two-blade rotor at the masthead. From there power is transmitted hydraulically to ground level. With the help of friction this power is easily converted into a flow of heat. Another idea that the engineers are considering for the greenhouse-heater is a heat pump, an energy-economizing device that works like a refrigerator.

Monster wind machines might not be out of place in the wide open spaces of land or sea, but they are less acceptable where space is precious and where amenity matters. Also, their power output is highly variable, making some form of energy storage even more desirable than it already is with most electrical generating machines. But there is a gigantic natural store for the purpose, constantly picking up wind energy and holding it in less perishable form for our consideration. It is the sea.

Ocean power

The oceans are vast reservoirs of energy. They stock the plainly physical forms of energy such as tides, waves and currents as well as the potential fuels hydrogen, deuterium, uranium, etc. Underneath the ocean beds there are also substantial resources of oil, gas and coal. So far man has scarcely licked the surface of what is on offer. In landlocked countries this is hardly surprising. But shorelined countries are looking seawards seriously only now that fuel supplies have become dear and insecure.

The form of ocean power that most takes most fancies is tidal. The rise and fall of ocean level, mainly under the alternating effect of the moon's gravitation, causes picturesque phenomena along the coast. It also accounts for power in a global amount that has been estimated at 3000 million kilowatts. But harnessing the energy of the tides does present difficulties. In mid-ocean the change in level is scarcely noticeable – perhaps a metre. This may seem an academic point, but it does highlight an important fact: tidal power, like most resources, is for those whom geography has favored.

Geography is favorable in estuaries whose natural frequency of rise and fall matches that of the tides, so that the difference between the highest and lowest levels is increased. One of the world's best estuaries from this point of view is that of the River Severn. With the Bay of Fundy in Canada and the Ile de Chausey in France, it is among the exceptionally good sites, with spring tides giving fourteen to fifteen metres between highest and lowest levels. The neap tides are only about half as good as this. The Bay of Fundy is being considered for a tidal scheme by the Canadians, and the Russians are reported to have some schemes too. The French have for some time had a scheme in mind for the Ile de Chausey, and they have carried through a small pilot project on the estuary at La Rance, near St. Malo. Technically this has been successful, but the main project has been shelved. Schemers have also studied the Severn Estuary and have produced different estimates of the amount of electricity that a barrage could generate there. Nobody questions that substantial power could be generated. But the ultimate deterrent, there as at other possible tidal sites, has been an economic one. Intermittent output means that there has either to be energy storage or conventional power stations have to be built to bridge the power gaps. Either way the costs are excessive at present. When fuel becomes dear enough the balance may change, but there are still environmental questions to consider. One of the Severn schemes, for instance, would present the population of Weston-super-Mare with a permanent three-kilometre mudscape.

For countries such as Australia, Britain, Canada, Japan and the USA there appears to be an attractive alternative that has been surprisingly neglected until recently : the waves that the wind whips up (by some still rather mysterious mechanism) and blows with great vigor towards certain shores.[7] Inventors in their hundreds have long been busy devising ways of harnessing this energy, as well as all the other natural energies known or guessed at, but there have been few to take notice. One man whose ideas have won commercial success in recent years is Yoshio Masuda in Japan. He has made bobbing buoys generate their own electricity for navigation lights and so forth. They work with an oscillating water column. The principle of this is usually explained in terms of an open can, held upside-down in the water so that the oscillating water level acts the role of an air-compressor piston inside. The compressed air thus producible (with the aid of valves to guide the flow) can drive a turbogenerator exhausting to the atmosphere. Devices working on this principle are being developed for bigger power outputs both by Masuda in Japan and the National Engineering Laboratory at East Kilbride.

The National Engineering Laboratory has assessed how much energy is flowing towards various shores. The lowest annual amount is in the Mediterranean – 275 megawatt-hours per metre – and the greatest in the Atlantic, off Scotland and Ireland – over 500 megawatt-hours per metre. On a 2700-kilometre contour around Britain, fifteen kilometres offshore, the cross-flow is about half a million million kilowatt-hours a year, or more than twice the output of the UK's electricity authorities in 1974. After recommendation in Lord Rothschild's energy report from the Government's 'Think Tank' in 1974 wave power was promoted from mad-invention status to researchworthiness, and over a million pounds were earmarked for chosen devices. At the time of writing there are five being studied with Government backing and there may be more to follow.

Four are floating devices. The National Engineering Laboratory is working towards a system of oscillating water columns in line abreast, virtually a floating breakwater that extracts instead of dissipating the energy of the waves. Sir Christopher Cockerell, inventor of the hovercraft, has produced an idea that is being developed by his firm, Wave-power Ltd. In one version of the idea pontoons are joined together like links in some giant's bicycle-chain. Moored head-on to the waves the pontoons pitch individually as the waves pass along the line, making the pontoons form a rough copy of the wave pattern. The pontoon's relative motion at their hinges drives generating machinery. Of course, in any major generating system there would be multitudes of such pontoon chains side by side. Another kind of device that would be moored in multitudes, head-on to the waves, is the invention of Prof. Michael French of Lancaster University. He is experimenting with a long rubberized-fabric tube divided along its length into sealed and inflated cells. As the waves pass along the flanks of the tube (which floats just submerged) the cells are alternately squeezed and released. Valved passages in the keel conduct the compressions and rarefactions to reservoirs from which an air turbine is driven, perhaps to generate direct current for shoreward transmission. The most efficient of the floating devices, at least under laboratory test, is the 'duck', invented by Stephen Salter of Edinburgh University. He has so cunningly contrived its shape that it can extract nearly all the energy from the waves. A line of such 'ducks' is hinged to a common 'backbone', looking like the keys of some enormous piano, played perpetually by the waves. And their playing drives the generators.

The only wave-converter in the program that does not float (and that therefore avoids mooring problems, enjoys a relatively cheap sea-bed foundation, and scoops up more accumulated wave-energy) is the one due to Robert Russell, Director of the Hydraulics Research Station, Wallingford. This is a two-basin scheme. A sea-wall has vertical slots, so valved that wave-crests fill one basin while wave-troughs drain the other through a low-head turbo-generator.

All these experiments are in their very early stages. There is talk of a pilot-scale power station in the North Atlantic off Scottish shores in the 1980s. But the Department of Energy's Chief Scientist, Dr. Walter Marshall, does not expect useful power from the waves before this century is out.

Ocean energy is much more than skin-deep. There are even greater resources under the wrinkles. Ocean currents are but another manifestation of solar energy. They are caused partly by the large proportion that must fall on water – 70% of the earth's surface is ocean. Surface water is heated by the sun and, in the tropics, stays fairly steady at 25°C. But a thousand metres down the temperature is kept at about 5°C by currents coming in from polar regions, balancing the poleward flow at the surface. The currents are not impelled by this convection alone. Wind plays a role (that is why wind and current directions often match). Just as power can be won from the wind by rotors, so might it be won from ocean currents. Proposals have been put for placing arrays of current-turned rotors 20 to 120 metres deep along a twenty-kilometre front near Florida, to draw on the Gulf Stream's energy and generate a million kilowatts. But the power potential that has drawn most research interest is in the temperature drop between the surface and the deep. The Caribbean and the Gulf Stream are among the waters offering opportunities not far from land, for utilization of the steady twenty-degree difference between currents about a thousand metres apart. The principle is to run a steam or other vapor engine, taking the hot current for boiling and the cold current for condensing. In early attempts sea-water itself was used. It was flash-evaporated in a chamber under partial vacuum (the boiling point of water falls with atmospheric pressure, as every mountaineer knows). Sea-water is too corrosive and needs too high a vacuum to be practical, so present thinking favours ammonia, propane and fluorocarbon refrigerants such as Freon.

It is estimated that temperature-falls between ocean currents make more power available altogether than is at present consumed world-wide, so even fairly inefficient submarine generators could be adequate. It is possible that, for countries such as the USA, which have suitable sites, power stations of the future will hum in buoyant reinforced-concrete hulls over a hundred metres long and thirty metres in diameter, moored to the sea bed and floating sixty or seventy metres under the surface, and communicating with the surface by means of towers. It is also possible that such ocean power stations will be combined with fish farms.

Sun and ocean together are responsible for another great power resource, which is already being tapped inland, generating electricity. Solar power amounting to perhaps forty million million kilowatts goes into working that great machine known as the hydrological cycle : water, evaporated from the oceans, makes a roundabout journey via the clouds and river basins back into the oceans. It is this great hydrological machine that provides the heads of water driving hydroelectric power generators.

Repeatedly we have found that solar power and its daughters (e.g. wind and water power) lead to the same energy end-product – electricity. This is for a great number of domestic and industrial purposes the most convenient and useful form of power. For some purposes, such as telecommunications, it is supreme. Yet there remains at least one substantial application for which electricity does not have universal utility : transport. Here the *mortal* solar daughter, fossil fuel, might still seem irreplaceable. Trains and trams can best be powered by electricity, but road vehicles, ships and aircraft stay obstinately fuel burners. This could change. Automotive heat engines will not necessarily be forced to run on fuel for all time. Mechanical engineers are experimenting with power units whose tanks are filled with liquid nitrogen. Atmospheric heat is enough to boil this, giving the vapour to drive a turbine. The exhausted nitrogen goes back into the atmosphere from which it was extracted in the first place – an environmentalist's dream! Solar-powered cars have been proposed, and experimental ones have actually been driven, but if such vehicles are to be available throughout the day they must be able to store energy. Batteries, as we shall see shortly, are not yet up to the task. Flywheels are a possibility raised by new materials of construction, but they too have a long development-road to travel, with perhaps even less chance of arriving anywhere.

The hard fact we end up with is that, to date, the best stored energy for vehicles, as for most other purposes, has been that held chemically in fuel. What are the prospects for fuel power?

FUEL POWER

Fossil fuels will not last for ever but they can be spun out by more efficient utilization.[8] This is most important in the case of oil and gas, the hydrocarbon fuels that are being most rapidly depleted. Coal, which has a smaller proportion of hydrogen, is in much more plentiful supply, but may turn out to be best used with hydrogen added, to convert it into gaseous or liquid form.

Gasification and liquefaction of coal are by no means new. Coal gas came with the Industrial Revolution. Hydrogenation, or building more hydrogen into coal, began on a large scale in Germany, where it was wanted to produce liquid fuel in advance of and during World War II. Oil is still being synthesized from coal by similar processes in South Africa. Generally, scientists are working to develop gasification and liquefaction processes for what is foreseen as 'The Coal Renaissance'. The Americans are particularly keen on gasification because of their pressing need to find a substitute for natural gas, and they are co-ordinating their research with Britain's to speed progress.

According to the National Coal Board's science chief, Leslie Grainger, substitute natural gas made from coal is a more efficient energy supplier than electricity because the consumer gets more than half of the coal's original heat (not discounting losses from gas appliances), whereas electricity, even with the efficient fluidized-bed-combustion boilers now under development, lets more than half the heat get away. Grainger dismisses nuclear power advocates' ideas of hydrocarbon fuel manufacture from hydrogen and carbon dioxide (the hydrogen from water electrolysis and the carbon dioxide from limestone). In his view coal will be for long the chief and cheapest source of carbon, most conveniently converted into more valuable forms with hydrogen from the coal itself and from reacting steam. Gasification can lead to production of various liquids, not only for fuel but for chemical feedstocks. As the prices of oil and natural gas go up, *coal* refineries may become a more prominent feature of the industrial landscape than are *oil* refineries today. Prophets foresee the 'coalplex', a plant in which different coal processes are operated to help each other, saving costs and producing various fuels, chemicals and other commodities.

In the much nearer future, liquid fuel derived from coal could well be driving spark-assisted diesel engines in road vehicles. Such engines are being developed because they should be less fastidious about fuel than gasoline engines. Diesels are anyway more efficient than gasoline engines, and with spark instead of compression ignition they become insensitive to fuel quality. Cutting short the route from the mine to the filling station even further is another potential fuel – pulverized coal. This has long been the form for power stations, but now engineers are driving experimental car-engines with it.

Radicalism is essential for the big innovations but conservative reform remains the realistic policy for all oil and gas users. It is the conservation of both fossil reserves and of the environment that is the objective of 'reformers' of engine design. In the world's development laboratories the engineers are refurbishing such ideas as the Stirling-cycle engine and the stratified-charge engine to supply the new necessities. For industrial power and heat yet other routes are being found to higher efficiency, as described later. Thus oil and gas deprivation can be postponed for as long as conservative measures permit and the earth's still plentiful coal can be mined. For the present, even with the profit margins imposed by the owners of easily extracted oil and natural gas, substitutes made from coal cost too much to market. Return to a coal economy would also be costly in human terms. Unless robots could take over the underground work (as, defying the mining engineers, Prof. Meredith Thring of Queen Mary College maintains they could) colliers' lives must continue to be more hazardous, unhealthy and uncomfortable than most. Also, in spite of the care and skill of modern mining engineers, there must be loss of amenity on the surface. Is a beauty spot like the Vale of Belvoir to be lightly sacrificed? And it is no minor matter to counter atmospheric pollution.

Coal can yield methanol but (as mentioned earlier) that liquid fuel can also be derived from wood. Advanced forestry has been put forward as an answer to the fossil-fuel problem, for example, by D. E. Earl the silviculturist.[9] Even biological gas – methane generated by fermentation from refuse – has been advocated as a universal fuel.[10] Such gas, from sludge, is already used for electrical generation in sewage works, where it is burnt with some oil in special engines. A small fleet of vehicles was run on the gas for many years at Mogden Sewage Works. Other carbonaceous refuse that could yield methane includes such farm waste as pig refuse. In some hot countries the gas is generated from sun-dried dung, which is anyway the principal fuel for domestic purposes. The inventor, Harold Bate, fuels his car with chicken-droppings.

Methanol liquid and methane gas are hydrocarbon fuels obtainable from vegetation and refuse instead of from mineral resources, and they may sound basic enough. But there is a school favoring a simpler and more basic fuel – hydrogen without the carbon. Hydrogen is advocated not only because it could drive engines without pollution, as it has already driven the engines of experimental cars and aircraft, but because it might with advantage be substituted for electricity. It is abundant. It can be transported cheaply by pipeline and stored compactly when liquid. A hydrogen pipeline one metre in diameter would transmit 8000 megawatts of power more simply and cheaply (it has been claimed) than an overhead transmission cable or a superconducting cable of comparable capacity. At the receiving-end electricity could be obtained either from engine-driven generators or from fuel cells.

Fuel cells are electrochemical cells that do the opposite of electrolysis: electricity comes out when hydrogen and oxygen are fed in and, in the presence of a platinum catalyst, combined. Fuel cells, possibly supplied with hydrocarbon fuel and air, are also being studied for automotive power, but they face the same problems of bulk, weight and operating risk that confront battery power generally. All batteries can be regarded as fuel-users, though the substances they 'consume' are no more or less 'burnt' than is the fuel of a nuclear reactor. The difference between a fuel cell so-called and other electrochemical cells is that the fuel can be fed into a fuel cell continuously and it is a permanently installed power unit with easily disposed-of waste products. Other cells have to be replaced when their charge has run down, and may have to be discarded altogether. A fuel cell is a relatively efficient power unit. Theoretically it can turn up to three-quarters of its chemical energy intake into electrical energy output. It has no moving parts to make a noise or break. Fuel cells power artificial satellites and experimental road vehicles today. In the future they could power your house and car.

Where it is not transport but transportability that is wanted, the 'primary' cell is still the most popular, and it is still being improved by research. But these cells become garbage once their electrical energy has been drawn off, and the electricity they supply is expensive. The cost is bearable only because the amounts are small and the convenience is great in transistor radios, pocket calculators and the like. More economical for higher-powered devices is the 'secondary' cell, which can be recharged simply by passing electricity back into it and thus restoring the inbuilt fuel material to its original state.

The trouble with the well-established lead-acid rechargeable battery is that there is so much battery for the energy it dispenses. There are however other electrodes and electrolytes, and some of them promise more energy for less weight. Research is being done to make some of them competitive with heat engines, at least for some types of vehicle. Advanced batteries of the same kind could also store generated electrical energy during troughs of demand and lessen the need for generating capacity at peaks of demand. High hopes rest on the sodium-sulphur cell, which has to be operated at 300–350°C. The Electricity Council and Chloride Ltd. have joined forces to develop this for the market. Another 'hot mix' is the lithium-iron sulphide cell being developed in the US. The molten salt electrolyte demands an even higher temperature, 375°C. Less exotic-sounding is the iron-air cell, which is attracting interest in Britain, Germany, Sweden and the US. One day, solar cells of some kind may be recharging such batteries as these on board the buses or trains they power – conceivably even the cars – but nobody imagines that the power-to-weight ratio can be lifted enough to fit them for aeroplane propulsion.

Fear has been expressed that aviation is threatened by the depletion of oil reserves. The fuel burnt in aero-turbines is generally kerosene (paraffin), which is a kind of petroleum distillate. It is ideal for the purpose, packing a lot of energy and letting itself be pumped easily even at the extremes of pressure and temperature reached at great altitudes and supersonic speeds. Kerosene is so good that, in the view of John Allen, Chief Future Projects Engineer at Hawker-Siddeley, change to liquid hydrogen should be postponed as long as possible. Kerosene can be synthesized from coal or oil shale – indeed, fuel made from oil shale has already been used in the US to fly aircraft. Aeronautical engineers, like many others, may prefer to link carbon to hydrogen for as long as possible.

Sea transport might be at least in part another story, it is being seriously suggested. There could be a return to sail, though on principles revised in the light of the aerodynamic knowledge that aeronautical engineering has brought with it. Some calculations indicate that, over long distances, for non-perishable bulk commodities, sailing vessels can already compete with oil-fuelled ships. In contrast, certain seagoing craft are already propelled by nuclear power, arousing the same sort of opposition as nuclear power on land.

Nuclear power being so beset on land and sea, economy with fossil fuel becomes the more imperative. In the generation of electricity, saving may become possible by the adoption of magneto-hydrodynamic and electro-gas-dynamic principles. Of more immediate practicability is fluidized-bed combustion as mentioned earlier. The bed is of some inert granular material such as sand or ash, and it is fluidized by blowing air up through it. Solid, liquid or gaseous fuel is fed into this foaming bed and burnt. The burning is very rapid. Through the bed may pass boiler or other tubing requiring the very intense heating that such beds make possible. Furnaces and combustion chambers can thus be made more compact, and more economical both to build and run. The bed may contain limestone or other material that will hold back sulphur and thus stop atmospheric pollution, so fluidized beds can burn sulphurous coal cleanly and might also utilize such otherwise intractable resources as oil sands and oil shale. What happens to the incombustible residues? Remember, the bed is fluidized. One just opens a tap at the side and runs them off.

Another unorthodox type of burner, the invention of Prof. Felix Weinberg at Imperial College, is known as the Swiss Roll because it has spiral passages corresponding to the cake and cream layers of that confection. Combustibles are led into the roll through the 'cream' passage while combustion products are led out through the 'cake'. The combustion chamber is down the middle of the roll. In ordinary burners there are limits of flammability, and if there is more or less than a certain proportion of air the fuel will not burn. Thus methane will not burn if it is more than 15%, volume for volume, in a mixture with air. Nor will it burn if it is less than 5·3%. In the Swiss Roll burner such limits disappear, so very weak gaseous mixtures can be burnt at very high temperatures in them (limited only by the material of which the burner is made) and at high efficiency. Low-grade solid fuels and liquids may be awkward because they break down or decompose on their inward spiral. But the researchers overcome this problem by putting a fluidized bed down the middle of the roll and counterflowing only the air and the gaseous combustion products. Thus the range of possible fuels is extended to virtually anything that will give out heat when, in a Swiss Roll burner, it is made to burn. Such burners would fit into a railway locomotive, say, but not into a car. For most industrial purposes their size would be no problem.

There is no space left to describe other new or newish ways to energy saving in industry. Without doing anything at all scientifically adventurous, however, some industrialists are already saving fuel and money by adopting 'total energy' systems. These are electrical generating sets whose exhaust heat is put to productive use. Similar economy may become possible in the home, too, following Fiat's development of one of their car-engines to generate both electricity and domestic heat from one methane- or natural-gas-burning unit.

REFERENCES AND FURTHER READING

1. Some general books to start off with are:
   British Industry Today: Energy (Central Office of Information).
   Gerald Foley et al., *The Energy Question* (Pelican).
   K. A. D. Inglis, ed., *Energy: From Surplus to Scarcity* (Applied Science Publishers).
   Energy, Europe and the 1980s (Meeting Report, Institution of Electrical Engineers).
   Michael Kenward, *Potential Energy* (Cambridge University Press.)
   Amory B. Lovins, *World Energy Strategies: Facts, Issues and Options* (Friends of the Earth).
   Energy in the 1980s (Meeting Report, The Royal Society).
2. Perhaps the best single up-to-date book on the nuclear power scene, in spite of its special-sounding title, is the 6th Report of the Royal Commission on Environmental Pollution, Chairman Sir Brian Flowers, *Nuclear Power and the Environment* (September, 1976, Her Majesty's Stationery Office, Cmnd 6618).
3. Michael Flood and Robin Grove-White, *Nuclear Prospects: A Comment on the Individual, the State and Nuclear Power* (Friends of the Earth).
4. B. J. Brinkworth, *Solar Energy for Man* (Compton Press). As we go to press the British Government is launching a four-year program offering £6 million of backing for commercially oriented research and development on solar energy. The program follows publication of the Department of Energy's

report, *Solar Energy: Its Potential Contribution Within the UK,* which contains an up-to-date introduction to the subject. The report is published by Her Majesty's Stationery Office.

5. A slender book encompassing several solar transformations is Godfrey Boyle's *Living on the Sun,* subtitled 'Harnessing Renewable Energy for an Equitable Society' (Calder & Boyars).

6. A classic of this subject has recently been reissued: *The Generation of Electricity by Wind Power,* by E. W. Golding, (E. & F. N. Spon Ltd). To get up to date see *Wind Energy Systems,* the report of an international symposium at Cambridge in September, 1976 (published by B.H.R.A. Fluid Engineering, Cranfield, Bedford MK43 OAJ).

7. *The Development of Wave Power,* by the Economic Assessment Unit, National Engineering Laboratory, East Kilbride, Glasgow.

8. *The Efficient Use of Energy* (IPC Science and Technology Press).

9. D. E. Earl, *Forest Energy and Economic Development* (Clarendon Press, Oxford).

10. C. Bell, S. Boulter, D. Dunlop and P. Keiller, *Methane: Fuel of the Future* (Andrew Singer).

FOOD RESOURCES AND POPULATION

Michael J. Pickstock, N.D.A.

For the past twelve years Mr Pickstock has been responsible for agricultural and natural history programmes for the BBC World Service. Now a freelance writer and broadcaster specializing in agricultural and natural history programs for the BBC World Service. Now a freelance writer and broadcaster specializing in

He has practical farming experience in England and a National Diploma in Agriculture from the Harper Adams Agricultural College, followed by farming in Canada and New Zealand, and travel in Australia and Sri Lanka.

Since joining the BBC in 1964 he has travelled widely in Africa, visiting or working in ten countries, including 18 months as broadcasting adviser in Lesotho.

Now living in Suffolk, his wife and two children enjoy joining him in farming in a small way.

What's the problem?

Every day people complain. Roads are crowded. Standing room only on trains. Price of food exorbitant. Government getting more and more remote from people. In another 25 years the complaints are likely to sound more shrill.

By the year 2000 there will be six people living for every four today. And for every individual living in towns and cities now there will be two. Population increase and urbanization have caused changes and set up stresses in modern society which will affect nearly everyone who is alive today. Not for 15,000 years has the human race faced a more critical time.

The last great turning point in man's history occurred when our ancestors chose to give up the free and easy life of hunting and gathering wild foods and settled to grow crops and domesticate animals. No doubt they thought that planted crops and herded animals would provide a more secure existence than constantly tramping the savannah. Their hope was only half realized : one set of problems was exchanged for another. For half a million years there had been little more than 3 million people following the herds and the ripening food plants as seasons changed around the world. For 500 centuries the population of man was stable, contact between families and tribes infrequent and friction and hostility rare. The development of agriculture at the end of the last ice-age 10,000–15,000 years ago set in motion the pendulum of population increase, food demand and need for technological improvements, which rocks us still. The danger lies in that the pendulum swings faster – and its ability to upturn society is threatening.

There have been so many forecasts, figures and discussions on food and population that many people call it 'apocalyptic thinking'. Warnings of impending disaster proliferate : if food shortage does not starve us, pollution will poison us or lack of mineral resources will reduce industrial society to chaos. Those who are sunnier personalities radiate optimism and feel that these are challenges to be overcome by man's proven ingenuity.

What are the facts? Should we be alarmed, concerned or exhilarated by these challenges?

Look at your watch for a minute – or count slowly to 60....

... So! During that minute 240 children were born somewhere on this planet. People died too but not 240. In fact the balance of arrival over departure will amount to 200,000 by this time tomorrow. By the end of the year there will be as many extra people as live now in the United Kingdom, Belgium, Denmark and Norway combined – about 75 millions. But the majority of new arrivals will be heard gurgling and crying, as babies do, a long way from western Europe : most will be entering a life of extreme poverty and need in the countries of Asia, Africa and Latin America. India alone can expect a population influx of 13 millions – equivalent to the population of Australia. Yet imagine the uproar if Australians, one and all, attempted to emigrate to India in a twelve-month!

But on their own statistics are meaningless. When William the Conqueror won the Battle of Hastings in 1066 he had less than 2 million new subjects as raw material for the national survey known as Domesday Book. Today there are nearly thirty times as many citizens of the UK. When the colonists entered North America there were perhaps a million Indians and Eskimos in residence on the continent. In 1976 the population of the USA and Canada amounted to some 235 million. Are the UK and US 'overcrowded' now? What is the 'ideal' population level for a country? Can we judge this on a basis of space, food producing capacity, raw materials, water ... or some other factor? For instance the African country Zaire is four times the size of France and yet only has half the population of France. Which of these two nations is the more 'over-populated'?

The answer to the question of 'over-population' lies in the

balance between a population and all the resources which are necessary to provide sufficient of the essentials of life to each and every one. And of these essentials food and water unquestionably are the most vital.

New directions in agriculture

During the last 20 years agricultural scientists have achieved a succession of major technological triumphs. Plant breeders have produced new high-yielding disease-resistant varieties and now are synthesizing entirely new plant species. Chemicals have taken over the control of weeds, pests and diseases – and are now being used to speed or slow the ripening of fruit, to shorten the length of straw. Irrigation is now possible using a minute proportion of the water used in the past, which is trickled through micro-bore plastic pipes direct to the roots of plants. Artificial fertilizer, long an essential but expensive necessity, is gradually being supplemented by nitrogen and phosphorus synthesized by soil bacteria. And animal science has so changed livestock production that first poultry, then pigs and now dairy cows have migrated increasingly from the sun and rain of farmyard and field into environment-controlled buildings where temperature, humidity and lighting are controlled so precisely that standards for domestic housing seem archaic. Animal nutrition is also monitored and adjusted more precisely than in any human situation. Breeding is carefully planned and the reproduction processes measured and manipulated with the help of electronics and hormones. And, as with plants, attempts are being made to create new animal species as successors to our established domesticates.

Growing food or concrete

The challenge to agriculture does not end with attempting to feed more and more people each year : farmers must do this from a steadily shrinking area of land.

More people need more houses, roads, airports, factories, shopping centers and recreation space. When great cities were born as small villages the first settlers inevitably chose a fertile place where their crops and animals would flourish. Inevitably, too, they chose a site which was well supplied with water. It is no accident that the population of Indonesia is so heavily concentrated on the island of Java, where the soils are rich, and that nearby Sumatra, larger but less fertile, has far fewer people. The bulk of Australia lies empty because the early colonist found a more kind and hospitable climate and farming land on the narrow edges of the country, where rivers flowed and the mountains precipitated rain. When the first Europeans landed in North America they sailed up the inlets and rivers to find sheltered, fertile and well-watered valleys and the population grew fastest from New Hampshire to Virginia. The long cold winters and thinner soils of Vermont and the swamps of Carolina did not appeal to the same extent.

Today, not only is the world population growing as a whole, people are flooding into the cities. In 1970 1400 million people lived an urban life and by 2000 the number will have more than doubled to 3300 million. This trend is universal and whether it is in Britain, Nigeria, the United States, Brazil or India, for one reason or another city life seems more appealing. There is the hope of work for the rural unemployed, the attractions of entertainment and social mixing for the adolescent, and the challenge of competition for the ambitious. That the majority of those already living in cities find them less appealing with each year that passes does not deter the hopeful new arrivals!

The United States, which has been the first to develop so many of our present trends, provides an example of how the ratio of land to people is shrinking. In 1853 there were 20 acres of cultivable land for every inhabitant. Today there are less than 3 acres per person. But which are the acres that have disappeared under concrete and tarmac, never to be ploughed again? The fertile belt of early colonial farm land, that succored the first arrivals, is now a vast conglomerate of cities, known by some as Megalopolis,[1] stretching from Boston down through Rhode Island, Connecticut, New York, New Jersey, Pennsylvania to Maryland and Virginia. The

THE EXPLOSIVE GROWTH OF URBAN POPULATION

THE WORLD'S 12 FASTEST GROWING CITIES

City	1970 Population (millions)	1985 Population (millions)	Growth rate %	Continent/Country
1. Bandung	1·2	4·1	242	Indonesia
2. Lagos	1·4	4·0	186	Africa
3. Karachi	3·5	9·2	163	Pakistan
4. Bogota	2·6	6·4	146	South America
5. Baghdad	2·0	4·9	145	Middle East
6. Bangkok	3·0	7·1	137	Indonesia
7. Teheran	3·4	7·9	132	Middle East
8. Seoul	4·6	10·3	124	Asia
9. Lima	2·8	6·2	121	South America
10. Sao Paulo	7·8	16·8	115	South America
11. Mexico City	8·4	17·9	113	Central America
12. Bombay	5·8	12·1	109	India

THE WORLD'S 12 LARGEST CITIES
(Rank Order)

City	1970 Population (millions)	City	1985 Population (millions)
1. New York	16·3	1. Tokyo	25·2
2. Tokyo	14·9	2. New York	18·8
3. London	10·5	3. Mexico City	17·9
4. Shanghai	10·0	4. Sao Paulo	16·8
5. Mexico City	8·4	5. Shanghai	14·3
6. Los Angeles	8·4	6. Los Angeles	13·7
7. Buenos Aires	8·4	7. Bombay	12·1
8. Paris	8·4	8. Calcutta	12·1
9. Sao Paulo	7·8	9. Peking	12·0
10. Osaka	7·6	10. Osaka	11·8
11. Moscow	7·1	11. Buenos Aires	11·7
12. Peking	7·0	12. Rio de Janeiro	11·4

In total the urban population of the world is *doubling* every 25 years.

city of Philadelphia spreads over soils of fertile coastal plain not up the thin soils of the Pocono Mountains. The best places for farms seem to be the best places for cities – whether the cities are Calcutta, Cape Town, Sydney, Tokyo, Djakarta, Manchester, Mannheim, Nairobi or Cairo. As Jean Gottman wrote in *Megalopolis*: 'Hay and trees can never pay as well per acre as motels, split-level or apartment houses. Dairy cattle and poultry may be kept profitably in urban areas only if they are as compactly housed as people and maintained in the same way, with feed brought to them from cheaper land.'

But the cheaper land is also the poorer land and land more remote from consumers. The farmers of today and tomorrow, with an army of scientists and technologists to help them, must somehow find ways around the 'Catch-22' of modern population growth and food resources. 'Produce more from less but mind the price.' That is the appeal from consumers to the food industry.

'All flesh is grass'

The prospect of continent-wide famine is nothing new. World population growth accelerated with the start of the Agricultural Revolution in Europe two hundred years ago and urban growth exploded a century later with the Industrial Age. Then the extra mouths were fed by opening up the virgin lands of North America Argentina, Australia and New Zealand. Today the answer lies in squeezing more from the same land, from less land, and the plant-breeders have led the way to show how we can extract a quart from a pint pot. Their greatest successes have been with cereals.

'By far the most important vegetable food are quite literally the grasses,' wrote Edward Hyams in *Plants in the Service of Man*.* And he continued: 'That family of green plants not only provides the bulk of pasture on which all kinds of cattle are fed; it also includes those noble species of grasses we call cereals. At the base of every one of mankind's great societies, of all the original civilizations, is a cereal

*Published by Dent.

grass.' Each year that passes we become more dependent on this family of plants and within the next two decades we will all increasingly become grass-eaters, relying on them for protein as well as carbohydrate. The cereals that we eat will be bred to contain more protein and the less noble cousins of cereals, the pasture grasses that have long been the exclusive preserve of grazing animals, these will be processed to yield juices rich in protein which can be used in numerous novel ways.

Cereal breeding has already had several dramatic successes. In the first half of this century yields of wheat, barley and maize were doubled in Europe and North America. The same techniques applied with greater urgency turned Mexico from a wheat importing nation in the '50s to self-sufficiency in the '60s; they resulted in a doubling of the Indian wheat yield in little more than seven years.

The latter day 'green revolution' had its origins in two Rockefeller and Ford Foundation funded research centers: the Center for the Improvement of Wheat and Maize (CIMMYT) in Mexico and the International Rice Research Institute (IRRI) in the Philippines. The 'miracle seeds' that they produced were of hybrid varieties of wheat and rice, which combined the capacity to respond to heavy doses of fertilizer and water with a short straw length that was able to carry the heavier seed heads without falling over.

Because Asia in particular was closest to the looming crisis of famine the new seeds were first launched there in a crash programme to avert disaster. In less than a decade they were being grown on close to 100 million acres.

*The introduction of these new varieties averted disaster in India, Pakistan, Bangladesh, Indonesia, the Philippines – but they did not resolve the underlying problems of argricultural insufficiency.

The new varieties were tailored for high-fertility conditions but since energy prices rocketed in 1973 fertilizers have become prohibitively expensive for peasant farmers. They grow best with ample water but land under irrigation has

*The Green Revolution by Stanley Johnson. Pub. H. Hamilton.

RAPID SPREAD OF HIGH-YIELDING
RICE AND WHEAT IN ASIA AND
NORTH AFRICA

Year	Acres
1965	200
1966	41,000
1967	4,047,000
1968	16,660,000
1969	31,319,000
1970	43,914,000
1971	50,549,000
1972	68,000,000
1973 (*est.*)	80,200,000

Source: U.S. Dept. of Agriculture

expanded three-fold in 30 years and there is little potentially irrigable land left to be utilized.

The new varieties were bred for high yield but their lush growth made many of them more susceptible to pest attack.

Traditional communities with conservative tastes found that the new rice varieties did not cook as well or taste as good as their old varieties.

Undaunted, the plant breeders have produced a new generation of seeds to attempt another 'miracle'. Attention has been focused on dry-land crops such as maize, millet, sorghum and rice, which can grow without flooding. Disease resistance and palatibility have been emphasized in the selection of characters and the growing season has been shortened for maize, sorghum and rice by two to three weeks. These are the crops of the next decade which will hopefully double yields in those extensive regions of the tropics where rainfall is concentrated to a few weeks or months and the rains can start late and finish early.

With irrigated rice the reduced days to harvest permits the growing of two crops per year where one was grown before, or three crops in place of two. In dryland conditions quick maturing varieties of maize and sorghum can be ready for

Variety	Disease resistance				Insect resistance			Growing season
	Blast fungus	Bacterial blight	Grassy stunt virus	Tungro virus	Green leaf-hopper	Brown plant-hopper	Stem borer	Growing season
IR 8	MR	S	S	S	R	S	MS	120 days
IR 20	MR	R	S	R	R	S	MR	120 days
IR 26	MR	R	MS	R	R	R	MR	120 days
IR 28	R	R	R	R	R	R	MR	105 days

S—Susceptible improving rice varieties
MS—Moderately susceptible
MR—Moderately resistant
R—resistant

harvest even if the rains end early whereas traditional varieties would die, their ears empty.

Towards man-made plants

The sexual breeding of plants (and animals) has serious limitations since sexual reproduction is possible usually only between individuals which belong to the same or a closely related species. By selecting, mating and selecting again varieties and breeds have been developed whose qualities would amaze our grandfathers. Unfortunately they will prove inadequate for our grandchildren.

Many modern plants and animals are so-called hybrids: their grandparents and parents were selected for particular qualities and lines showing very different qualities and were crossed to provide progeny which are superior to both parents. But there is a limit to genetic improvement by this process and that limit is the variety and quality of characters which can be found among the chromosomes of the whole species. Where a crop, such as the cultivated potato, has been developed from a handful of original seed, the genetic variability is very small and the scope for permutations of characters limited. If a character for disease resistance, hardiness or food quality does not exist in a population, how can it be introduced?

The food plants and animals of the past have served us well but even if we can grow four tons of wheat per acre in 1985 where only one ton was the average in 1900; milk 2000 gallons per cow per year where the average was 500; produce one pound of poultry carcase for 2 pounds of food, in a matter of weeks, where the same bird used to take twice as much food and time to fatten; these standards of excellence could still leave many hungry in the two decades to come.

To break through the ceiling of species improvement plant breeders have produced the first commercial man-made plant. Called triticale, it is quite a modest example of the possibilities that lie ahead.

As the two equines – the horse and donkey – were crossed to produce the mule, so two cereals – wheat (genus *Triticum*) and rye (genus *Secale*) – have been crossed to produce triticale. The two hybrids have many parallels : as the mule combines the size of the horse and the endurance of the donkey, so the new cereal can yield as heavily as wheat and yet grow in sandy, acid, infertile soils in the coolness of mountains and in more arid conditions as can rye.

Another, unfortunate, parallel is that like the mule the progeny from a simple crossing of wheat and rye are sterile. The reason is simple: wheat contains 28 chromosomes, rye 14. When the sex cells are formed the number of chromosomes is halved and when the sex cells of two individuals fuse the progeny inherit the sum of the two halves. Triticale inherited 21 chromosomes and so could not divide to form sex-cells with an equal number of chromosomes.

The key to unlocking inter-specific sterility has been provided in some measure by the chemical colchicine – an extract of the autumn crocus. Colchicine has the remarkable and useful property of making plant chromosomes double their number and the cells of triticale treated with colchicine, duly 'doubled-up' to become sexually viable. But even this was not enough; it took several years of selection a chance mating between a triticale and a wheat plant, further cross-

ing between triticale lines and back-crossing with wheat to
produce a commercial crop.

With strains now available, which out-yield wheat, triticale will undoubtedly become a crop of great value, sup-
planting the low yielding wheats which struggle to survive
in the poor soils and cool climates of northern India, and
the uplands of Kenya, Ethiopia, Mexico and central Spain.
The discovery in Texas of triticale strains with 18% protein
(wheat has 12%) puts the crop as close to the peas and beans
for protein value as to the cereals.

Radiation treatment of seeds was once thought to hold
great promise for plant-breeding but it has proved of limited
benefit. The proportion of useful changes which are produced by irradiation are minute compared with the great
majority of unviable mutants and mortalities that result. Perhaps finer control of dosages will change the situation or
perhaps radiation will be used in conjunction with other
plant-breeding techniques, as occurred with a species of wild
and a cultivated oat.

Many species of cereals and other crops, such as potatoes
and cabbages, have relatives living wild, often as weeds,
which plague the cultivated species. The wild species, long
outstripped in yield and most other characters favored by
grower and consumer, often retain an enviable capacity to
resist pests and diseases. Among the most devastating scourges
of both cereals and potatoes are fungal diseases, variously
known as mildews, rusts, blights and blasts. Their danger lies
in the speed with which fungi can produce new strains to
circumvent the characters of resistance so laboriously bred
into cultivated varieties. The resistance in cultivated species
is not a defence in depth but often depends on a few genes;
what is required is the multi-gene resistance of some wild un-
improved plants.

But to cross two species, however closely related, can prove
as problematic as crossing more distant 'cousins'. This was
the problem that faced scientists at the Welsh Plant Breeding
Station, Aberystwyth when they tried to introduce the mildew
resistance of a wild oat to a cultivated variety 'Manod'.

The wild and cultivated oats produced sterile progeny because of unequal division of chromosomes at sex-cell formation. But again the use of the drug colchicine induced the doubling of chromosomes and the seeds of the now fertile progeny then could be repeatedly back-crossed to the cultivated oat parent to reduce the unwanted wild oat genes while retaining the genes for mildew-resistance. Finally the seeds were irradiated to cause the chromosomes to fragment and rejoin and the result was a high yielding oat with the desired mildew resistance. The use of a drug and radiation had produced another man-made hybrid !

New hybrids of wheat, sorghum, rice, sugar cane and potatoes can follow, since wild relatives exist in profusion, and the genes of many of these wild plants will help fill 'the hunger gap' in the decades ahead.

4000% yield increase?

Crop yields in the tropics could be increased by 4000% ! That startling claim was made at a conference in London in 1975 when a Dutch scientist, W. van Monsjou, reported on a survey of world soil resources, which had just been completed by the University of Wageningen. His point was that the greatest potential for crop growth is where the sun shines brightest. But to complement solar energy plants also need mineral salts, mainly nitrates, water and carbon dioxide. The carbon dioxide present in the atmosphere is readily available to plants, the nitrogen much less so. Plants can use nitrogen only as nitrate salts and to boost crop growth farmers have been pouring nitrogen fertilizers into the soil at a considerable rate. But most of this fertilizer is used in the more developed countries where industry is available for manufacture and farmers have sufficient wealth or credit to buy it.

Transfer the bulk of this fertilizer-use to the tropics and the total world food output could be boosted forty-fold. Yet this is not likely to happen : economics dictate that fertilizer, like any other commodity, is sold where people can afford to pay most for it.

In 1974 world use of nitrogen fertilizers was 40 million
tons and it is estimated that this will quadruple to 160 million
tons by 2000. The consumption of fossil fuels in the manufacturing processes will be between 250–300 million tons – 1·75
tons of fossil fuel to produce 1 ton of fertilizer nitrogen.[2] At
1976 prices 1 ton of nitrogen fertilizer cost $200–250 per ton
and the cost in 2000 can only be an inspired guess. What is
certain is that the most needy farmers will be least able to
afford to buy what they need.

Similarly plants in the tropics respond most to moisture
and yet irrigation potential is limited by the need to buy
energy to pump the water where it is available and where it
is not, farmers must be content with the fickle rains.

Yet how can responsible scientists remain optimistic about
the future? Their faith centers on two very recent discoveries: that many plants are capable of tapping supplies
of natural nitrogen; and some plants have developed more
than usually efficient systems of photosynthesis to make better
use of carbon dioxide and water when both are in limited
supply. If these qualities can be exploited in the species concerned, and more especially if these abilities can be transferred to other less well-endowed crops, food production
could be dramatically increased at minimum cost.

Free nitrogen

It has long been known that certain plants, the peas, beans
and clovers (legumes), have developed a mutually beneficial
arrangement with soil bacteria whereby the bacteria take
sugars from the plant and provide in return nitrates, which
they are capable of synthesizing from atmospheric nitrogen.
Now soil chemists have revealed that one teaspoon of fertile
soil can contain upwards of 10 million bacteria, one million
fungal cells and tens of thousands of minute animals and
algae. Not only do these teeming inhabitants of 'Soil City'
break down organic matter and release soil minerals such
as phosphates for plants to absorb, many of them are capable
of 'fixing' nitrogen into nitrates. Whereas earlier it was
thought that only the legumes could benefit in this way

through the bacteria (*Rhizobia*) that live in the tiny nodules on their roots, scientists now realize that many other bacteria (and also fungi) 'fix' nitrogen and manage to trade what they do not require for sugars produced by the plants around them.

In the air over every acre of soil there is the vast resource of 80,000 tonnes of nitrogen and having established that there are at least 150 non-leguminous species capable of having a mutually beneficial relationship with soil bacteria and fungi, every attempt will be made to capitalize on this very recent discovery.

Only a year ago an Anglo-Brazilian team, working in Brazil, announced that they had found that at least 40 tropical grasses and cereals have this association and among the favored ones are millet, sorghum, maize and rice – the most important cereals in the developing world. Probing deeper they discovered that the nitrogen-fixing bacteria traded their nitrates most willingly with those plants that have 'leaky' roots – dribbling small quantities of sugar-rich root sap into the soil – and it is in the vicinity of the roots of these plants that the bacteria and fungi cluster most closely

Can plant breeders select strains of these cereals to ensure 'leaky' roots? Why not? Could the ability to 'leak' root-sap and encourage a nitrate-for-sugar swap be transferred to related cereals? Undoubtedly it will be attempted.

In parallel with this work are attempts by scientists at the University of Sussex to mimic the reactions of soil bacteria in the laboratory and so to develop a process for synthesizing nitrogen without the high temperature and pressures usually required in industrial processes. In 1975, at the Agricultural Research Council's Unit of Nitrogen Fixation, the Sussex team joined molecules of nitrogen to molecules of the metal molybdenum and held them in an acid solution. 30% to 40% of the nitrogen was converted to ammonia – the first stage of 'nitrification'. Replacing the molybdenum with tungsten, 90% of the nitrogen was converted. They had succeeded. But a snag remained : in the reaction the molybdenum and tungsten were destroyed by the acid and without the acid the conversion would not take place. If the molybdenum or tungsten can be sustained to be recycled continuously a simple low-cost system for nitrification would result and the availability of nitrogen fertilizer would benefit millions of farmers, whose crops presently lack sufficient of this vital plant food.

Making the most of sunlight

As some plants have developed to become more than unusually adept at securing their needs of mineral salts such as nitrates, others have only recently been discovered to be far more efficient at turning carbon dioxide and water into carbohydrates through photosynthesis.

In hot dry conditions plants are faced with a choice : keep the leaf ventilators, the stomata pores, open to absorb carbon dioxide (essential for photosynthesis) and water is lost through the stomata. Since plants under these conditions cannot afford to lose water their second option is to reduce water loss by closing their stomata and accept a much reduced intake of carbon dioxide. The latter choice would seem to be equally inhibiting to rapid plant growth.

Some tropical plants, including maize and sorghum, have adjusted to this dilemma and are able to grow with a remarkably low carbon dioxide intake and this they do by making much more efficient use of the carbon dioxide that they do absorb. The efficient plants have been labelled C_4 plants because when their carbon dioxide is absorbed and fixed into a stable organic compound the product is C_4 dicarboxylic acid, acetic acid. As research reveals more about C_4 plants it should be possible to make better use of them and also to better understand the processes by which other plants absorb their carbon dioxide so that they too may be made more efficient and less demanding of water.

The C_4 process occurs in at least eight separate families of plants, although the advantage may show in only a single genus or species in the family, and scientists are now endeavoring to transfer the genes responsible for the more efficient photosynthesis to other plants less fortunate in their adaptation.

What future for animals?

A broiler fowl eats 2–3 pounds of grain for each 1 pound of usable meat produced; a pig consumes 5–6 pounds, a beef steer 8–9 pounds per pound of meat. Do such inefficient converters of scarce grain and protein have a place on the farms of the future?

Many people think not and they back their argument by pointing to such facts as these : the farm animals of the EEC consume more protein in a year than the human populations of India and China combined; that British farmers grow less than one-fifth of the protein that they feed to their animals and have to import the rest; that animals, of necessity kept more and more intensively, produce vast mires of effluent that become an embarrassment to agriculture and a pollution-hazard to the environment. Do away with animals, let us become vegetarians and in so doing release millions of acres of food production to direct human use!

The argument has merit – but it is over-simple. Without the grazing animals (ruminants) Man cannot, in the foreseeable future, crop the extensive pastures that grow on hillsides and moorland and the vast dry savannahs that stretch almost unbroken on both sides of the equator. Land such as this, too high or steep or dry to be cropped with food plants, can only grow herbage. To benefit from this we must utilize the ruminants and let them process the cellulose, indigestible to Man, into meat, milk and wool.

But if ranching continues do we need the chicken, pig and dairy cow? The answer is a qualified 'Yes'. Milk is an important part of the diet of many cultures, although many Asians and some Africans have learned to live without it and are even incapable of digesting milk. Moreover the cow can digest many waste or by-products of crops. In the past 20 years, during an era of relatively cheap cereal and protein, dairy cattle have been fed extravagantly on grains but already it is obvious how diets will change for the next two decades. Our milk will be got from feeding straw, molasses, urea, the trimmings from vegetable-canning and freezing plants, manure and even sawdust!

The Danes have developed systems for treating cereal straw with strong alkalis such as sodium hydroxide (caustic soda). One large straw-processing plant has been commissioned in Britain by BOCM Ltd., part of Unilever, and smaller equipment for on-farm treatment is now available. Untreated straw is not totally digestible because of the coating of lignin that surrounds the useful cellulose. Chopped and soaked in heated alkali the lignin dissolves and treated straw can be incorporated in cattle feed as a substitute for cereals. Molasses, a by-product of the sugar industry, provides additional carbohydrate and urea, a simple nitrogen salt (which is also a common fertilizer) can be fed to supply protein. The bacteria present in the rumen stomach of cattle and sheep can synthesize urea into a protein that ruminants then use.

Manure from poultry, pigs and cattle is rich in food materials: since no animal manages to extract all the goodness from its diet a proportion inevitably passes into the dung. Dried, sterilized to kill salmonella and shredded, the dung can be included in cattle feed without harm. Poultry manure when dried, contains an average of 30% protein and 4 pounds of it can replace 1 pound of barley and $1\frac{1}{2}$ pounds of soyabean or groundnut meal.

Distillery wastes in the form of brewers grains and 'draff' from whisky distilling have long been valued as cattle feed but in India where silk culture rather than brewing is the major industry cattle are being fed the mulberry leaf stalks which have been stripped by silkworms. The stalks contain 11% protein and useful quantities of calcium, phosphorus and copper.

In many parts of the tropics, rivers, reservoirs and irrigation canals become choked by the prolific water hyacinth. The weed is a great nuisance but the cost of clearing it can be saved by feeding it. On its own the hyacinth is scarcely palatable but washed, dried, chopped and mixed with chopped rice straw and molasses it can be made into a silage.

In Japan sawdust has been mixed with sediments from whisky production and waste tomato juice and pulp and the 'cocktail' fermented with yeast. The cost of this feed is one-twentieth of grain.

Sawdust treated with sulphite has proved digestible enough to feed cattle in the US, the sulphite boosting the digestibility of beech sawdust from 5% to 59% and this material has been included instead of alfalfa pellets at up to 31% of the daily ration with no loss in animal production.

Wherever sugar cane can be grown this fast-growing and carbohydrate-rich crop could be used extensively for cattle feeding. The tough outer rind removed and the white pith chopped, cattle in Mexico and the sugar islands of the Caribbean have yielded well on this mix of 'comfith' and urea. There are many areas where cane grows prolifically and yet there is no adequate market for sugar. Indeed as sweeteners are extracted increasingly from maize (corn sugar) and from supersweet fruits of certain West African shrubs (some are 2000 times as sweet as sugar) the future of cane growing may, of necessity, have to switch from feeding people to cattle.

To feed pigs and poultry on wastes is not so easy since they have simple stomachs and cannot digest coarse fibrous foods. Yet having started their careers in agriculture as the scavengers of the farmyard and risen to the position of eating only the best (and most expensive!) foods, pigs and poultry will have to adjust their tastes. Inevitably the human demand for wheat, maize, soya and fish proteins will deprive them of the diet to which they have become accustomed. But they too can continue to play a useful role as converters of food which is in less demand.

Cassava, also known as manioc and tapioca, is a tropical tuber capable of yielding over 20 tons per acre – with little fertilizer or water. Although it has been used as a staple in many parts of the developing world, more and more people are changing their tastes, preferring bread, the other cereals and even the 'European' potato. Yet plantations of cassava could, and already do, provide a carbohydrate-rich substitute for cereals in pig and poultry feeds.

Protein for feeding these animals is being made from kerosene by British Petroleum and methane (natural gas) by ICI and Shell. At present the manufacturing plants are concentrated in Western Europe : BP have small plants in Britain and France and have recently completed a 100,000-ton-a-year plant in Sardinia; ICI are to build a plant in NE England to produce 75,000 tons a year; Shell hope to have a commercial unit on-stream in the mid-1980s. The processes use yeasts (kerosene) and bacteria (methane) to convert the hydrocarbons to protein by fermentation. The material that they produce is 70–75% protein (soya = 28%, fish meal = 55%) and these single-cell proteins (SCP) will undoubtedly become the important sources of animal feed in the near future.

And if animals prove that they can thrive on these microbially produced proteins the next step would be to incorporate them directly into human foods!

Feeding and breeding are the sole preoccupations of the lifetimes of most domesticated species and to ensure their future they must reproduce with unfailing precision and prolificacy. Electronics and hormones make certain that there is no undue 'resting' between pregnancies.

Artificial insemination of all livestock is common practice, but the farmer can now manipulate the breeding cycles of cattle, sheep and pigs so that any number of a herd or flock are at their peak for mating on a given day. It is achieved by injecting hormones, known as prostaglandins, and once 'in step' batches of mothers can be moved through farm maternity units so maximizing the use of equipment and buildings and simplifying management. Where a pig farmer is doubtful about the readiness of sows to be inseminated he can use a probe, which measures the acidity of the vagina and displays the degree of receptivity of the animal as a meter-reading. Since boars are often impetuous this electronic aid also enables the stockman to use his boars at the optimum period of oestrous and not before or after the maximum of ova are ready to be fertilized.

Just as AI was developed to spread the genes of a super-sire over a much greater number of females than he could possibly serve naturally, now animal breeders can use the technique of 'ova transplant' to multiply the progeny produced by 'super-dams'. The technique is most useful with cattle and sheep, which produce only one or two young each year; with ova-transplants they can conceive many dozens or hundreds but leave the nurturing of them to foster-mothers. Prostaglandin may be used to bring the cow or ewe in season for insemination naturally or artificially. The fertilized eggs can then be removed surgically or by flushing them from the uterus and the tiny embryos are then ready for insertion in the cows or ewes which will carry them in their wombs for the rest of their foetal development. But the breeding cycles of donor and foster-parent must be synchronized so that the recipient females will be at precisely the right stage to accept the ova-transplant and injections of prostaglandins will therefore be given to donor aid recipient animals simultaneously.

Using this technique a bull in Texas can be mated with a cow in Scotland and the several progeny flown (incubated in the wombs of rabbits!) to be implanted in cows in Australia, Argentina or Kenya. The schoolchildren of Iran will be able to benefit from families of cattle whose parents have never set foot in their country and the expense, risk and problems of quarantine will be largely eliminated once ova-transplants are in general use.

This technique has, in fact, been used for exporting sheep to South America. The embryo sheep were implanted into the wombs of a rabbit which was flown from England to South America. On arrival the embryos were removed and reimplanted in the wombs of sheep – to grow normally.

New animals for old – why Guyana is eating mermaids!

Sixteen species. They are the total of Man's efforts to domesticate mammals during the past 15,000 years. Yet there are 4500 species to choose from – why stop at sixteen?

Few farms number the eland among their livestock yet this large-bodied antelope of southern Africa has proved that even when removed far from its native home it can be tamed and farmed. A century ago they were taken to Russia, where they adapted very well to confinement and the Russians developed milking strains of eland which outyielded those in the wild by 400% !

More recently the musk-ox of the Arctic tundra, renowned for its beligerence, has been tamed so that the Eskimo might have access to the fine wool of its undercoat, similar to that of the alpaca of South America.

Could the gazelles, antelopes and even the giraffe, hippopotamus or water-living manatee be domesticated? If they were through them we could utilize the grasses, shrubs and trees on which they graze and browse and so harness vast areas to feeding the world, which are presently unfit for cropping or grazing by domesticates. It has been suggested that the manatee, that soulful-looking creature sometimes called the sea-cow (and the origin of the mermaid legend), could be grazed with double benefit on the water-weeds which choke so many tropical waterways. The manatee itself could be 'cropped' to yield meat and a tough hide and to attempt this 'domestication' a research station has been established in Guyana.

Against the bulk of the manatee the so-called 'grass-cutter' or cane rat of West Africa, seems insignificant, yet this vegetation-eating rodent has long been a favorite food among the people of Nigeria, Ghana and their neighbors. Now in Ghana they are trying to domesticate the animal as a local 'rabbit', which small farmers or even town-dwellers may keep as a source of cheap meat fattened largely on vegetable wastes from roadside, garden and kitchen.

What are the chances of success in developing new 'domesticates'? Obviously it will depend on the character of the species. The eland is naturally placid, other antelopes and gazelles less so. Yet the horse, camel and the giant primitive longhorned cattle of past ages were hardly passive, yet their descendants were pressed into our service.

Steal their genes!

While some scientists are hopeful about taming wild species others are not. Dr. Roger Short of the Medical Research Council's Reproductive Biology Unit in Edinburgh has a more original idea : he proposed that animal breeders should appropriate the useful genes of certain wild species but leave the wild animal wild – or in a zoo.

He points to the Marco Polo sheep of central Asia : 1·5 metres high at the shoulder, it is twice the size of any domesticated breed. Living in the sparse harsh country that borders the high Gobi Desert, it is capable of surviving tough living conditions and growing rapidly on indifferent pasture. To try and domesticate the Marco Polo would take years and even if successful they would be almost impossible to fence.

Artificial insemination techniques, now a commonplace in livestock breeding, could provide semen from captive species held in zoos. Other giants in the sheep family are the Argali of Afghanistan, Pakistan and USSR; the Bighorn of the Rockies; the Urial of Asia; Moufflon of Sardinia. Their genes could boost the performance of present domestic breeds, although not before careful matching and genetic manipulation has ensured that chromosome compatibility will result in fertile offspring.

Already inter-specific hybrids of cattle have been produced : the 'Yacow', combining the hardiness of Himalayan Yak with the milking ability of European or tropical cow; and in North America the bison (or buffalo) and cattle have been mated to produce the 'Cattalo'. But both Yacow and Cattalo are sterile or produce progeny of very low fertility. To break this sterility a group of US ranching and business interests have backed the breeding work of California rancher Bud Basolo who has crossed and back-crossed two species and now claims to have produced a fertile hybrid $\frac{3}{8}$th bison and $\frac{5}{8}$th cattle. Called the 'Beefalo', this hybrid

has been widely publicized but geneticists are suspicious:
to guard a potential fortune the sponsors of the 'Beefalo'
refuse to disclose how a fertile cross was achieved but are
promoting the animal as a potential new species for crossing
on ordinary cattle. They grow faster, live on poorer forage,
provide a higher proportion of meat on slaughter and are
hardier and more disease-resistant than cattle. So it is
claimed, but most cattlemen are watching and waiting before committing themselves.

Whether or not the 'Beefalo' is the first successful inter-
specific bovine hybrid, other attempts will follow.

With ducks and geese the introduction of 'wild' genes
could be equally dramatic. Both are to a large extent grazers
and utilize grass where chickens cannot. Dr. Short has suggested that since many Arctic species of water-fowl can in-
cubate their eggs in a week or ten days less than domestic
species this advantageous character should be put to good
use for food production. The Red-breasted goose hatches
in 23 days whereas the domestic goose takes 28–31 days. The
young of Arctic breeding species also have the ability to grow
extremely quickly on little more than grass. It is essential
for them to reach a size and strength to fly out before the
short summer is over and the sea and the land freeze. Within
the first three weeks of life the Red-breasted goose increases
its hatching weight by 17·7 times!

Genetic engineering

It is difficult to visualize what new plants and animals will be
created to boost food supplies by 2000 : some may have the
above ground appearance of a heavy yielding cereal with
the tuberous roots of a potato or parsnip; others may be roots
or cereals capable of fixing nitrogen. If that sounds improbable, nature has already produced such a plant – the
winged bean of New Guinea (*Psophocarpus tetragonolobus*)
which : '... produces seeds, pods and leaves (all edible by
humans or livestock) with unusually high protein levels;
tuberous roots with exceptional amounts of protein; and an
edible seed oil'.[3] And the winged bean is no 'freak' – it has

yielded over 2 tons of seed per hectare with 34% protein and 18% oil which compares very well with the top-yielding grain legumes such as soya.

Using techniques that totally by-pass normal sexual reproduction, scientists in Britain and the US are attempting to combine such various and useful characters, as appear in the winged bean, in man-made plant creations.

At Nottingham University Prof. E. C. Cocking has shown that cells of tobacco (*Nicotiana tabacum*) can be fused with the cells of petunia hybrids – two species which are impossible to cross-breed sexually. Sex-cells are invariably extremely choosy with whom they will mate so to avoid these inhibitions Prof. Cocking has worked with somatic cells taken from leaf and stem tissue. Using enzymes to strip the outer cell walls he has brought the cell contents (protoplasts) of tobacco and petunia together in a solution of plant growth chemicals. 50% of the protoplasts have fused and grown on to form new cell clusters.

So far this has been the limit – with cells of widely differing species. Yet research workers in the US at the Brookhaven National Laboratory, have used the technique to fuse protoplasts of two species of nicotiana and the cell-clusters have grown into plants – hybrids which were no different from those produced by sexual propagation. 'Parasexual hybridization', as it is termed, can produce viable plants, but work continues to find which species will prove compatible to fusing into on-growing cell clusters and which plant-growth chemicals will provide the best media for the process to achieve fruition.

Once the techniques of parasexual hybridization are perfected our present crop and animal species could be relegated to the status of curiosities and genetic reservoirs. This will not happen in 10 or 20 years but early next century cattle, sheep and pigs could become rarities seen only in zoos and animal breeding stations, where they might keep company with bison, giraffe or kangaroo! A livestock specialist once complained about the waste in a beef carcase, commenting that to satisfy public demand a steer should be 'all

rump and no front' – like a kangaroo. Unwittingly he may
have sited a future scientific possibility!

Chemical agriculture

The unthinking use of chemicals, which provoked Rachel
Carson to write *Silent Spring*, is hopefully past: if all
farmers are not yet equally conscious of their responsibili-
ties to the environment they are very much aware that
chemicals – pesticides and fertilizers – are increasingly expensive and bear using judiciously.

In the last 50 years manpower has drained away from the
land and muscle has had to be replaced by more mechanical
and chemical power. United States agriculture which dominates world food exports, directly employs only 5% of the
country's labour force. In the UK little more than 3% are
employed in farming. But these figures do not include the
army of people in the manufacturing and processing industries, who lend their support through the products that
they manufacture – machinery, chemicals, pharmaceuticals,
plastics – and the services that they provide – research,
technical advice, veterinary care, mechanical repair, transport and food processing.

We all depend on agriculture for food and a surprising
number of us depend on it for income. In the less developed
country's labor force. In the UK little more than 3% are
employed in farming. But these figures do not include the
must be developed to absorb the upwelling mass of school
leavers. In the next decade alone 300 million more people
will want work. In the words of Prof. Iskander, Director of
the Demographic Institute at the University of Indonesia,
'Four people are being born every second throughout the
world and nearly 3 of them are born in Asia. The Asian/
Pacific area contains more than half the world's people, with
a population density that is already 3 times higher than the
world's average.' In Asia, for reasons of population increase,
in Europe and Japan because urban sprawl is destroying fertile land, the pressure will be greatest to intensify and use
'chemical agriculture' to the limit. In the US also, but there

not so much due to desperate need, but because of burgeoning American research and inventiveness.

Food production units in the future will tend to be very large, or small but very intensive. Field crops such as cereals, peas, beans, potatoes and sugar beet will be tended by giant tractors pulling equipment many metres wide and balloon tyres will be introduced to spread the heavy loads and reduce damage to the soil. Already tractors approaching 200 horsepower are on the market in the US and they and much smaller machines are fitted with sound-proof dust-proof cabs where operators can sit in air-conditioned comfort and relax to the sound of radio and cassette tapes.

A control panel of meters enable drivers to monitor all vital functions remotely from the driving cab : cultivator depth, volume of seed in the drill, spray-tank pressure and even the extent of grain-loss through the inefficient operation of combine-harvester.

Laser beams are already being used to guide direction and depth of operation of drainage machines and in the US research is proceeding on developing a 'robot' carrier to replace the tractor and trailer. On the open fields of the American corn-belt, the plains of northern France or the fens of East Anglia such machines would be guided by laser or electronic beacons from harvest field to silo and back again, sensors guiding them around obstructions and helping them to 'dock' and unload without human help.

In situations of potential hazard such as spraying pesticides or working in the dust and often lethal gases of storage silos farm workers will wear suits and air-conditioned helmets powered by re-chargeable power packs : their appearance remarkably akin to an astronaut.

Intensive horticultural techniques have already passed beyond the simple market-gardening and glasshouse production of the past. In modern greenhouses small computers are programmed to adjust automatically screening for less light, extra artificial light, more ventilation to reduce temperature and humidity, heating and irrigation to increase them; the monitoring and adjustment of plant nutrients in the irrigation water. Plastic tunnel structures have extended the acreage of crops grown under environment-controlled conditions and techniques for using solar energy and even wind power for heating plastic and greenhouses are well advanced.

One complex called 'Environmental Farms Incorporated' of Tucson, Arizona, grows tomatoes under 10 acres of glass. Greenhouses, some larger than a football field, contain desert sand, which has been washed and screened. Eighteen miles of underground pipe and 6 miles of plastic tubing trickle nutrient-enriched water to the plants. Similar complexes are appearing in the energy-endowed but water-scarce states of the Middle-East and there it is economic to desalinate sea water and to use the carbon dioxide-rich exhaust gases from desalination plants to enrich the atmosphere of the growing houses.

The latest development to revolutionize intensive food production uses neither soil nor sand : plants are grown in flat plastic tubes, their stems supported by metal clips and their roots constantly bathed by a slow flow of liquid nutrients. Called the Nutrient Film Technique (NFT) and developed by Dr. Alan Cooper of Britain's Glasshouse Crops Research Institute, it has been available to growers for less than five years. Already NFT has proved itself in the US and Australia : one farmer in California is able to grow seven crops of cucumbers a year with a yield of 400 tons per acre (1000 tons/hectare). He has now expanded his NFT culture to 40 acres. In Australia a lettuce grower has used NFT outdoors, without plastic or glass cover, and has managed to raise 10 crops per year of very high quality lettuce.

The secret of NFT is that the tubes of plastic are almost closed, losing little moisture in the heat and avoiding dilution in the heaviest tropical downpours. Soil quality is irrelevant since none is required. Prof. Cooper is now trying to develop the technique to grow potatoes. He says : 'The potato plants develop a root-mat, which is common to all plants in the row. This root-mat has a very high tensile strength so that it would be possible to withdraw the potato crop from one end of the row and if equipment were developed it would be possible to strip the potatoes from the remainder of the plants automatically.'

So far the potatoes have only grown marble-size tubers but the development of such a technique would eliminate the energy-intensive deep ploughing, ridging and harvesting processes by which tons of soil are moved and compacted for every acre of potatoes grown today.

The vegetable foods of the future will largely be products of scientifically precise matching of seed, nutrient, water and sunlight. The intensive techniques for growing them will enable their production in the heart of cities and rising generations will be all too thankful to be eating to complain that tomatoes grown in culture do not have the piquancy of those grown on sewage sludge or animal manure.

Bacon that never said 'oink'

Despite advances in animal science and the more efficient use of feedstuffs, including wastes (pigs at Michigan State University have been shown to thrive on a diet containing their own dung and urine which has been digested by suitable bacteria!), the demand for meat will outstrip supply. Real meat will become a luxury, eaten on special occasions, while ordinarily the meat-hunger of our taste-buds will be part-satisfied with meat-substitutes.

Already the protein of soya beans and field beans has been used to 'fill' meat products such as sausages, pies, meat loaf, mince and hamburgers. The next stage has been to make meat chunks, bacon rashers and even steaks. Dr. Magnus Pyke, speaking to British butchers assembled for a celebratory luncheon in London, provided a glimpse of the coming age of 'meat'. 'There are two processes currently in use. The first requires protein to be extracted from soya or field beans, purified to remove raffinose and stachyose – these have no effect on the technical excellence but produce flatulence in the eater – and dissolved in alkali to form a kind of "dope" which is pumped through fine spinnerets. The delicate threads are drawn through an acid bath, stretched and worked to give any required consistency from the toughness of simulated bacon rind to the delicacy of tender veal, passed through a bath of fat, another of coloring matter and flavor and the whole wound up into a hank, compressed and fixed with a binder. Cut across the grain, this has the exact structure of meat.'

'A less expensive process than that based, as it is, on nylon spinning, owes its origin to a method once used for making buttons and umbrella handles out of the protein of milk. This time the protein is extruded under high pressure through a die. The trick is to use an appropriate machine exerting the proper pressure and combine the (vegetable) protein with starch, fat and other ingredients to obtain the desired result.'

And perhaps some of the 'desired results' of color, flavor and texture may be achieved by including, as 'other ingredients', the meat fragments and suspended solids that now disappear down the drain in abattoir effluent. New Zealanders, with their extensive meat exporting industry, have begun to filter and process the millions of gallons of liquid which has run to waste from their slaughter and packing plants. Now they are using the salvaged high-grade meat protein for animal feed ... but in the future?

If consuming abbatoir effluence nauseates it should be remembered that early efforts to make margarine, the first 'substitute' food, were based on chopped cow's udder, whale blubber and vegetable oil. Many nutritionists now consider the substitute a more healthy food than the original : margarine is supplemented with vitamins A and D and because it is produced from vegetable oils it is low in the saturated fats common to animal products, which are said to be a contributing cause of heart disease.

Not to be out-done, Australian scientists have bred a strain of cow capable of yielding milk containing polyunsaturated cream and, when killed, polyunsaturated beef. But brilliant though the achievement, animal sources of food must give ground to vegetable.

Conclusion

So many patterns of modern society first developed in the United States that the development of 'Megalopolis'[1] is likely to foreshadow similar urban growth elsewhere.

Just as super-highways and railways connected the cities from New England to Virginia and catalyzed a boom in industrial growth and population concentration, so the same pattern is repeating itself. The Great Lake cities of Milwaukee, Chicago, Detroit and Cleveland are growing into a 'Megalopolis' that could join the giant coastal connurbation at Pittsburgh. Meanwhile another arm of growth moves south from Toronto to Niagara and Buffalo. In California a 'Megalopolis' is developing from San Francisco to Los Angeles and San Diego and in time this could join with the growth of Vancouver south to Seattle and Portland, Oregon.

In colonial America the average density of population was one person per square mile and well into this century over 50% of Americans were still farmers and most of the rest lived in well-dispersed small uncrowded towns.[4] By the mid-1960s 70% of Americans lived on 1% of the land and by 1985 as many people could be living in cities as occupied the entire nation in 1960.

In Europe the extensive network of motorways, autobahns and autostrada is only 25 years old and already the American pattern is apparent and the natural consequence will be a 'Megalopolis' reaching from the industrial north of England through the Midlands to the Thames Valley and London. Onwards it will march (with ferry and hovercraft connections) through Kent, Essex and Suffolk ports to Holland, Belgium and France. Here it will join the industrial belt that already extends from Holland, Belgium and northern France along the German Rhine to Cologne, Koblenz, Frankfurt, Mannheim and Strasbourg to the Swiss border and Basle.

In the Far East, Tokyo and Osaka will form the basis of a Japanese 'Megalopolis' while Hong Kong spills onto the mainland New Territories. And in India Calcutta sprawls outwards to the very borders of Bengal, towards the Bangladesh capital city of Dacca.

Everywhere the growth of 'Megalopolis' will present problems of organization and communication, but few of these can be more vital than ensuring food and water.

Food supply will probably be maintained – provided that sufficient investment and incentive are given to implement the varied marvels of modern agricultural science. But the development of agriculture and food resources, in the ways described, could present a 'Pandora's Box' of unfortunate consequences. The 'biological back-lash' could threaten environment and most especially water supplies.

The success of the Roman Empire was as much due to their skills as water engineers and the building of aqueducts as to the prowess of their legionaries and the construction of military roads. The demise of the earlier empires of Sumer and Babylon can be blamed on the salt-pollution of soils through over-irrigation. The ancient Greeks cropped and grazed the vegetation of their home-land beyond its capacity to regenerate in such a dry climate and the eroded soils of modern Greece bear witness to their folly.

In the decades that are to come it will be relatively easy to provide fertilizers and pesticides, plant and animal hormones; to feed, manipulate and protect our food sources. But in the process we may destroy our soils and poison water supplies. 'In antiquity, leather workers and dyers were usually banished to quarters on the outskirts of town because of the offence given by urine and other smelly chemicals used in their workshops.'[4] Modern industry is learning to clean its gaseous and liquid wastes but the problems of similarly treating all agricultural wastes is not so easy. Where agricultural units are compact the problem is simplified – drainage water can be recycled, chemical pesticides can be used more precisely and inside glass or plastic houses. Natural pest predators can be used to reduce the dependence on chemical insecticides. But how to cope with the nitrogen fertilizer run-off in the valleys of the Ganges, Mekong or Nile? How to ensure that soils of only marginal cropability do not turn into dust-bowls as the pressure of food-demands presses far-mers to 'eat the seed-corn' of fertility?[5] Forests are rapidly diminishing throughout the tropics, giving ground to agricul-ture, which in many instances cannot be sustained. The underlying soils can support trees but they cannot sustain cropping which takes away already scarce nutrients to distant urban consumers, who void them through their sewage to the sea. The cut timber provides instant income as firewood or for lumber but the soil erodes, rivers silt up and wide-spread flooding brings disaster and death in heavily popu-lated plains and deltas.

Science in agriculture, as in any other sphere of life, is a tool : to be practised or ignored, to be used or abused. There is little doubt that every scientific 'straw' will be clutched at to ensure that the race between human popula-tion growth and Earth's ability to produce nutriment is not won by population growth. If it is, mankind is doomed.

Hope lies in the first hint that birth rates in most in-dustrialized regions of the world are now stable and that in some less developed regions the percentage annual increase of population may be slowing. Statistics published in October 1976 by the 'Worldwatch Institute', Washington DC, sug-gest a trend that is encouraging but their conclusions are based on government figures and their facts of life and death have been gathered in remote areas, dense shanty-towns and among largely illiterate and suspicious people. We can only hope that the trend is accurate and optimism not misplaced.

Population growth appears to slow naturally where in-comes and social expectations rise, infant mortality falls and there is improved health and standards of living. These changes in society can be achieved by government decisions regarding the optimum use of all resources. Ultimately the race between food and population will be decided by politics and not by agriculture and food technology.

WORLD POPULATION GROWTH 1970 AND 1975

		Crude birth rate (per thousand)	Crude death rate	Natural increase (per cent)	Population (millions)	Natural increase (millions)
North America	1970	18·2	9·2	0·90	226	2·04
	1975	14·8	8·8	0·60	236	1·42
Western Europe	1970	16·2	10·6	0·56	333	1·89
	1975	13·7	10·5	0·32	343	1·12
Eastern Europe	1970	17·4	9·1	0·84	368	3·14
	1975	18·0	9·4	0·86	384	3·31
East Asia	1970	30·6	12·1	1·85	941	17·43
	1975	19·6	7·8	1·18	1005	11·91
Southeast Asia	1970	42·1	15·5	2·66	278	7·40
	1975	38·6	15·3	2·33	317	7·37
South Asia	1970	40·8	15·9	2·48	709	17·57
	1975	37·1	15·8	2·13	791	16·89
Middle East	1970	44·3	15·5	2·88	136	3·91
	1975	41·7	14·5	2·72	155	4·22
Africa	1970	47·1	21·0	2·61	312	8·16
	1975	47·1	20·0	2·71	355	9·65
Latin America	1970	37·4	9·7	2·77	276	7·64
	1975	35·5	9·0	2·65	317	8·39
Oceania	1970	20·9	9·0	1·19	15	0·18
	1975	17·4	8·1	0·93	17	0·16
World	1970	32·2	13·2	1·90	3594	69·36
	1975	28·3	11·9	1·64	3920	64·44

Source: Worldwatch Paper No. 8

REFERENCES

1. Jean Gottman, *Megalopolis* (MIT Press).
2. *Resources Available for Agriculture* (Scientific American, September, 1976).
3. *Under-exploited Tropical Plants with Promising Economic Value* (Nat. Academy of Sciences, Washington, D.C.).
4. R. J. Forbes, *The Conquest of Nature* (Pall Mall Press, 1968).
5. Erik P. Eckholm, *Losing Ground: Environmental Stress and World Food Prospects* (W. W. Norton, New York).

RAW MATERIAL RESOURCES AND CONSERVATION

Robert Waller

Robert Waller worked at the BBC for the Third Programme and later as an agricultural producer. He turned freelance and worked on television documentaries. He now spends his time writing and speaking on human ecology. He is the author of a number of books – including a novel and poems – and the editor of Sir George Stapledon's classic *Human Ecology*. He is also the biographer of Stapledon, who was a distinguished agricultural research worker and agrarian philosopher. Robert Waller's last book, *Be Human or Die*, is a study of the impact of ecology on the humanist tradition.

Industrial civilization depends on energy slaves

The energy slaves on which Western civilization depends eat fuel, devour raw materials and spew out wastes. This is the basic reason why ecology, conservation and resources have become a world-wide concern.

How many energy slaves can we maintain more or less indefinitely? Energy slaves are, of course, substitutes for human labor; societies using them are called capital intensive, while pre-industrial civilizations are labor intensive. When, for example, we in the West discover how to synthesize a natural material artificially and put our energy slaves to manufacturing it, we cause widespread unemployment in other societies that were exporting the natural material. This process of beggaring your neighbor to increase your own prosperity has widened the gap between the poor and rich nations. But as we shall see as this essay proceeds energy slaves and the technology that they employ create environmental problems that are certain to limit their activities. They represent but one phase in the continuous history of mankind. How clearly can we anticipate the next stage?

First let us try to understand the problems of an energy slave civilization.

How durable is an energy slave civilization?

In September 1965 the British Association discussed the theme *World Fuel and Power Resources and Needs.* It is interesting to compare the views of scientists over eleven years ago with their views today. What is remarkable is that, although the speakers accepted that the world demand for energy was rising without obvious limits, only one contributor anticipated a decline in material progress. The paper of the dissenter, Prof. A. R. Ubbelohde, FRS, provides us today with a model of how to reason on the relation of an energy slave society to its environment.[1]

He began by outlining the growth of our dependence on energy-slaves. As early as 1911 Sir William Hartley had estimated that each British family had an average of 20 energy 'helots' in its service. Another way of putting this is to say that each one of us has between 100 and 300 wheels turning on our behalf, all of which need energy to move them – there have been civilizations that had no wheels at all! Sir William warned us that known stores of energy were not unlimited, so that we might find our slaves unable to work because we could not feed them.[2] Prof. Ubbelohde went on : 'This dependence on energy slaves becomes keenly felt only in unforseen interruptions of energy supplies.... Adequate safeguards against all major sources of breakdown of energy supplies should be given overriding national priority, next in importance only to maintaining adequate food supplies....'

Eight years after this warning our energy supplies were interrupted by the Israeli–Arab war. Prices of oil soared and we were confronting the situation the professor had predicted. He might have added that energy supplies and food production are linked when agriculture is energy-intensive. The rise in the price of fuel raised the price of food. We have become uneasily aware that an industrial society is twice as vulnerable as a pre-industrial one because it needs food for humans and fuel for machines. We could starve because our energy slaves do not have the food they need – fuel – to work the land for us or to supply us with nitrogen fertilizer.

What are the prospects for keeping our energy slaves at work?

What did Prof. Ubbelohde have to say about our prospects? The optimism of the other speakers was based on their theoretical estimates of the world's unexplored reserves of minerals. If these can be recovered at an economic cost, they are indeed immense. Many physicists, such as Alwin Weinberg, believe that there need be no resource scarcity if our scientific knowledge is consistently applied and supported by economic incentive. This might be so, it is impossible to prove one way or the other, though it seems to me that phosphatic fertilizer, a limited resource that has no substitute, must ultimately limit food production and hence population.* However, the approach taken by Prof. Ubbelohde suggests that it is not the actual availability of resources that is crucial but the impact of their use on the environment, that is, on natural systems. To many ecologists this idea is a platitude. They know that every animal species and most living organisms could breed *ad infinitum*, if there were no restraints. But beyond a certain point inhibitions start to come into play that gradually slow the rate of increase. The growth curve must be shaped like a bell with a building-up period, accelerated growth, approach to saturation and then either a levelling off or a drop down to a much lower level. This bell-shaped curve is as familiar to a biologist as an exponential upward line to an economist. He assumes that nature cannot tolerate indefinite growth in a finite world.

'In following the growth of energy consumption of individual communities, well-known features of biological growth may provide suggestive parallels.' The professor then went on to point out that in Western Europe the accelerated curve might now be expected to head for saturation and that the steep growth was only to be expected after a build-up of industrial potential.

'What restraints are we likely to experience? In a general answer to this question, there seems little doubt that inhibiting factors must grow in strength as the total number of energy slaves grows. At present evidence that this is happening is difficult to gather. But they are likely to arise through difficulty of control and to a lesser extent difficulty of supply.' The professor then pointed out that our political system is organized in such a way that it is very difficult to get information about what is happening environmentally and that when we do find out it is impossible to adjust and adapt owing to our short term political policies. 'The feedback of information to those who decide policy for the future

*See p. 473.

may become increasingly inadequate as the population of energy slaves outstrips the humans they serve.* As a consequence clear views are rare and the public are kept in ignorance, and we find ourselves,' the professor said, 'unable to improvise in an emergency.'

Does nature show us the way to invent our machines?

Prof. Ubbelohde gave as his opinion that in the future most of our energy slaves would be thrown away and we should return to a simpler style of living. I would put this by saying that we shall discover that we are over-industrialized and that we have yet to explore the degree of industrialization that nature will tolerate and that we essentially need. But in the meantime we must continue to find out how we can keep our energy slaves at work since we cannot suddenly change from an industrial to a post-industrial society. We must begin to plan the transitional period now. The professor suggested that we should adopt the principle of studying natural organisms in order to learn how they manage to use energy so efficiently : this principle has indeed become popular among many modern inventors and has been beautifully illustrated by a German computer expert, Felix Paturi, in *Nature the Mother of Invention* (1976). Ubbelohde cites as an instance the electro-mechanical processes utilized by the hopping flea or the buzzing mosquito 'which are immeasurably ahead of any modern contrivances in the range of human inventions.... Power engineers of the future may conceivably achieve unorthodox micro-engines based on useful clues from biological models....' Paturi calls his book a study of the engineering of plant life and confines himself to plants. The sun is the greatest of all nuclear fusion stations and plants convert sunlight into energy : sunlight is a constant flow, while fossil fuels are a limited store. Many research scientists all over the world are today studying how to reproduce the photosynthetic process artificially. Such a

*As an example of the population of energy slaves outstripping human beings, in 1968 one car was born every five seconds in the US and one baby every twelve seconds.

discovery would do far more to revolutionize our energy problem than building nuclear power stations which require such gigantic concentrations of energy, as well as being dangerous.

In the rest of this essay we shall discuss the points raised by the British Association Congress in 1965 in the light of what has happened since.

Examples of the rise and fall of growth: cars and electricity

Peter Chapman, Director of the Energy Research Group, T.V. University, has plotted the growth of automobiles and electricity supply against the time between 1960 and 1975. The graphs made show how after a steep rise that had been building up since 1900 or so demand begins to falter and the phase of exponential rise comes to an end. Chapman points out that for social reasons, in particular maintaining employment, the car industry is given priority and artificially stimulated. The installed capacity for electricity, based on projecting exponential demand – in the same way that we plan our motorways – now greatly exceeds actual demand. Peak demand may only occur on one day in a year and yet we build generating stations whose capacity is in excess even of that. The production curve began to drop well before the oil crisis, suggesting that it is saturation rather than shortage of fuel that is holding down further growth in output.

Rise and fall of North Sea oil

Another example of the bell curve is the predicted output of North Sea oil. Here again the curve follows a building-up period as the oil gradually comes on flow, an accelerated growth rising exponentially and an approach to saturation followed by a rapid fall. When this happens around 1990, the British will have to resort to the world market again. But as the world's oil supply is likely to follow the same curve, peaking at roughly the same time, the competition for oil will have intensified, prices will have risen still further and many of the present exporters, such as Iran and Venezuela, will be limiting themselves to their own home market. Our situation from 1990 onwards will be much worse than our situation in 1977, unless we have progressively brought on stream other sources of energy to take the place of oil when its output rapidly falls. The Norwegians are being more far-sighted and are planning to keep a reserve of 20% for as long as possible, while Britain intends to use up its as fast as possible in order to pay its foreign debts and rid itself of the burden of interest. But equilibrium will be shortlived. The policy of preparing for famine by feasting well illustrates Prof. Ubbelohde's criticism of Britain's short-term political policies allied with public ignorance.

It is highly improbable that in thirty years' time Britain will have alternative sources of energy ready to take the place of oil. Even if we go over to coal this will mean an expensive change-over to coal-burning equipment and if new mines are to be brought into operation and a large force of miners mobilized, we must prepare for that now. But coal is still relatively expensive compared to oil and mining is an unpopular occupation. Burning coal is so polluting that at a scientific conference in Germany it was said that we could not burn all the coal that exists without suffocating ourselves and changing the climate by overheating the atmosphere.[3] If we expect to substitute nuclear energy for oil, we shall have to make gigantic sacrifices of other industrial needs to pay for the nuclear power stations we should require. Such a prospect can be written off, not on account of the dangers, which are bad enough, but on economic grounds. The amount of energy required to build nuclear stations that only have a working life of thirty years, together with the cost of monitoring their discarded hulks and their radio-active wastes, must surely suggest, if one takes everything into account and not just running costs, that nuclear energy is the most expensive form of energy ever discovered! Merely to consider running costs once the station has been erected and to ignore all the rest is extremely misleading but a frequent practice wherever energy is involved.

The energy statistics and prospects lead to one conclusion:

that if we are to meet the situation that will confront us after 1990 we must recognize that we are an over-industrialized society and that we must start to economize in the use of energy at once and explore alternative energy slaves. Japan is already spending $54 million annually on this, France $18 million and West Germany $12 million while Britain, it seems, is spending about $36,000.

The interpretation of exponental growth curves

The biological experience of growth, which takes into account the restraining environmental factors, teaches us that statistics are abstracted over a limited period of time and that they are meaningless as a guide to the future unless we know what precedes that period and is likely to follow it, in other words, from what part of the curve they have been abstracted. We cannot, of course, know exactly what will happen in the future and the curve invariably zig-zags, but we can make a pretty good guess at the general trend. The popular way of explaining exponential growth is to imagine a lily pond, that is, a finite habitat, and assume that the lilies keep doubling their numbers annually. At first no one will be concerned because the number is still small; even when the pond is half full it may not cause undue alarm. For what is not realized is that in *one* more year, however many years it has taken to become half full, in *one* more year, it will be completely full; after that it will suffocate and die.

The alarming aspect of exponential growth, if it is unrestrained, is the suddenness with which a safe situation can become an uncontrollable one; how rapidly a resurgent and pulsating life can turn into its opposite once the capacity of the natural habitat is exceeded. A cause for rejoicing at first, vigorous growth turns before people realize what has happened, into a cause of suffering and mourning. Once a society is seriously overcrowded, everyone has to be regulated in order to ration food, space and work. Liberty is sacrificed. While the world as a whole has a 2% to 3% birthrate this is the situation that must confront much of mankind.

Natural systems, as the bell curve shows, have inbuilt safeguards against exponential increases continuing beyond the capacity of the environment to sustain them. The commonest safeguard is a dwindling food supply; but predation and disease and mere accident usually keep down a population rise before the food runs out. Man has endangered his survival by his attempts to insure it; by breaking through the natural barriers he has been able to increase his numbers as no other species can do, eliminating thousands of other living species to make way for his own expansion and thereby undermining his own base in nature. This free field for exponential expansion can only be temporary and must bring about its own nemesis.

What do the demographers have to say about it?

The rise and fall of populations

In *Population and Social Change* (edited by D. V. Glass and Roger Revelle) the few professional demographers in our universities collaborated in a study of the population problems of our time. Demographers have learned to be cautious about making forecasts on the assumption that the present trends will continue even in the short term into the future. Our demographers believe that the best way to study what is likely to happen in the future is to study what happened in the past. The estimated world population in 8000 B.C. was 5 million; then population doubled in 1500 years; between A.D. 1650 when it had reached 500 million and 1850 when it was a thousand million, it had doubled in 200 years. It doubled again in eighty years then in forty-five years and it is heading for doubling yet again in 37 years; so we have the lily pond sequence. If the annual increase continues at 2% it will be 8 billion in the first decade of the next century. Although some countries of Europe have population densities greater than those of Asia at present, they constitute only a small part of the globe. If such densities become world-wide, it will be an ecological catastrophe.

In 1970 when the world total was 3607 million, 2 billion of

these, 56·8%, inhabited Asia and Australasia. By the year
2000 this number is expected to have risen to over 60% of
the world population and to exceed the present population
of the whole world. A glance at the table (p.541) shows that
the populations of the Western countries will have dropped
percentagewise and if there should be a fast fertility decline
will level out at zero growth, i.e. replacement level. The
question is, at what point on the bell curve are all these different populations?

The general view of the demographers is that the populations of all the industrialized countries will begin falling.
It took eleven thousand years for Britain to reach a population of ten million in 1800 : but the last nine million of these
needed only a thousand years. Then forty-six million were
added in one hundred and seventy years. We had more than
quadrupled our population in 1·5% of the total time. The
slow build-up followed by the accelerated surge forward has
been completed; the rate of increase of a hundred years ago
has almost halved from 1% to 0·6% evidence that saturation
is approaching. The trouble is that we now have such a large
base that the actual number of people, despite the falling
rate of increase, has not dropped at all. Nevertheless the
downward trend is appearing and if the increase of 0·6%
were to become a decrease of 0·6% by, say, the year 2000,
a population that had then risen to 60 million would fall to
ten million in three hundred years' time. This is by no means
an absurd speculation, for the object of the industrialized
countries will be to maintain their level of prosperity so far
as they can with scarcer resources; a lower population is the
sole way they can do this. For example, we do not only have
to consider the increasing environmental restraints that will
be coming into play, but the fact that when the British
population doubled in under a hundred years, the nation
was living off food produced overseas, much of it from the
colonies; in addition to that emigration was at a high level.
One of the basic contributing causes of the two great wars
in Europe was that the central powers were hemmed in and
could only expand within Europe itself, unlike the maritime

powers which acquired overseas territory. Now the sources of imported food over which we had administative control have gone and the scope for emigration has shrunk, we are back once again in a situation that resembles the pre-industrial eras. The historical researches of the demographers reveal that pre-industrial civilizations, recognizing the limits to expansion, never did procreate without some form of family limitation, even in Catholic countries. English village communities seem to have been particularly skilled in limiting birthrate.

Perhaps for the industrialized countries, then, we do not need to be too alarmed about population increase and all our errors may arise from assuming that the increases of the past hundred years will continue both in population and economic production. We should plan for saturation levels followed by a fall; this would entirely change our planning of motorways, airports, car production and so on. We have to become conscious that we are at the top of the accelerating curve and that we are, consequently, over-industrialized relative to our future; that we shall only be able to adapt to this future provided we do not continue on an out-of-date momentum, a momentum that is not matched either to reality or natural systems. At present we are increasing or maintaining the production of many of our goods for the sake of keeping people employed rather than because the goods are needed. We are persuaded they are essential by massive advertising campaigns. This is not the way to plan for full employment, but can only end in a sudden massive increase of unemployment. We have to design new technologies.

The population puzzle lies in the under-industrialized countries, the poor countries. It has been assumed that they will steadily become industrialized and that their growth will follow the usual curve of build-up to the point of take-off and acceleration. As a result their population curves would follow the same trends as the West and begin to level off toward zero growth. What has happened is that the populations have gone into acceleration phase without industry and agriculture keeping pace. Population increases of 2% to 3% are, however, only a phenomenon of the last three decades. Although they would seem to be explicable with hindsight, they certainly have taken us by surprise. A large fraction of the population as a result of the suddenness of the jump forward is under fifteen years of age. These young people are the gunpowder of the population explosion; they mean that it will be thirty years at least before birth control can significantly slow up population growth.

The explanation would seem to be that Western science has created the conditions for exponential population increase when people simply followed their traditional pattern of family life, breeding in the hope that the survivors would keep them in their old age. There are now too many survivors and instead of making old age secure, they make it less secure by creating a scarcity of traditional resources. As we shall see when we discuss food, the gap has been made up by American food imports. There are several reasons why there has not been an economic take-off, one of which is that people living in hot climates do not have the same incentive to employ energy slaves as people in Northern regions where cold and hunger are more keenly felt. The cultures of Asia and Africa are different from ours and to build up a Western industrial society rapidly requires a Western outlook, a Western education and Western technical training. The greatest mistake of all has been to start to industrialize before a larger agricultural base has been secured. In other words the new nations have not developed in the right order. But even if they had, the natural systems of the world could not sustain a Western expenditure of energy on a world scale. So the present tragedy has its brighter side, even if that is cold comfort for the poor countries.

It would seem to be inevitable that by the year 2000 there will be too many people in the world and that it will not be possible to sustain such numbers for very long. Pressure of people on land and resources, the overburdening of natural systems, the need for authoritarian government and so on must lead to a decline from the peak. National wars are quite likely as boxed-in communities struggle to secure the basics of existence for themselves. It is not difficult to predict that the world will never again carry such a large population. This population, as we have seen, has been created by freak conditions and the belief that Western civilization can spread all over the world. As soon as this error is recognized, as soon as the facts permeate people's consciousness, as they must do over the next fifty to a hundred years, the human communities will begin to plan for equilibrium between people and nature. Certainly it will be a race against time but evolution, natural and human, has seen some astonishing changes.

In the meantime surely we in the West must begin to sacrifice some of our affluence, both for the sake of our own survival and the welfare of poorer nations. We cannot predict how the material wealth of the world will finally be distributed : this will be partly a matter of cultural preferences and traditions; nevertheless we can be sure that there will be a considerable levelling out, so that the over-industrialized nations find a new stability and the under-industrialized ones seek their stability by a rise in wealth.

'The limits to growth' – a commentary on the Club of Rome report

People ask if the resource predictions made in *Limits to Growth* are now out of date. *Limits to Growth* did not make that kind of prediction. The authors only showed that, if present trends continued, oil and many of our most essential minerals would be largely exhausted by the end of the century. Nobody has been able to refute that. In 1971 both people and industrial capital were growing exponentially, with industrial capital growing faster than people. The present 'recession' must indicate that the rate of growth of industrial capital has slowed down; but all the industrial countries are scheming to get it rising again. The statistics of mineral reserves and resources – that is, confirmed reserves and unexplored resources – were the most authoritative that could be found, many of them provided by the US Bureau of Mines.

The critics object that we cannot know what will happen in the future in the way of changing demand, substitute materials, new methods of exploration and less wasteful technologies and so on; but they are only saying what the authors themselves said. They stressed that their world model was a crude first attempt towards constructing a system that could be used to program a computer; it was a model using the principles of systems dynamics. This dynamic thinking includes as many as possible of all the factors at work, often simultaneously, that determine the resources available to society. The difficulty of using such a complex model is that all these factors interact on each other, as they do in a natural ecological system that follows the bell-curve. No model could accurately reproduce the future if only because we cannot know in advance what decisions people will make about what kind of a civilization they prefer. Consequently such a model can never do more than help us to sharpen our intuitive judgements, which for so long have been wrong. Anyone who can improve upon the model by providing evidence of changes in our present system will be doing what the Limits' Team invited people to do. Some sourer critics give the impression that they are indignant because their intelligence was not included in the feedback loops.

Limits to Growth popularized the importance of feedback. We are using the idea of a feedback when we talk of a 'vicious circle'. The wages/prices spiral is a vicious circle. The rise in wages is fedback through increased costs and prices, so that wages have to rise again. This is positive feedback as it augments the rate of growth. Negative feedback works the other way as when the death rate cuts down the base from which the birthrate is determined. Birth control is a negative feedback; as birth control depends upon increasing social services, these services act as a negative feedback; they also divert resources of skilled manpower and may be opposed by religious traditions.

To leave these feedbacks out of account in our estimates

of the future rate of growth of people and resources must give a misleading picture. By leaving out of acount feedbacks that oppose a future development that we favor, we can 'prove' almost anything we like.

I believe that the most important thing that *Limits to Growth* taught us is that the rate of growth of industrial capital can be compared to biological growth. It 'breeds' goods and services. It tends to multiply itself up to the limits of available demand and resources. If we think of capital as 'means of production' rather than just as money, we can picture what is happening in the world. Every industrial 'plant', like a natural plant, consumes energy and resources. So far as it is using non-renewable materials, there must come a time when it cannot regenerate itself; also, unlike a natural plant, many of its waste products are not bio-degradable, hence they are polluting. Industrial capital does not last for ever; everything wears out and has to be replaced. This is known as depreciation; even before death there is the constant cost of maintenance. Although tax allowances are made for depreciation, the real cost is under-estimated in all the planning of industrial societies, as in the way their plans for motorways and power stations are presented to the public, for example.

The growth rate of industrial capital is implicit in the popular term 'economic take-off'. Once it is well established, an economic system is supposed to take-off the ground like an aircraft and soar exponentially. As this has happened in the Western nations everyone has been waiting to see it happen in the rest of the world. But the gap between the West and the rest has widened not shrunk. The rich have goods and the poor have children. In the US the economy took-off with such success that in 1968 it was breeding cars faster than people were breeding children! When we pause for a moment to consider the materials and energy used in the production of a car and the energy that it will use up in its working life, we have some idea of the drain on the world's resources of an exponential increase in the industrial growth rate. It becomes clear that it will be more difficult to feed our voracious machines than to feed people; although the two are related in our economy. Tractors, fertilizers and sprays are the product of energy-intensive industries and if they fail to be fed, then the farmer will lack his means of production and the food production will fall drastically. Less dependent systems of farming would be unaffected.

Can we, then, take it for granted that industrial expansion can provide for our increasing numbers, if it devours resources so extravagantly? Mass production industries, indeed, demand that we buy more than we want in order to keep the price down! Economies of scale are economic economies, they do not economize in resources. Far from it; they waste them. Older methods of production, however limited in what they did, provided our necessities with a much greater thrift. On a livestock factory farm, for example, the animals do not graze the land but are fed feeding stuffs imported from countries where people are undernourished. This is one example of how it is that although the rate of industrial growth has kept ahead of population growth, nevertheless the world is still divided between rich and poor countries. The exponential increase in industrial growth has been confined to a few countries. Whether they are fortunate remains to be seen. So far they have been able to blackmail the countries with the materials devoured by their machines into selling their resources at bargain prices – because they dominate the capital market. But now the ports of the oil exporting countries are being surrounded by vast industrial belts to which the resources are being diverted. The older industrial nations with few natural resources (or in Japan's case almost none at all) may be left with nothing to sell but their professional skills.

Limits to Growth has taught us above all to look at every resource problem as the centre of an intricate dynamic mass of forces. We no longer just look at population statistics but at industrial growth rates as well. How we interpret these depends upon our ecological knowledge. Capital growth may make our problems worse. A country with a falling birthrate and a rising industrial growth rate could threaten the future prosperity of mankind more than a country with a high birth rate and a low rate of industrial expansion. We should also remember that automation 'eats people' – as sheep runs did when arable farms were destroyed to make way for them because wool had such a high price. Automation may in this way affect the birthrate as people will be reluctant to have children when there are no jobs for them.

Looked at from this perspective it could happen that industrial civilization must come to an end because its energy requirements could only be satisfied by nuclear energy. The supporters of nuclear energy are probably right about that. Many of the opponents of nuclear energy also recognize that. A nuclear power station has a life-time of about thirty years. Before it goes into operation, while it is being born, so to speak, it devours quite a high fraction of the amount of energy it will ultimately produce; it also consumes a huge amount of expensive materials in its construction. While it is working it creates radioactive pollution, both long-term (250,000 years) and short-term (pollutants with a short half-life of only a few years). When it dies it remains a useless radioactive hulk on a country landscape.

To meet the energy requirements of an industrialized world would require at least 3000 nuclear power stations, each one being replaced every thirty years. The cost would be prohibitive even for America.

Many of the critics of *The Limits of Growth* assert that it is wrong, even wicked, to depress people with prophecies of doom. Man, they say, is so clever that he will always find a way of solving his problems provided he is not discouraged. This, of course, is not true of human experience; men have only overcome their problems when they acknowledged what they were. If an industrial society is impossible to sustain, all right, let's work for a post-industrial one.

The influence of American food exports on world development

Studies carried out recently reveal the important fact about the world's food supply. It is that the export of grain has been effectively reduced to North America and that North America is becoming increasingly dependent on imports of raw materials: consequently the Americans are going to be increasingly pressed to sell their grain for these raw materials instead of giving it as aid. The United States has increased its export of grain from 5 million tonnes to 94 million within a single decade. Its exports are now sufficient to feed 600 million people, yet this only amounts to 10% of the total US cereal crop, so much of the rest is fed to animals to maintain the affluent meat diet of the West. It is reckoned that one-third of the world's grain production each year is used in livestock production with the result that, while poor countries are undernourished, rich countries are over-nourished, to the detriment of the health of both. A great part of the food of the West is a luxury and should be taxed as such.

Even so, this ten per cent of its grain production gives the US an almost monopolistic control of the world's exportable grain supplies. The dumping of food as aid or at concessionary prices so depresses the farm economy of the recipient nation that its farmers lose their incentive; if food aid were given constantly the receiver would become dependent on it. If this were to happen and the US no longer had surpluses to dispose of, the result would be famine for millions of people. In one year in the 1960s, the US supplied India with enough grain to feed 60 million Indians. One of the effects of this grain aid to India has been to delay the reform of land ownership in India, which is probably the most important means of giving the Indian land worker the incentive to improve his farming.

The question is, for how long will the US continue to grow these food surpluses? Are they part of an accelerating trend that is now beginning to level off as the conditions of production change? The answer to that question is, yes, they are. There are many reasons why the US farmers will not be able to keep up these yields indefinitely. Firstly we have to remember that the American farmer has to import great quantities of energy on to his farm to grow his crops. The energy that he feeds into his production does not give as great a return proportionately in food calories as the labors of an Asiatic peasant who only uses his own strength. But as the American uses his energy slaves, his yield per man is vastly greater. Nevertheless he does not provide these food surpluses with his own muscle, but because he can purchase so much of the world's fuels and other resources necessary to manufacture and use machines. If he could not do this, his productivity would fall below that of the Asiatic peasant.

The rising price of all the inputs on which the American farmer depends to cultivate his 450 million acres of arable land will make production far more expensive and there will be sharpening rival demands for the energy and machinery from other industries. Ten per cent is not a great deal of the total output and the rising American population will in time want it for themselves. America is one of the least densely populated countries in the world with only 55 people to the square mile as compared to Japan's 770. Consequently, although the American birthrate is easing off, it will continue to maintain a higher rate than more densely populated industrial countries like the Netherlands and Belgium whose densities are greater even than Japan's. The American pressure on land is already taking a million acres of the best farm land every year. Lastly, surveys have revealed that the US has lost about one-third of its topsoil; the amount of fertilizer used has reached saturation in the corn growing areas and diminishing returns have set in. All these losses to farm production should reduce its yields by the 10% on which the world relies to buffer itself against famine.

A consideration of this probability is enough to make any nation dependent on American food supplies take steps to make itself self-sufficient. But there is yet another factor, briefly mentioned above, that is likely to withold American supplies from the poor nations very soon. That is the fact that America is running short of raw materials, including oil, and must buy them on the world market. Therefore it must use its food exports in exchange for fossil fuels and primary commodities. Obviously America will confine its exports to the nations that can pay most for them. Until the poor nations can be sure of growing their own, the United Nations will have to set up an agency for buying the world's food surpluses and distributing them where they are most needed. Otherwise millions are going to die of famine.

Other Western nations can help the world food situation by changing their diet. Only 8·3% of the land in the UK provides food directly to Man. And even then there is not enough to feed our livestock; so much of it must be indoors all the year round and fed imported grain. Our animal factory farming, about which we boast so much, is the most inefficient farming in terms of land and energy that we have. The fifteen million tons of feedstock that we import to feed animals every year should be fed directly to human beings in the land where it was produced. Western nations should confine their animal husbandry to grasslands in upland areas or integrate their animal husbandry with mixed farming and rotations. The technology used in agricultural monoculture is wasteful of energy and resources and ecologically destructive. It has been calculated that farm animals in the UK monopolize land which could theoretically feed over 200 million people on an all-vegetable and cereal diet.[4] Although it is not necessary to go to such extremes, the evidence is clear that one of the reasons that the world is short of food is the greediness of modern high farming technology and not simply the backwardness of pre-industrial agriculture. A study of energy in relation to agriculture, *Energy and Food Production* (Gerald Leach, 1976), gives us all the statistics in painstaking detail. The final picture of the wastefulness of the UK food producing system as a whole is shattering. As Leach says, 'Words almost fail.' To quote one paragraph :

'UK agriculture, employing 413,000 in 1972, or only 1·9% of the total workface of 22·12 millions, is highly labor efficient (that is, in terms of output per man). Each of these farm workers feeds some 66 people on a food energy basis and rather more counting proteins. But if one counts up all the direct and indirect labor employed in the food system from farm to shop, the total is probably close to 3 million. To this must be added the workforce abroad that supplies imported animal feedstuffs and the 40–50% of food that is imported – perhaps one million or more allowing for the greater labour intensity of imported foods. Thus each person in food-related employment in the UK or abroad, feeds 'only' 13 to 14 people. Hence on a total population basis, 7–8% of the working week is spent in food provision : a performance only two or three times better than pre-industrial farmers and hunter-gatherers.'

The Institute for Food and Development Policy (US) has made a study of world food production (reported in the United Nations publication *Development*, November 1976) which has led them to state categorically that while we have the present system of food distribution 'More food means more hunger'. Their argument is briefly that while food is grown for profit it will be grown in the wrong place, by the wrong people and eaten only by those who can afford it however much of it there is. 'The lure of great profits tempts large landlords to take back land they formerly rented out. Many use their now higher profits (from Western farming methods) to buy out small neighboring farmers. Throughout the under-developed world, the landless now comprise 30 to 60 per cent of the agrarian population.'

Even in the UK the monopolization of farm land continues so that 40% of the land is farmed by 10% of our farmers.[5] But large farms do not produce the highest yields per acre; they make the most profit per acre. As this profit has depended upon cheap fuel, fertilizers and machinery, the large farm is going to push up the price of food.

The world food shortage is not then entirely due to increasing populations; it is also due to the economic system by which food is grown and distributed, the husbandry employed and the system of land ownership. The Chinese have shown the way in feeding a large population with limited energy resources. Where land is privately owned, when population increases, the richer must become richer as the poor become poorer, because the pressure of increasing population on land pushes up rents. It is only where men can move on to land of their own when rent and wages are at subsistence level that this 'iron law of wages' does not apply. The only economist I know that has seen the answer to this problem that is in harmony with the liberty of the individual is Henry George. George lived in America at the turn of the century when the iron law of wages was just beginning to bite as it was realized that land was not as infinite as it seemed. He suggested that the land should not belong to anyone, neither to the state nor the private capitalist, but that every occupier of land should pay a graded land tax to the community as this would prevent speculation in land. Doubtless this system has its defects and loopholes but in principle I believe it to be right. Anyone interested should read George's great book, *Progress and Poverty*, which, as Maynard Keynes said, deserves to be as influential as *Das Kapital*.

Pollution and modern technology

Following the lead given us by Prof. Ubbelohde, let us search for some of the restraining factors on growth arising from our technology. Our energy slaves are working modern technology and the technology introduced since the last war is fouling our own nest. Automobiles, pesticides, herbicides, synthetic textiles, packaging, detergents and aerosols are all the products of a new technology that disregards the need to protect natural systems. They are all new and they have all increased in use since the nineteen-fifties. To these we can add that major source of pollution, sewage disposal, which has intensified its burden with the growth of connurbations. As we ourselves are self-regulating and self-cleansing organisms, we must include among pollutants those which injure us directly such as tobacco,* chemical additives to foods that our body's metabolic system cannot cope with and the medical drugs with their unpredictable side effects. All these injuries to natural systems, our own and that of external nature, add up to a formidable assault on the tree of life. The noise of traffic, aeroplanes and machinery, the monotony of working at the pace set by the assembly line and the machine, all oppose the rhythms of nature and irritate or damage the nervous system. Consequently they are all becoming restraints on uninhibited growth; in addition to their physical injury, they arouse psychological revolt; the age of progress was intended to relieve people of this kind of suffering and they are becoming disillusioned. Man is an adaptable creature and no doubt he can adapt to much of this. But like everything else in nature, he has levels of tolerance and these are being grossly exceeded 'for the sake of progress'.

How the natural systems work

Plants have a period of growth that leads to flower and seed; the vegetation then dies down and falls on to the soil or we put it on the compost heap. This 'waste' then disappears from sight decomposed by bacteria without whose labors death would be incomplete. The 'dead' substances have disappeared because they have been transformed into humus and plant nutrients. This process has released energy by

---

*Tobacco and lung cancer. Between 1945–65 there was a phenomenal rise of 800% in the death rate of lung cancer. This corresponded with the decline in the amount of naturally cured tobacco (air cured) used in cigarettes. Oven-drying of tobaccos, according to the researches of Dr Beffinger, a Pole who has spent most of his life in tobacco research, does not allow for the full enzymatic fermentation responsible for the carcinogenic activity of tobacco smoke. Dr Beffinger's work was described by Professor Eysenck in his book *Smoking, Personality and Health*. Professor Eysenck wrote to Dr Beffinger 'It is a scandal that thousands have to suffer because medical research will not take the obvious seriously.' Beffinger claims that the tobacco industry prevented him from doing research in this country.

breaking down cell structures, energy equivalent to that needed to build and maintain the cell structure. It is this energy that is locked in fossil fuels; fossilization has prevented them from completing the process of decomposition because the bacteria could not get at them. The active compost heap warms up. If we have been foolish enough to burn our wastes because they looked unsightly, we have broken the cycle of nature, impoverished the nutrients in the soil, weakened soil structure and polluted the atmosphere.

Life and death in the cycle of natural systems is one self-renewing process, following the curve of build-up, acceleration of growth and decline. But the decline is an essential preparation for the renewal of life. In this process no energy or matter is lost : it is transformed. The seed must fall into a fertile ground created by the death of its parents.

On the vegetable patch, however, we harvest the flower or the seed or the greater part of the plant to eat ourselves, thus taking in as fuel the energy in the plant which is oxidized and burned in our bodies. But what we extract from nature's cycle must be replaced; this is partly achieved by returning our own wastes (sewage) to the land and partly by the decomposition of our own bodies when we die. Animal wastes play the same part. It has not always been understood that harvesting must slowly starve the body of nature on account of the gap between what the plant or farm animal has taken from the land and what the farmer puts back. The first farmers moved from one site to another, so nature had time to regenerate the soil they had exploited. As late as Roman times, there was a great debate about whether mother nature was growing too old to continue giving birth. A great Roman writer on agriculture, Columella (*circa* 117–27 B.C.), argued that by good husbandry nature could be kept fertile indefinitely provided that what is taken from it is given back in some equivalent form. The Romans were already experiencing the way large-scale civilization denudes the fertility of the land. Victor Hugo in *Les Miserables* wrote that the fertility of Sicily went down the sewers of Rome. The fertility of the whole world goes down the sewers of the great industrial capitals of today. How often do we reflect when we throw away the kitchen refuse that we are stealing fertility from the land, perhaps from the farm of some peasant thousands of miles away overseas?

Relation of commerce to the natural systems

From the commercial point of view the first half of the natural cycle is productive and profitable, while the second half, the stage of decomposition, is not. After the farmer has sold off the produce from his land, he wants to spend as little time, labor and money, as possible in preparation for next season's sowing. In the whole history of mankind we have never taken so much from the land as we are doing now; and yet, for the most part, we have abandoned the husbandry, learned through trial and error, that matched productive methods to the natural cycle. This departure was made possible by scientists studying how plants feed and concluding that they required nitrogen, potash and phosphate. This crude model of the nutrient requirements of plants has become the basis of modern husbandry and the foundation of the great industry that provides these three nutrients as artificial fertilizers. The use of these fertilizers has enabled us to feed millions of people who might otherwise have starved (or, who, more probably, would never have been born) as the example of the American grain trade illustrates. If fertilizers can no longer be produced, then these people will starve. Apart from that gamble with nature, the defects of this husbandry are now becoming apparent in the form of intensive pollution of surface waters, diminishing crop yields per acre of fertilizer used, deterioration of flavour and loss of topsoil. As Western husbandry is locked into the high energy industrial system, the energy crisis has caused the cost of food to rise as dramatically as everything else. The essential fertilizer phosphate is almost a monopoly of Morocco, so far as supplies for Europe are concerned. Horses might become more economic than tractors costing £7000 each or combine harvesters costing £30,000. We can see why economics dictate larger and larger farm structures, contrary to the welfare of man and land.

According to Leach in the book already referred to, between 1900 and 1972 inputs of nitrogen rose eightfold and phosphate and potash some thirtyfold, while yields of crops receiving the increase barely doubled. Leach considers that increase in yields of animal products over this period was only 50%, a rather remarkable case of diminishing returns. Nevertheless at the prices then ruling the bargain was not a bad one. Our agriculture like our industry settled into its present form when oil was 1·5 dollars a barrel. But when a barrel of oil is 12 dollars and phosphate has quadrupled in price, how much longer will high energy technology continue to be a good bargain? The small farmers produce as much or slightly more than the large farmers from the same amount of land, but the economic structure of farming prevents their numbers from increasing, indeed tends to eliminate them. Yet the small farmer is far more likely to adapt to the emerging situation.

Why artificial fertilizers and sewage pollute[6]

Artificial fertilizers do not maintain the humus content of the soil which depends upon roughage, fibre and trace elements from plant remains. Humus holds the plant nutrients so they are gradually released as the plant calls for them. Half the artificial nitrogen applied to the soil trickles out of it when there is insufficient humus to hold it in place; a good deal of the other applied plant food is also lost. All these escaping nutrients join the natural losses from the soil to over-fertilize the plants growing in surface waters. So much oxygen is then absorbed by the affluent vegetation which has had an exponential growth that the fish die. The movement of the water is obstructed and the whole system stagnates instead of cleansing itself. If this overfeeding (eutrophication) accumulates, the natural system must finally die. Thus, by depriving the soil of humus which serves to control drainage as well as hold nutrients, we overload the surface waters. The *deprivation* of one system and the *overloading* of another are seen to be related and a negative dynamism is set up. Nature never stands still. The Broads in Norfolk are a good example of this; like many lakes all over the world they are dying as a consequence of draining vast areas for farmland.

The other contributor to this death by over-eating is sewage. As we have seen sewage is a plant food. Traditional farmers had a slogan, 'Feed the soil and the plant will look after itself.' But, once we have built big cities, feeding our wastes back to the soil becomes a difficult technical problem. Now that we house large numbers of farm animals all the year round indoors, many farm-factories (which ought to be highly rated as factories) disgorge as much sewage as a town. Is it not incredible that we should have been so foolish, guided by false economics, as to add agricultural pollution to our own! When we turn animals into machines, they become polluting and wasteful like everything productive that is not matched to natural systems. If we had designed our sewage disposal system in the 19th century on the model of nature, we should have forestalled the problem of sewage pollution by returning our wastes straight to the land instead of carrying them to the rivers, lakes, estuaries and the sea. No doubt the Victorians thought God had kindly provided them with rivers as drains, so the industrial revolution should triumph – 'and restore the lost property of the Fall' – as Francis Bacon put it. But a beneficient Providence did not give us rivers so that we might become rich on the cheap; on the contrary, by taking this short cut, we solved an immediate problem by creating one that is more expensive to solve. It is seldom recognized how stupid our alienation from nature has made us.

Other high energy polluting processes

The wastefulness of the processes that produce our synthetic textiles have not received as much attention as fertilizers and sewage. This is revealed by applying an energy analysis; energy has been the invisible cost in our industry, as Leach's work shows. Energy analysis is far more comprehensive than the traditional economic analysis; it penetrates a step further back in the productive chain and includes the energy resources required, for example, to build the plant that produces the product. It is energy analysis that puts such a different complexion on the economics of nuclear power stations. Now that energy has become so dear we have begun to take account of the loss to nature of the substances that provide us with it.

Since the war natural fibres, such as wool and cotton, have been increasingly displaced by synthetic ones. Natural fibres decompose: they do not become waste. Try putting your wife's nylon stockings on the compost heap; when burnt they pollute the air. Natural fibres are better adapted to the body's needs. In the summer heat they absorb sweat and in the winter they keep us warm by trapping air. People are rediscovering the pleasure of natural fibres and their healthiness at the same time as they are becoming aware that synthetic textiles are a source of pollution, both in themselves and in their productive process, and that they waste energy. They also cause widespread unemployment in the poor countries where the natural products are usually found.

Nylon requires as many as ten steps of chemical synthesis, each needing considerable energy to overcome the entropy associated with the chemical mixtures used and to work the reaction apparatus. By contrast the energy required for the natural synthesis of cotton is obtained free from a renewable resource – sunlight. The natural process does not pollute, while the artificial one does. Moreover, the raw material for nylon synthesis is petroleum or a similar hydrocarbon, non-renewable resources. The natural process draws upon a constant flow of natural energy, while the artificial one reduces a limited store.

Nylon production (unlike cotton) requires chlorinated intermediaries as a reagent. A considerable proportion of chlorine production is carried out in electrolytic mercury cells: at present the manufacture of chlorine releases mercury into the environment. Thus, if we follow the chain of production through, we find that the substitution of nylon for cotton has increased the pollution of the environment as well as reducing energy resources.

Detergents, aluminium, plastic bottles, aerosol propellants, cars

We can list many other postwar technologies that have similar effects. Detergents generate a more intensive environmental impact than soap, since soap is wholly degradable to carbon dioxide; even the new degradable detergents leave a residue of phenol which is a toxic substance alien to the natural aquatic system. The phosphate in detergents increases the growth of vegetation in surface waters, but soap is free of phosphates.

The displacement of steel and timber by aluminium adds to air pollution, for 29,860 BTUs of power are needed to generate sufficient electricity to produce a pound of aluminium as compared to 4615 BTUs per pound of steel. The substitution of disposible containers for reusable bottles is another example of polluting waste, and one that is impossible to justify. It cannot be said that it is strictly affluence that has led to this additional burden of litter, but the convenience of industry. Aerosols are a further instance of pollution that has no justification but convenience. These propellants are now banned in the US because of their destructive effects on the ozone layer,[7] a layer of atmosphere that shields us from lethal radiations of the sun. All these destructive impacts on the environment by modern technological processes are often described as the technosphere destroying the biosphere. Man cannot survive without the biosphere. The extent to which the technosphere increases our wealth has to be set off against the extent to which it damages the biosphere. Whether supersonic aircraft and aerosols cause chemical reactions in the ozone layer that dangerously weaken it, we don't know. But there can be no doubt that any process which did so, would have to be absolutely banned or human existence would be imperilled. In this way the biosphere imposes absolute restraints on human activities. But at what degree of pollution this happens it is often difficult to estimate. But as Prof. Ubbelohde said, when that point is reached, then the saturation level is reached and the development must start to decline.

The pollution caused by the emission of chemical wastes, especially lead, through car exhausts has been increased for no better reason than to make automobiles more prestigious for the buyer. The horsepower and compression ratio have been steadily raised since the war, so the owner can boast of increased speed (which exceeds the legal speed limit) and a more silent, though powerful, engine. The compression ratio has augmented the amount of tetraethyl lead needed as an additive in order to suppress the engine knock that occurs at high ratios. Nearly all this lead is emitted through the exhaust into the air. Lead is not a functioning element in any organism, and is toxic. There is considerable evidence that it causes brain damage to infants and retards the mental growth of school children. In adults it causes severe depression.

It is clear from the above examples that a lot could be done to reduce pollution without a radical threat to our way of life. Indeed we should improve it. Taking action is a social problem and nowhere is this clearer than in the obstinate determination of successive governments to give roads and road traffic outrageous advantages over rail and water traffic. For example, roads are financed out of revenue whereas almost all other social necessities such as houses, hospitals, schools and railways must raise their capital by means of loans that have to be repaid at high interest rates. It is true that railways are expected to finance themselves by fares; but how many of our feeder roads would survive if they were financed by tolls on motorists? As motorways, costing up to $7.6 million a mile to construct and then requiring heavy maintenance costs, do not have to repay loans, let alone the interest on loans, this acts as a massive hidden subsidy. It has been calculated that if the interest on the money spent on roads was given to the railways, we should be able to travel free.

Industrial pollution and climate

On what evidence do scientists believe that we could not burn all the coal reserves without suffocating ourselves?

Unfortunately the long term effects of inadvertently changing the climate by our industrial activities have not been a fashionable subject for our scientists and engineers. According to one of the leading research workers on this theme, Professor Gordon Macdonald of the University of California, there are even in the US only a handful of small research groups studying in a professional way the influence of man's activities on climate. Considering that the alteration of climate, inadvertently induced, could make the planet uninhabitable, encouragement for research in weather modification would seem to be a priority for survival.

What are Professor Macdonald's own views?

In an article written for *ENVIRONMENT, pollution, resources and society* (Sinauer. Stamford, Connecticut) in 1971, he says that the magnitude of the changes produced by man are of the same order as those produced by nature. He gives as an example that the carban dioxide added to the atmosphere by the burning of fossil fuels can bring about a change of several tenths of a degree in the average temperature; and a change of this magnitude has been observed naturally. Such alterations produced by man, he says, can no longer be regarded as local. Direct heat input by a city changes the mircoclimate of that city. The combined effect of many cities can change the climate of the planet.

'Our understanding of the physical environment is sufficient to identify inadvertent modification, but it is far too primitive to predict confidently all the consequences of man's unwise uses of his resources.'

Theoretical predictions based upon observed changes in temperature related to man's industrial activities indicate that we might boil, freeze or drown. Few of us think very often of the fact that both ends of our planet are packed with sufficient ice to form a layer fifty feet thick over the entire surface of the earth! Ice has played a dominant part in the evolution and development of nature and human society. Polar ice might either extend into the continents again as it has in the past or it might melt and inundate the continents. We should respect the Poles for we might place our fate in their hands!

The climate of the world is an integral system largely governed by its heat balance. Heat flows as we know from a higher temperature to a lower one as if endeavoring to equalize temperatures. Differential heating is then obviously very important, especially the degree of contrast between the equator and the Poles. Raising the overall temperature of the planet a degree or two would speed up the circulation patterns. It is possible that this has increased the violence of some hurricanes in parts of America. Contrary to commonsense one effect might be to increase the intensity of winter by blowing the arctic cold further south and antarctic cold further north. On the other hand if the rise in temperature were sufficient it might melt the poles and raise water levels throughout the world. This uncertainty points to the need for increased research.

If the temperature fell, as a result of the particles and dust cast into the atmosphere by man's industrial and agricultural activities, then the ice might extend. We know that dust and particles can dramatically effect climate by the consequences of the eruption of volcanoes. The Krakatoa eruption in 1883 lowered the temperature on a global scale. The ash in the atmosphere joined with the clouds to reflect back the sun's radiations before they reached the earth. (This reflection back is known as the albedo fraction.) The dust thrown into the atmosphere by eroded soils can lower the temperature over a wide region and decrease crop yields in fertile areas. At a time when large regions of the world are threatened with famine this increase of the albedo fraction would be catastrophic.

According to Professor Macdonald fluctutations of temperature between 4–6 degrees centigrade either way can result in the melting of the ice caps or a new ice age. How possible is this? We have to remember that while man's contri-

bution to natural change may be compa
though it would seem that in this case it is
nature's – it can tip the balance and overthro
regulation, as in the case of the pollution of l
inforcing a natural trend, as in temperature (
hasten the final result.

To quote Professor Macdonald: 'If the present rate of energy increase of 4% per year is maintained (a doubling time of seventeen years) then in 200 years artificial energy input into the atmosphere would equal one half of the radiation balance [the heat from solar radiation reflected off the earth's surface.] This level would be reached in only a hundred years with a ten percent yearly increase of energy. As Budyko [a famous Russian meteorologist] has argued, an increase of only a few tenths of one percent in the radiation balance (0.2 to 0.6 watts/m²) would be sufficient to cause the melting of the polar ice. With a doubling time of seventeen years, we could increase energy production 25-fold in about eighty years to reach an artificial energy input of 0.2 watts/m².

The combined effect of carbon dioxide pollution [which creates the so-called greenhouse effect by trapping the heat reflected back from the earth's surface so that it cannot escape into space] and direct heat pollution is strongly in the direction of warming the earth's atmosphere. On the other hand, urban and agricultural pollution tend to lower the earth's atmosphere by casting dust and particles into space. Which pollution will win in the end? Will we drown or freeze? These will become critical questions in the not too distant future.'

Changes in climate are one more of the good reasons why we should restrain economic growth, so far as growth depends upon our present industrial technology. To quote President Carter: 'More is not necessarily best.'

Philosophical considerations raised by modern technology

The experience of biologists shows that no living organism can expect to increase its numbers indefinitely without re-

ints being imposed by the environment. This generalization is illustrated, as we have seen, in the bell shaped curve.
The trouble is that science does not give us 'facts' in the hard
material sense : it gives us 'laws', but they are not laws in our
legal sense. These laws are assumptions that have a high
degree of probability and since they are created by nature we
cannot repeal them. If a scientific generalization has a high
degree of verification it is futile to oppose it. For example,
when the laws of thermodynamics demonstrated the impossibility of perpetual motion, invention took a leap forward,
though, on the face of it, this was a discouraging discovery.
By accepting it, inventors did not waste more time trying to
design the impossible. The laws of thermodynamics still have
important lessons to teach politicians and industrialists,
namely that unlimited growth produced by machines, our
energy slaves, is impossible. It runs up against entropy. So
that biological generalizations and the laws of physics alike
support the contention that development must have carefully
estimated limits.

Until we accept this we shall not start to design sustainable
societies. It is doubtless a nice thought that we are evolving
from ape to superman, but it is nature's game and if we don't
accept nature's laws, which no legislative act can repeal,
we shall have perished before we have attained our evolution.
ary potential. It is a highly dangerous assumption that nature
has been designed to raise man up to God by an evolutionary
process with which we don't have to concern ourselves !

Entropy and resources

Thermodynamics is the study of the general laws of processes which involve heat exchanges and the conservation of
energy. The first law is that energy can be neither created nor
destroyed and this suggests at first glance that there is a constant amount of it on which mankind can draw for ever. Our
laws of economics were formulated when this was vaguely
considered to be the case. But it is nonsense as anyone can see
for himself when he puts a lump of coal on the fire. The same
lump of coal or gallon of oil cannot be burned twice and

though fossil fuels may be still forming it takes millions of years to make them. Once disintegrated into its chemical parts by burning, the energy that has been used by man to warm himself or drive his machine-slaves cannot be used again, although there has been no loss to the terrestrial system.

If it were true that machine energy could be perpetually recycled without loss of material substance and without pollution of the environment, there might be some plausibility in the idea of perpetual economic growth. The second law of thermodynamics explains why you can't burn the same lump of coal twice, have your coal and burn it. Burning substances to release heat breaks down their biological structure and the exchanges between the burning body and the chemistry of the atmosphere must involve loss of energy in terms of work and cause pollution as well. When burning, fuel is disintegrated into random disorder that makes it impossible to reconstruct the original structures so as to break them down again; the process is irreversible. If the substances burned to produce heat are renewable, such as wood and plant products, this would not matter so far as energy loss is concerned, provided no more substance was destroyed than nature was able to regenerate. But however great the renewable resources available for combustion, it is pollution that must impose the ultimate restraint.

Life counters entropy. Life strives to reduce the loss of energy to a minimum; obviously if all energy becomes random and is no longer contained within biological structures, there could be no life. Processes that match natural systems are therefore the least entropic. Our use of energy makes us an increasingly entropic force in the world. Hence the slogan DOWN WITH ENTROPY is used by the protestors against the expansion of our economic growth.

Principles for redesigning our society

The principles of ecology are becoming political principles; beneath the surface of orthodox political life, new groups are forming with new revolutionary ideas. The situation in which we now find ourselves demands a revolution in the idea of revolution. We know that the Faustian bargain with which the industrial revolution began cannot be won; the environment is more powerful than Mephistopheles. H. G. Wells suddenly appreciated this fact with overwhelming horror when he wrote one of his last books, *The Mind at the End of its Tether*. Wells had been a great enthusiast for modern technology. But in this book he said that if he concentrated his attention on the relation between industrial production and nature he knew that our civilization was doomed. Nature could never sustain the demands we make on it. The only reason everybody did not share his views, he said, was that few people are capable of concentrating their thoughts : they live by their natural animal optimism. In the second part of his book he admitted that he too was only able to continue with his life because he had pleasant dreams and irrational hopefulness.

Like many of his generation who trouble to think Wells found himself in a hopeless *impasse*. Was he right? From the purely rational point of view I think he was. If we cannot admit any new dimension or factor emerging from below into our lives I do not see how we can change course, even when we see disaster ahead. We have no motive for doing so that is sufficiently compelling to change our daily actions. We shall go as blindly into ecological disaster, led by mediocre politicians, as we muddled into the two world wars. But this next crisis is far greater than any that has confronted industrial civilization so far. We have reached one of the great turning points in world history.

The great changes in civilization have always been unpredictable. Those who lived before them could not have anticipated them. Take farming, for example, perhaps the greatest of all human revolutions, for without it we would not have civilizations at all. Have you ever noticed that it is Cain, the first grain grower, who was also the first builder of a city? And his descendants have continued to murder their brothers, the more primitive societies, ever since. Perhaps they will realize at last that what these societies know about living with nature is more valuable to us than seizing their land. But how is it that Adam, who was condemned to agriculture, according to the myth of *Genesis*, for eating of the tree of the knowledge of good and evil, did not appear on the scene earlier? There had been a million and a half years of hunter gatherer societies. Why should man suddenly discover farming after all that time? We don't know. But it was a totally unpredictable change in human existence. The first digging stick in the ground was practically simultaneous with the explosion of the atom bomb on such a time scale.

Nearer our own time who would have predicted in the Middle Ages that the wars between the religious sects that, after the Reformation, looked as if they might tear Western societies into pieces, would have been calmed by the emergence of natural science? Science distracted men's attention from the competition between creeds and made them more interested in the nature of nature. This in its turn led to the industrial revolution and new threats to Western life. What will emerge to save us now?

For my part I think that the saving dimension will be an expansion of consciousness and self-knowledge. This expansion will have several sources, but one of them will arise from the application of the scientific attitude of mind to our own psyche, including the interpretation of dreams, extra sensory perception and a clearer understanding of the psychic unity of man and nature. It is this that will help to provide the motive for the application of ecological principles to industry and development. As these discoveries spread, people will become so excited by them they will tend to forget about material standards of living and seek new satisfactions. The material changes required, which frighten us now with our conviction that only money can save us, will not then seem insoluble. In his Introduction to *The Secret of the Golden Flower* Carl Jung wrote :

> The greatest and most important problems of life are all in a certain sense insoluble. They must be so because they express the necessary polarity in every self-regulating system. They can never be solved, but only outgrown ... This 'outgrowing' ... on further

experience is seen to consist of a new level of consciousness. Some higher or wider interest arose on the person's horizon, and through this widening of his view the insoluble problem lost its urgency. It was not solved logically in its own terms but faded out when confronted with a new and stronger life tendency.[8]

What happens to the individual can happen to society. The change in the idea of revolution will be that the next revolution, the second revolution, will be to foster the new life tendencies, some of which have been described in this book, now emerging into our social consciousness. The old sterile problems of economic man will be left behind and forgotten simply because they are too boring for people who have drunk the new wine.

REFERENCES

1. His paper was reprinted in the *New Scientist* September 9th, 1965.
2. The wheel illustration is advanced by Harry Walters in *Ecology Food and Civilisation* (Charles Knight, 1973).
3. Reported in *New Scientist*, Vol. 72, No. 1032, p. 750.
4. G. Mellanby, *Can Britain Feed Itself?* (1976).
5. On the theme of the UK farm structure see my *Taxation and the Future of the Farm Structure* (*Ecologist*, June, 1976).
6. The examples of pollution by modern technology are taken from *The Origins of the Environmental Crisis*, an address given by Barry Commoner, Director of the Center for the Biology of Natural Systems (US) to Council of Europe, July, 1971.
7. *New Scientist*, Vol. 68, No. 696, p. 7 et seq, contains a detailed discussion of this hazard.
8. Quoted in *The Dream Game*, Ann Faraday (Penguin, 1976).

CONCLUSIONS

I INTRODUCTION

Before we discuss the implications of all that has been said in this book, it is vital to take a deep breath and ask ourselves in all sorts of terms what all this is likely to mean for us all in the future. Even to decide upon the appropriate parameters in which our thoughts should be incubated taxes the imagination.

We know, before we start, that from our analysis of the ideas and the forecasts we have heard there will be a very large number of priorities involved, some for us to decide and some decided for us. If we do X we cannot do Y, and we can easily caricature such a situation : should a nuclear super-bomb be developed and exploded then all other forecasts would cease to be worth considering. We have to accept the fact that one invention or development may well eliminate another. A possible example may be one from the field of transportation. We might expect at the moment that hypersonic jets, nuclear powered high speed trains and the like, may cover the globe and provide our main means of travel. But if the complete 'sensory telephone' were developed, then we could, for all practical purposes, be with our (business) friend in Australia without leaving our study in Oxford. This would mean that one of the otherwise practical needs that we would remove is the further development of supersonic jets.

So priorities of one kind or another will influence our ideas about what will actually happen and what we will help to happen; these are the vital matters we need to decide, and decide quickly, in view of our limited resources. One thing which will certainly occur is that communication's huge strides forward should help the dissemination of knowledge. Such a dissemination of knowledge must inevitably help in turn to dissipate some of the alienation and misunderstanding that occurs between people, or peoples. This is true of those many who have failed to understand each other because of the lack of a common language – either literally or *conceptually*. It is lack of understanding, through lack of knowledge that is the basis for suspicion which in turn is the basis for so much confusion : people fear what they do not understand.

We are concerned with the development of science and this means with prediction, understanding and control. It is the last, from a social point of view, that especially concerns us. We are involved with control in all its aspects : *how* we control, *what* we control and *how much* we control. This control is invested in (or is acquired by) individuals, groups, whole countries and continents and by all people, eventually on an international basis – or will be some day. This we must bear in mind all the time we try to assess the work of our contributors in this book.

II THE CONTRIBUTORS AND THEIR WORK

Let us now, before we attempt any overall assessment or evaluation, consider the particular contributions. Let me take them in the order in which they occur in the book. An order that is not accidental and progresses from computers and the information sciences to medicine, surgery, genetics and then on to space research. An application in defence, with reminders of war, takes us on to the psychological fields which leads us, via a detailed study of energy sources and foodstuffs, to the very practical problems of transportation and finally a more philosophical reflection on the most basic of subjects – our resources.

At first blush one might expect the world of computers and artificial intelligence to present the most devastating, even horrifying, of all the advances in modern science. Perhaps few will immediately think of it as horrifying – although it has its potential horrors – but whatever one's view of the future, some of the other articles in this book are certainly more hair-raising in their forecasts.

The next stage in computing

Dr. Christopher Evans starts with a discussion of the history of computing which serves admirably to remind us that the digital computer, which is having such a dramatic impact on our present world, is something that was conceived of, and could reasonably have been made, long before it was. Charles Babbage designed the essential features of the modern digital computer well over a hundred years ago, and the calculator goes back well before to Pascal and Leibniz.

It goes without saying that the abacus and the very process of counting on fingers and toes that preceded arithmetic, held implicit in them all the essential ingredients of the modern computer. The lesson here is, and we meet it more than once in this collection of essays, that there are a large number of modern scientific inventions and developments which were dependent upon basic – often simple – facts that we knew a very long time ago. This helps to remind us that the actual development of science – now in a state of flood – will often depend on apparently incidental things. In the case of computers, this dependence was, among other things, upon the development of precision engineering.

The speed of development of the modern digital computer has been astronomical. Speeds of operation have increased exponentially and size and cost have come down in much the same way.

The problems of computing were once mainly in the software – data preparation and the like – but now with the development of so many computer languages – often of a universal kind – the problem of speeding up the slowest ship in the convoy has been solved : the whole convoy is now enormously fast.

One of the main implications of computing is that it opens up so many new vistas. We are able today to tackle problems that even yesterday were utterly unthinkable. Indeed much that has been said in the rest of this book owes its existence to the possibilities of computer-usage. Science has many bases, but common sense perhaps still takes priority, even over electronics, sophisticated apparatus of every kind and financial investment; the computer though must be near the apex of the whole edifice.

But all aspects of computing, as Dr. Evans points out, are not equally beneficial, some are distinctly sinister. The 'data bank society' is a phrase which has become very familiar. When we learn, that with the aid of modern data processing, we can keep a summary of the details on every living person and that that summary can be made available to a data bank user, more or less instantaneously, we begin to glimpse the enormity of the problem.

Methods of detection, credit-worthiness, reliability, suitability for jobs, health record, etc., are all potentially part of the universally available data bank summary. Such information lends itself to devastating abuse. It must be made possible to control and canalise such information to make sure it is available only to the correct sources. Even to try to decide when or what are the correct sources is itself a daunting problem.

There is a clear cut threat to humanity from what has been called the Information Explosion, especially in the form of the resultant Information Pollution. We have first to fully realize the dangers and second to try to sensibly 'defuse' the threat. The difficulty is the usual one : 'power corrupts and absolute power corrupts absolutely' – control is power, and the very people who should control the information explosion are the very people who could use it for their own power. It is precisely because of this that the threat of totalitarianism lies over us : but more of this later.

Artificial intelligence is the other large arm of computer development. We have learned how to synthesize and simulate human capabilities. We can make machines that see, hear, think, learn and speak. They can make judgments, formulate plans and generally could *now* do the job of the most intelligent human being.

What started out, in effect, as the search for a blueprint for organisms – including human beings – has become vastly different. The accumulating knowledge in the fields of cybernetics, systems engineering and automation, make clear that we may well speed up enormously the development of biology and medical science, but at the same time have other less obvious but even more important impacts.

Ths 'spin off' for biology and medicine links in well with the work described in this book under the titles of medicine, surgery, and genetic engineering. To this extent they represent the best use of science, even if they hold out the hope (or threat) of immortality at the end of the road.

The other side of the coin is that of manufacturing an artificially intelligent species which will take advantage of the data bank society, of automation and our gratuitous gifts of knowledge and intelligence. Unhampered by the size of its skull – it can be as large as it pleases – it may, since it understands *the principles*, easily surpass us in magnitude and therefore in ability – perhaps not just a little more intelligent, but more intelligent by a factor of 10 or even 100 or more : our fate would then lie in the hands of such a species.

Computing is seen as fundamental to almost every aspect of scientific development, in making 'astronomical' problems feasible to tackle. It has also, by the same token, contributed to the social threat, both through data banks and the machine species (UIMs).

Electronics and automation

Before we return, as we must eventually to consider this matter further, let us look at Electronics (this second article is the work of Brian Jennings).

The most important development in recent years has been that of the microprocessor, because it lends a new dimension to the whole field of computing which is the power behind automation. We have developed, in the past, so called analog computers (servo systems) which work on the basis of negative feedback. They were used for GEORGE – the automatic pilot – and were used for automatic controls of all kinds, including simple thermostats which controlled the air and water temperature in the home. From such beginnings automation started and has grown and grown. The digital computer became more and more important and now increasingly tends to replace the analog computer.

The limits on the field of automation are – like with most of our subjects – virtually unknown. The field is apparently limitless. We see the household being automated; the possibility of all heating, cooking, entertainment, even sleeping and resting, being automatically controlled is now obvious. Only the financial consideration stops it becoming an immediate reality. This is a barrier we shall meet again and again.

The significance of such domestic automation is that we may have to ask ourselves how we spend our time. It will no longer be necessary to be domesticated, since the foodstuffs will be delivered into the back store input (like a more sophisticated fuel dump) and the food will be processed and cooked automatically. It could even be that the choice of food will be automated, thanks to market research, or even extra-sensory perception.

If our jobs – factories and offices – are also automated together with transportation then we may be hard put to find anything at all to do, except for professionals – lawyers, accountants, scientists and the like. But even these jobs could be automated.

One of the most important joint developments of computers, cybernetics and automation would be to produce the automated thinker who could be any sort of professional you want; here yet again we see the emergence of the new all-powerful machine species.

On the shorter term we may expect to find machines capable of automated entertainment, with or without television. This development could facilitate automated crime as easily as a music hall performance : the sword as ever remains two-edged.

By automated entertainment, we mean supplying methods of satisfying our needs as soon as those needs arise and automatically initiated by those needs. Aldous Huxley suggested the 'feelies' and Dr. Jim Olds has shown those parts of the brains than can be stimulated electronically to provide such pleasureable feelings : all we need is to be 'wired up'.

Automated crime is largely crime by computers. This is now so well-organized that even the plans for a large-scale crime can be simulated and fed into the computer and tried out and tested before being put into operation !

Equally the enormous growth of credit-cards has already resulted in multi-million pound frauds and the major computer companies are having to devote very large budgets to making their hardware and software secure.

Automated education already exists in the form of teaching machines and programmed books. The present developments in this field, where all the teaching processes are under computer control, could radically change our whole educational system.

The need for security is also evident. Electronics can be used both in the need for security and by providing the source of that security. In the broadest terms we need to worry about automation in just the same way as we worry about data banks, cybernation and the emergence of the machine species. The key is control and the problem is as to who controls the controller and how much control should be exercised and where ! Electronics is handmaiden to automation and both are handmaiden to cybernetics and play a fundamental role in buildings, among a host of other things, UIMs.

The medical world

The next two articles which are on Medicine and Surgery (by John Newell) and Genetic Engineering (by Susan Goodman) can be taken together, because of their obviously interlocking nature.

The idea of genetic engineering is well conveyed by the title. We can interfere with genes and their structure, reconstruct genetic forms and eventually construct them *ab initio*. The names DNA and RNA have become – thanks in some measure to Crick and Watson's discovery of the double helix structure of DNA – bywords of the language of science.

The individual man-made gene first saw the light of day in 1976, and from it we may expect to see big advances in the control of disease – cancer is especially the target – and of course linking up with the work on artificial intelligence, in the eventual construction of artificially intelligent systems dressed in the same colloidal protoplasmic fabric as man.

The disease control and pain control aspects of genetic engineering are especially important on the short term and the difficulties encountered in the use of enkephalin are exactly the type of thing one might expect to meet in these sort of developments. The advantages of an anaesthetic of enormous strength, the benefits from the use of antibodies from immunization are obviously immense and could clearly be achieved on the short term. Less obvious and yet in keeping with the whole feeling of our entry into *Brave New World* is the use of genetic engineering to multiply the production of livestock and crops. The most important aspect of all is that, with the ability to control disease, there soon looms up again the possibility of immortality, something that threatens also from the development of artificial or man-made artefacts and also linking with John Newell's paper *Medicine and Surgery to 2001*, in particular with spare-part surgery.

Heart and other organ transplants are now thoroughly familiar, and in a way, less dramatic, is the capacity to remove an organ from the body, keep it alive – by freezing say – clean it, or repair it and then sew it back into place to carry on with its job. Then we can transplant intestines into an artificial kidney, we can transplant kidneys themselves and can of course substitute increasingly efficient artificial kidneys for the real thing.

What can be done for the kidney can be done for any organ, although when we come to the brain and the nervous system the difficulties are far greater than they are for the other organs. The question of memory arises because at the point where you substitute one brain for another you have to ask whether you any longer have the *same* person. There are enormous moral difficulties with brain transplants which do not arise with other organs.

The question of drugs and chemicals plays a vital point in transplant work but is also featured in behavioral matters involving the cure of psychotics (such as schizophrenics) and the improvement of learning and memory – nootrophyl is one of the drugs which could play a prominent part here. The whole question of making better use of our brains is a matter that is bound to come under intensive study.

Blood swaps prevent heart attacks, jabs prevent tooth-aches, drugs can stop infections – you name it, it can be done: sooner or later. The question is not only one of time and priorities, but, as John Newell points out, also one of money. It would be very sad if the best that could be done to alleviate the medical problems besetting people failed be-cause of lack of funds. To know how to save someone's life and not be able to do it for lack of money is the ultimate frustration.

Dr. Paul Berg's appeal in 1974 for a moratorium on dan-gerous genetic experiments is a matter of great significance. It represents one of those rare flashes of realization – for-tunately increasing–that the implications of scientific research for society are enormous. We have already met this aware-ness in the case of the dangers of data banks to society and shall meet it again in a variety of different ways. The logic of the situation is quite simple and can be illustrated by the example of inoculations. If you inoculate someone against a disease you normally stop him getting it and expose him to a minimal risk. If that situation changes and the likelihood of acquiring the disease is slight and the risk in the inocula-tion high, then, on balance, you stop using the inoculation. Exactly this sort of situation has arisen with whooping cough. The inoculation against whooping cough carries a small but definite risk of brain damage, but the chances of getting whooping cough are now almost zero – the disease has been virtually eliminated – so the balance of advantage has changed and the exercise of sensible medical control becomes obvious. The short-term impact of these medical advances is obvious and disease is clearly on the run, but the longer term link, with longevity, the population increase and more leisure, as well as supplying the flesh for our UIMs!

Flight into outer space

Our next topic, astronomy and space science is another of those which has had a good press for several years now. There can be few people in the world who are not aware of Moon landings and Mars probes.

For obvious reasons astronomy is one of the oldest, if not the oldest, science. Looking at the moon and the stars, and being even more intimately involved with the sun, goes back to the very beginnings of human awareness.

In modern radio-astronomy, as Peter Beer describes in *Space Research and Astronomy*, nothing more kindles the imagination than quasars – sources of radio waves that are moving away from us at 90% of the speed of light. Pulsars, small pulsating stars with their emissions, were an even more immediately exciting phenomen, because of the possibility of their emanation from an 'intelligent source'. Now we know that pulsars have natural sources and so the main focus of interest has switched back to quasars.

The importance of quasars lies in their 'hint' – it can be no more – as to the size of the Universe. This raises again the whole question of dimensions – in this case distances. The moon is 240,000 miles away. Mars is 200,000,000 miles away, and the Barnard's Star (a recent acquisition!) is 36 billion miles away – that is 36,000,000,000,000 miles! Light waves from quasars started 20,000 million years ago and since light travels at 186,000 miles per second, they emanated from a source that is approximately 120 thousand million billion miles away – that is 120,000,000,000,000,000,000,000 miles. To put this in perspective, if that is remotely possible we say that if the one step towards the wall is the distance to the moon, Mars is about half a mile away, Barnard's Star would take us about four times around the globe and the furthest quasars would take us 16,000,000,000 times around the globe.

If, on the other hand, the quasars took us just once around the globe then the proportionate distance to the moon would involve a movement too small to measure.

We are a mere puff of smoke in the Universe and there can presently be absolutely no way of knowing where human intelligence stands in such a vast league of possibilities. The colonization of Venus may come next and this will be a huge, almost unbelievable step, but in the light of the known minimum size of our Universe, it makes it sound like going down the road for a drink.

War in the future

It takes time to recover from such astronomical dealings and come back to earth and deal with weapon research, although even here, as Denis Archer effectively shows in *Defence and Weapon Research*, the bulk of events are likely to be conducted in space. Intercontinental ballistic missiles (ICBMs), with the explosive power of 50 million tons of TNT and more, are enough to make one pause, and fortunately also make the potential user pause. The advantage of such deadly – near terminal – weapons is precisely that they are not used because everyone (including the user) will probably be destroyed.

In more earthy terms, we expect further progress in chemical and biological warfare and here we may hope for the setting-up of effective international controls, since such warfare could be just as 'terminal' in its effects and more likely to be used because the dangers are less explicitly recognized. One can hope here that the same principle of abstention, because of its chance of rebounding on the perpetrator, may also apply. This would place it in the same category as the use of gas in the last world war.

Money is a huge factor in defence science and perhaps the financial limitation could be a basis for international agreement and therefore control. At the moment it is significant that total military spending is of the order of $300 billion annually and research and development represents about one tenth of this total. The most interesting point is that some

80% of this is accounted for by the USA and the USSR and about 10% by France and Britain.

Denis Archer concludes that, even though the threat of nuclear war remains a real one, should such a war occur, the human race will survive. I find this only slightly encouraging because in some ways destruction through a nuclear war could be more final and more painless than the other forms of death and oblivion which haunt us. But the importance of space research and defence weaponry I see as being closely bound up with our future colonization in space.

The mental world of tomorrow

Our next two articles, *Psychology and Mind Control* (by Martyn Partridge) and *The Paranormal* (by Roy Stemman), also, at least to some extent, go together. Psychology clearly has brain and nervous system at one end of a continuum and the 'less material' fields of extra-sensory perception at the other. Psychology as the study of behaviour provides the link.

In his analysis of modern behaviorism Martyn Partridge talks of the unedifying comparison between Skinnerian theory and the idea of humans as robots who are at the beck and call of external circumstances. It is ironic that humans should be likened to inert robots at the same time as cybernetics is constructing highly flexible robots which make active human beings seem inert. Some hope for 'free will' is though still maintained, in whatever guise it appears.

The most important aspects of mind control lie in the short term possibility of the rehabilitation of the in·habitants of our prisons and others who are at odds with society. The work on sensory deprivation is now well known to be one of the main platforms on which brainwashing is based and brainwashing is, or can be regarded as being, a means of worthwhile change, even therapy, as opposed to the more sinister aspect of such work, which specializes in control of the individual.

Such research is inevitably first cousin to more neurophysiologically orientated research, where for example, it is possible to implant a miniaturized transmitter-receiver in the skull and have a monitoring control as remote as you please from the individuals so monitored.

The problem then becomes one of selectivity. Can you select out exactly the information you want to enable brain control of a positive kind as opposed to the easier negative control by blanket effects and destruction? The experimental work of Wilder Penfield, which is quoted in the text, provides an exciting reminder of what is potentially possible but has not yet been realized. This is clearly a field where huge advances are imminent. Hypnosis and crypto-amnesia are other techniques which are being developed and it is here that we start to rub shoulders with the paranormal.

The paranormal I described in the Introduction as being an outsider in terms of horse-racing odds. But as we said there 'if the horse does come in it will cause a sort of sensation.' The very fact that the Soviet Union seem to be placing such weight on paranormal research must provide a gathering weight of positive evidence. This is all the more so because of the very strong, even parochial, materialistic and behavioristic tradition which has so far pervaded Soviet science.

The modern equivalent of the famous Duke experiments conducted by J. B. Rhine are very much more content-orientated and less statistically designed. Even the work done by Dr. Lozanov in Russia, which was apparently well controlled, achieved about 70% efficiency in telepathic communication. The odds here against it happening by chance are less than one in a million. But if these odds are impressive, those derived from Helmut Schmidt's results are more so. The chances of his telepathic results being achieved by chance are one in 500,000,000.

One thing is certain and that is the communication involved in telepathic-type experiments are not a matter of coincidence. They are either 'rigged' in some way (even unconsciously – one remembers Hans the 'counting horse') or they represent an unidentified means of communication.

Eysenck's comment on the implausibility of the idea of 'rigging' so many experiments is much to the point : the evidence in favor of certain paranormal phenomena certainly seems to be growing.

Kirlian photography is now well known and all that is uncertain is its significance. We should note that this is an apparently separate issue from the question of telepathy and a failure to confirm these results would in no way impair the authenticity or otherwise of telepathy. Water divining, pendulum swinging and other such techniques have the same independent logical standing as Kirlian photography : we must simply wait and see. But positive results could dramatically change our other forecasts.

One type of activity dealt with here under the paranormal, which is perhaps more easily explicable than most, is that of faith-healing. It is not that faith-healers and spirit-healers provide *explanations* themselves that are necessarily correct (nor indeed always even plausible) but the fact that the effects occur seems very likely to be true, and the implausibility of some of the explanations should not cloud the issue as to the facts. Roy Stemman talks of the reasons for success lying in self-suggestion and talks of this as a 'skeptical' explanation. The plain fact is that if this is achieved by either suggestion or self-suggestion, the result is of the utmost importance.

One of the reasons for the relative slowness to recognize possible physiological or physical explanations or 'psychic phenomena' is the failure to appreciate that when we talk of moods, emotional states and the like, we are specifying a particular physico-chemical state of the body. There is a clear possibility that laser-type beams or other forms of radiation lie at the foundation of the phenomena. Indeed it seems that our basic ideas of 'action by contact' bred largely from the work of classical physics may itself have to be revised, or at least generalized, perhaps in the way the Newtonian Physics was generalized to Einsteinian physics. All that we have glimpsed about the nature of reality makes one increasingly sceptical as to whether or not we can say that all the explanatory concepts are already known to us. In physiological terms, we could argue that lie-detectors and even EEG traces would once have been laughed to scorn and yet they have become explicable and clearly understood in physiological terms – so it is that the Occam's razor approach is still correct : we must not multiply our entities (concepts) needlessly. In other words we should always seek the simplest explanation, even if we do not (or cannot) find it, nor indeed always decide what is 'simplest'.

If science has detected the soul, and some people view the paranormal in that light, then all the ideas of the conceptual framework of modern science will have to be revised, or at least the total context in which they occur. I do not expect this to be the case, but at the same time we do not rule out the need to explain the paranormal, nor indeed do we wish to rule out the possibility of souls. There is no need to emphasize the vital influence that the paranormal could have on our future, nor for that matter the psychological research. This last is two-edged : it helps supply 'mind' for UIMs, and helps – as do UIMs – to enable us to increase our own 'mental' control.

The vital role of raw materials

From the spiritual, or at least the mental and (in general) the more functional type of scientific study, we now move on to a group of three fascinating articles that deal very much with the material side of our lives. The first is on marine research by Carolyn Roberts and she calls it *The Living Sea*. The second is on energy by Arthur Conway and the third on food and population resources by Michael Pickstock. These all hang together as a story in so far as they reflect above all else, humanity's need for fuel sources for bodily survival in all its various ways.

Carolyn Roberts has us all on the edge of a different sort of chair with what, by comparison with much we have read, is highly plausible and therefore not difficult to accept. But few of us remember that 71% of the earth's surface is water and few of us know (so I suspect) that its average depth is about two miles.

The importance of studying the oceans is obvious, since the sea is indeed a 'treasure trove' and one we shall need in the gathering exponential demand for resources. The seas of the world are made up of 350 million cubic miles of sea-water and this *could* all be for drinking. More important is the fact that every cubic kilometre contains about 39 million metric tons of dissolved solids.

The fact that by 1980 one third of the world's oil and gas will come from offshore wells is probably not the surprise it would have been only a few years ago, but its importance and its significance for future marine mining is obvious.

Problems of pollution are now of course well known. We have already mentioned information pollution but physical pollution is even better recognized. The sea with oil slicks and radioactive wastes reminds us that we have too often used the sea as a huge garbage dump, and have also, all too often totally failed to realize the dangers of doing so. The general attitude seems to be – as unfortunately in most other aspects of scientific development – that we'll worry about it all tomorrow when it comes : the plain fact is that tomorrow *has* come.

The living sea promises enormous possibilities for the future and the manner in which it kindles the imagination is bound up with the wonders of nature and the novelties it presents. There may be more mysteries bound up in the oceans than in the whole of outer space – and, perhaps, a more immediate and obvious pay-off for the human race.

To get the maximum value from our marine research requires a great deal of investment in the various aspects of marine life, including the use of laboratory ships and expeditions, involving deep-sea diving etc. The notion of expeditions is part of the process which we so much wish to encourage – a process of exploration of new horizons which can provide the much needed motivation necessary for human evolution. It is the adventurous aspect of life that war used to provide and must now be provided in a more constructive way.

Arthur Conway writing on *Energy* gives us the picture which stretches our minds from idyllic dreams to the desperate search for survival.

It is of interest that he immediately strikes the planning and architectural note, describing the possibilities which range from the autonomous house to the fully integrated city. He poses them as either dreams or nightmares and in doing so he is right. People vary enormously among themselves; we have the full gamut of introvert to extrovert and differences of character and personality on other dimensions besides. These differences are reflected in different sorts of families with different ways of life. It would be an outrage if in the world of tomorrow we paid even less attention to individual differences : so by implication we may expect to pay more, much more !

It seems likely though that there will be enormous architectural advances, but within the compass of these we may expect to cater for the intentionally-isolated as well as the gregarious. 'Away from it all' could come to mean living on another planet, since Tahiti and Samoa are now on the jet airlines of the world. Escape would be away from the huge linear integrated city built around communication links – although the isolated house could also be *inside* the city where circumstances demand attendance for business purposes.

The power aspect of the accommodation is a different matter, but solar energy must be a front runner. Arthur Conway points out the possible need to go out into space and collect our solar energy and the Glaser island some 40,000 kilometres away in space (and fifty square kilometres in size) is an exciting possibility in keeping with our gradually emerging picture of the future.

Perhaps more effort – it is tempting to say energy – has been given to the tapping of natural sources of energy than any other single aspect of human life – from wind to water, from sun to moon. Here the Swithenbank idea of a floating chemical works over the sea-wells is another exciting possibility.

Most stimulating of all are thoughts about the great

'Hydrological machine' which is the cycle that is made up of sun and ocean and is driven by some 40 billion kilowatts – dimensions that confound the imagination, but make us keen to tap their source.

One conclusion that *does* come through strongly is that we would – in the light of so many possible energy sources – be extremely foolish to embark on Nuclear Fission Projects without the (currently missing) safeguards.

We must always eat

The importance of foodstuffs in the direct and most obvious sense is the subject of *Food and Population Resources* (by Michael Pickstock). The question of population is clearly paramount. In an age where the demands for higher standards is steadily increasing and the expectation of life is steadily rising, it is adding salt to the wound to find the population of the world will be half as large again by the end of the century – a mere 20 years away.

A glimpse into one aspect alone of our total problem is well put in Michael Pickstock's own words :

> Population increase and urbanization have caused changes and set up stresses in modern society which will affect nearly everyone who is alive today. Not for 15,000 years has the human race faced a more critical time.

The fact that 20,000 additional people are created EACH DAY is a very good example of a scientific triumph replacing one huge problem – death from disease, starvation and childbirth – by an even larger problem. This larger problem infiltrates all aspects of the search implied by this book, but the main and most immediate task is that of feeding everybody.

The agriculturalists are certainly making their contribution. They can perform magic, including making high-yielding disease-resisting varieties of plants and they have developed pesticides allowing greater control of our agrarian environment. The range of animals which are important sources of foodstuffs is better controlled. All this has not happened without bringing about a new sort of threat. The threat of loss of health through inappropriate diet is symptomatic. So, as usual, the context in which scientific development takes place, turns out to be all important. It is the Robert Waller warning which we have on our plate again. We shall not attempt to deal with this here in any detail, but we will issue the general warning that the interrelatedness of life – a multiplicity of casual issues – is such that whenever we introduce a change we must also look at all the other things we change at the same time. It is this sort of problem that makes it so difficult to see the implications of our plans and decisions. We need to worry about the 'side effects'. All these articles on the sea, foodstuffs and raw materials are interrelated and, with their complicated feedbacks, contribute to the uncertainty of the future : they are, above all else, problems which we *must* solve in order to survive.

The future of communications

The next article is on *Transport and Communications* and is written by Arthur Garratt. It is clear that transportation could be regarded as a link in the communication chain.

The ergonomic bicycle and the high speed trains are natural developments – as are larger jumbo-jets and hypersonic aircraft. Lunar probes and space flight provide the dramatic jump from speed around (or upon) the earth to transportation to the stars and planets.

No discussion of space-time travel would be complete without some reference to extra-terrestrial communication and this would not be complete without reference to Pioneer 10, launched in 1972, it will not enter the planetary system of any other star for over 9 billion years, so the Cari Sagan message it takes with it had better be of lasting value. The whole problem of transportation relates to all that we have said so far in that it is mostly electronically dependent, automated, consumes raw materials and on the other side of the coin makes the future exploration of the universe, as well as our own earth, possible and this is vital to survival.

These are hugh problems to solve in terms of structures and fuels but the conceptual problem is solved in the sense that, provided the incentive and the money are both available, the technical ability will quickly emerge to make such changes possible. This in turn leaves us with the vast problem of deciding how to reconstruct our cities, countries and policies around the development of such enormous dimensions.

The fact that communication is more than language, and also involves gesture and manner, for effective understanding has always meant that physical transport was vital and the only question is whether or not that importance could diminish in the light of other forms of communication – is telepathy too long a shot?

Thinking about resources

Now we pass to the last of our exciting series of articles. We go to Robert Waller on *Physical Raw Material Resources.*

Robert Waller deals, in a philosophical vein, with his analysis of what amounts to a warning of what faces us if we exploit nature and risk disturbing our ecological norms. His thesis is as follows :

> When, for example, we in the West discover how to synthesize a natural material artificially and put our energy slaves to manufacturing it, we cause widespread unemployment in other societies that were exporting the material wealth.

The argument is that such a step widens the gap between the 'haves' and the 'have-nots'. The question is whether or not we can offset the unfortunate (to some) effects of such a step, since to synthesize 'artificially' is exactly what we are inevitably going to do.

The first point here is to consider how we can acquire the resources necessary to keep alive (and fit) an exponentially increasing population. Naturally, population control both through laws and by use of contraceptives – another field for extensive research – will help. But we ought to face the facts that the population will grow exponentially for at least as long as our forecasts extend in time. If the problem disappears, then so be it, but we must expect to take such steps as farming the seas and synthesizing resources and, as a result, our plans must be laid now.

The argument is that we depend upon 'energy-slaves' and we may not have the necessary 'food' to feed them and as the slaves die off so do we. This would almost certainly be true if we relied on natural resources, and is precisely why we are involved in synthesis. We know well the material risks that this entails, since our evolution is bound up with processing those features that are opposed to our survival : the idea is quite familiar.

If we have no exposure to disease, we do not build the necessary anti-bodies to protect us from disease, so a vicious circle could arise if we did not take steps to inoculate or in other ways avoid the situation which is typified by the synthesis of 'too pure' foodstuffs. Seemingly the flavour and the 'goodness' are inextricably bound up with bacteria and threats to our survival. It seems that at all levels we evolve by opposition and if science removes such opposition it spells the doom of the human species.

Professor Ubbelohde's view that we will throw away our energy slaves and return to a simpler style of living is probably quite right, and we shall argue that it is good advice, not perhaps so much for physical or biological reasons as for psychological ones. More of this later.

There is no doubt whatever that Robert Waller has touched on another vital truth when he says :

> The alarming aspect of exponential growth, if it is unrestrained, is the suddenness with which a safe situation can become an uncontrollable one ...

It is for this reason that we will synthesize foodstuffs and energy sources, and must try to do so within the control of some international body. It is precisely this sort of international control – only eligible to deal with *vital* matters – to which national private enterprise should be subject. National control of the 'welfare state' kind spells doom for psychological and political reasons. The problem of deciding what is 'vital' should occupy a lot of people for a long time.

The problem of living space – of sorts – can be exaggerated. It is theoretically possible that every living person, standing shoulder-to-shoulder can be put on the Isle of Wight, or very nearly. While admitting that we would not wish to survive in such cramped circumstances, it does look as if, particularly using that extra 71% of the earth's surface in the form of sea, that we might be all right for space for quite a while yet. It is difficult to believe that every person needs so much space to survive in a highly technological age. The uninhabited parts of Russia, Australia and America alone provide ample space for the time being – the cloud that hangs over us though, it must be admitted, is of the exponential variety.

We can, as we know, by having marine cities and marine communication links, provide a huge increase in living space, but again it is not space *per se* that provides the problem – it is the provision of energy and food for an exponentially increasing population. It is not only that there are more people but those who already exist live longer and want more and more, whereas the likelihood is that there will be less and less!

The time has come now to look beyond these remarkable projected developments and consider their implications for society. The social implications are not only in terms of the sort of world we must envisage but the sort of life we are to live in it and the goals we are to set ourselves.

If we start at the level of nuclear war we have a well known yardstick. If a war is avoided it will be because a common awareness of the horrors makes international agreement possible. The sort of conditions which might cause it, in spite of all, are those where one group is forced by circumstances to use nuclear weapons (which may escalate) because they have 'their backs to the wall' and have nothing to lose, so that if they have the means they destroy the world. It is questionable, as has been suggested, whether civilization as we know it would survive.

The immediate effects of the bomb may not be to obliterate every person, but the aftermath of radiation could eventually do so and make our earth uninhabitable. The story is well known and should it happen then many other forecasts are irrelevant. On the positive side we may still manage to make nuclear energy a really worthwhile source of energy on an economic scale.

Granted the degree of international control necessary to stop nuclear war and to avoid a major disaster with the emergence of the currently emerging nations, then we can see more clearly the sort of changes science and technology will bring us.

III THE ECONOMIC AND SOCIOLOGICAL IMPLICATIONS OF SCIENTIFIC PROGRESS

Science and technology are the driving forces of our Western civilization. The explosive growth of knowledge, wealth and population all stem from developments and inventions of the modern world.

Looked at in isolation, most scientific developments are justifiable and most certainly the benefits of science are legion. However, the cumulative impact of science on our economy and society is by no means all good.

Our technically based society has been dramatically successful in creating wealth and increased living standards. Western populations have now an expectation of continuous growth in income, and there is certainly nothing wrong with that. However, the achievement has been at the cost of exploiting and using up irreplaceable raw materials at an ever increasing rate. Industrial production is increasing so fast that between 1977 and 1984 we expect to produce as much as man produced in the whole of his history up to 1945!

It is obvious that in an apparently finite world there is an *ultimate* limit to reserves of raw materials. At some time they will be used up or at least become so prohibitively expensive that the whole basis of production is changed. Fossil fuels, such as coal and oil especially, will disappear unless we utilize solar energy and hydrogen energy sources and do so quickly, and with the exhaustion of resources we as a civilization will grind to a halt. The debate is only as to *when* it will happen and *what* we should do about it.

We reproduce below the key tables from the *Report of the Club of Rome* and *The Year 2000*. The model used has been often severely questioned but to us the important issue is not whether the Club of Rome Report is absolutely accurate in its projections but that the speed at which we will run out of resources is much faster than we appear to recognize – even on the most optimistic guesses of yet unknown, undiscovered reserves.

Why is all this happening? The answer is that advanced technocracies use up materials at an increasing rate (as more and more inventions demand more and more materials). But this is only the tip of the iceberg. There are only one billion people living in the 'developed' countries – imagine the explo-

FORECAST WORLD RESERVES OF SELECTED NON-RENEWABLE NATURAL RESOURCES

(Taken from 'The Limits to Growth' published by Pan Books and based on the Report of the Club of Rome)

	Number of years that known reserves will last at forecast usage rates (with exponential growth)	*Number of years resources will last if unknown discoveries boost reserves 5 times*
Aluminium	35 years supply	55 years
Copper	21 ,, ,,	48 ,,
Gold	9 ,, ,,	29 ,,
Iron	93 ,, ,,	173 ,,
Lead	21 ,, ,,	64 ,,
Mercury	13 ,, ,,	41 ,,
Coal	111 ,, ,,	150 ,,
Natural gas	22 ,, ,,	49 ,,
Petroleum	20 ,, ,,	50 ,,
Silver	13 ,, ,,	42 ,,
Tin	15 ,, ,,	61 ,,
Tungsten	28 ,, ,,	72 ,,
Zinc	18 ,, ,,	50 ,,

From 'The Year 2000' by Herman Khan

	Population, Continents (Millions)			
	1965	*1975*	*1985*	*2000*
Africa	310·7	398	520	779
Asia	1,889·0	2,343	2,863	3,701
Europe	674·7	732	792	886
Oceania	14·0	16	20	25
North America	294·2	354	431	578
South America	166·2	221	291	420
World	3,348·8	4,064	4,917	6,389

	GNP, Continents (Billion 1965 United States dollars)			
	1965	*1975*	*1985*	*2000*
Africa	43·9	69	109	216
Asia	287·4	501	883	2137
Europe	923·9	1,447	2,271	4,476
Oceania	28·0	41	60	107
North America	774·2	1,203	1,865	3,620
South America	59·4	91	144	292
World	2,116·8	3,352	5,332	10,848

	GNP Per Capita, Continents (1965 United States Dollars)			
	1965	*1975*	*1985*	*2000*
Africa	141	174	209	277
Asia	152	214	308	577
Europe	1,369	1,976	2,867	5,055
Oceania	2,000	2,510	3,080	4,310
North America	2,632	3,403	4,329	6,255
South America	357	413	496	695
World	632	825	1,085	1,700

sion in usage if the four billion underdeveloped peoples, whose usage *per capita* is currently less than 10 per cent of the West's, achieved the Western standard of living, which virtually every government is trying to achieve, and which means four times as many people demanding ten times the amount of goods. Add to this potential pressure on our resources, the fact that over all population is increasing, and we have an exponential *decrease* in resources – and these are the reasons the environment is being plundered so swiftly.

It is clear that no such system producing this situation can be described as stable, and the root problems seem to be :

1. *The God of G.N.P.* (Gross National Product) has conditioned people to expect a continuously increasing material standard of living (Gandhi succinctly commented 'Earth provides enough to satisfy every man's need but not every man's greed').

2. *The Population Explosion* which results from the triumph of medicine in cutting death rates, is now compounded by the impending success and threat of longevity.

Allied to total population growth is the gross overcrowding in urban areas as the lure of the city apparently offering high income potential concentrates population even further. One single complex of apartments in Hong Kong built by the Mobil Oil Company as a more profitable alternative to using land as an oil terminal now houses 70,000 people in an area of little more than a thousand square yards.

It has been understood at least that for most people over-concentrated urban life causes a high degree of loneliness and psychological stress. A less obvious consequence, as yet to be understood, is that urban dwellers who have little day to day contact with immediate neighbors cannot judge or be judged on the value of their actual contribution at work or within the society. This in turn has meant that an urban dweller tends to demonstrate his prestige to others not in terms of his intrinsic ability but in terms of more indirect clues about his value – we usually call such clues 'status symbols'. The acquisition of status symbols requires wealth and money which in many cases has now become not just an indicator of prestige but an end in itself.

3. *Excess novelty is disturbing.* Although the majority of readers of *Science Fact* will find the developments which are surveyed exciting and challenging – at least intellectually,

coping day to day with a flood of novelty is and will be a major strain. For many, if not most, it can become a nightmare – as bad as the cultural shock of being thrown into a totally new society. In his book Alvin Toffler calls this 'future shock'.

To get some idea of how novelty is likely to make its impact on us it is worth repeating some of the more fundamental changes predicted in our pages.

Imagine a world in which man will be moving on to and living in the sea – possibly as the only practical way left of obtaining continuing supplies of raw materials; imagine a world in which weather is totally controlled; imagine a world in which animals in the form we know them will be fundamentally changed through genetic engineering and in which the IQs of 'useful animals' will be genetically and chemically vastly increased to press them into the service of man.

Imagine above all a world in which we have created biological machines with artificial intelligences several times greater than man's own; imagine the actual application of cloning – Dr. Ladeberg predicted in 1972 that a cloned man would exist within 15 years, i.e. by 1987. Dr. Hafez predicts a 'biological Hiroshima' and raises the moral and philosophical issues of the cyborgs – half man half machines. How can we categorize them? Dr. Hafez also envisages an imaginative solution to the huge time periods necessary for space travel – why not pack frozen fertilized human eggs in space craft which will, on a predetermined schedule, be developed to become the future populations of distant planets? Everything else is miniaturized in a space craft argues Dr. Hafez, why not the passengers? Imagine a world where we have the ability to breed super individuals with supernormal hearing, sight, muscular skills, where we could for example breed men with implanted gills for underwater living, where we could realistically ask the question 'how long do I want to live?'.

Dr. Hafez's predictions are of extreme significance because the history of science shows that when it is thought that something *can* be done it inevitably *will* be done. It will be done. It will be done partly at the simplest level to satisfy the researcher's own curiosity, partly because of the 'arms race syndrome'. If a nation or bloc sees an advantage in genetic engineering or creating a pool of super brains there is an enormous impetus to do so, because 'if we don't, they will'.

Probably no area is more likely to come under this type of pressure than the developments of Dr. Hafez's cyborgs. The symbiosis of man and machine is already with us. We already have natural language computers. We already have detectors which can pick up signals from the nerve ends of the stubs of amputated limbs which are then amplified to activate artificial limbs. In other words, we already have machinery directly, sensitively and automatically responsive to the nervous system of a human being. The possibility of linking a computer directly to the human brain in an organic fashion is currently being studied at various institutions

Thus Professor Robert White, Director of Neuro-Surgery at the Metropolitan General Hospital in Cleveland, has already detached the brain of a monkey from its body and hooked the brain alone up to the carotid arteries of another monkey, thereby keeping it alive. The brain actually survived for 5 hours under these conditions and could have survived a lot longer. This same Professor White is on record as saying :

> 'To me there is no longer any gap between science fiction and science. We could keep an Einstein's brain alive and make it function normally.'

According to Professor White, at any rate, the possibility of transplanting a head to another body could be done with 'existing techniques' (although it must be admitted that this assertion is queried). He even specifically predicts the Japanese will be the first to keep an isolated human brain alive. He himself would not, because it seems that he could not resolve the moral dilemma implied.

We have gone into some considerable detail on the cyborg development because it not only dramatizes how far science is creating totally new situations but because it illustrates another potential danger – the danger that when a development is as fantastic as that of the cyborgs and progressing so fast, people generally will just not accept that it *can* be true. The (natural) tendency to reject some current developments as outlandish and fictional – particularly when they are unpalatable – itself opens the way for the more knowledgeable and unscrupulous to exploit such developments. This above all is the reason why knowledge of all areas of scientific developments should be open and public – and is one of the motives for producing this book.

4. *The speed of scientific change is traumatic.* Allied to the shock of novelty, Alvin Toffler in *Future Shock* persuasively argues that the sheer speed of change is itself a major stress-producing factor in our society.

Two quotations, the first from John Diebold, the American automation expert, and the second from social psychologist Warren Bennis, put the position starkly, if slightly dramatically Diebold says:

> The effects of the technological revolution we are now living through will be deeper than any social change experienced before.

Bennis says:

> No exaggeration, no hyperbole, no outrage can realistically describe the pace of change. In fact only the exaggerations appear to be true.

We would agree. We have already read described in the pages of *Science Fact* a world population doubling every 21 years – or to put it another way, every single city would need to be duplicated – every 21 years we will need to build a new Rome, a new London, a new New York, a new Tokyo! Already urban planners are drawing up subterranean cities, submarine cities, supermarine cities. We have already seen the phenomenal growth of production. It is important to realize that the whole exponential growth of knowledge and invention occurs precisely because each new development itself multiplies the possibilities for further developments.

We do not wish to repeat unnecessarily the results of the combination of over-stimulation through speed, novelty and

transience on society. The symptoms are all too obvious –
anxiety, senseless violence, depression, apathy, hostility.
What is interesting is that recent research in America has
been able to identify the very real and measurable physical
strain involved in adapting to new situations. The survey was
conducted by Dr. Holmes and Dr. Rahe of the University of
Washington School of Medicine.

The originators of the measurements of the effects of
changes in life on bodily health, in particular Commander
Ransom J. Arthur, Head of the United States Navy Medical
Neuro-Psychiatric Research Unit at Santiago, has claimed
that :

> for the first time we have an index of change. If you have had
> many changes in your life within a short time this places a great
> challenge on your body ... an enormous number of changes
> within a short period might overwhelm its coping mechanisms'
> Certain viruses live in the body and cause disease only when
> the defences of the body wear down. There may well be gen-
> eralized body defence systems that prove inadequate to cope
> with the flood of demands for change that come pulsing through
> the nervous and endocrine systems.

In other words, change carries a psychological and physiological price tag on it.

The actual physical responses that come into play when a
human being is subjected to new stimuli are categorized as
'orientation responses'. To a significant stimulus the pupils of
the eye will dilate, photochemical changes appear in the
retina of the eye, hearing becomes momentarily more acute,
and our muscles are unconsciously directed against the incoming stimuli. The pattern of brain wave changes, fingers
and toes grow cold as the veins and arteries in them dilate,
palms sweat, blood rate increases, and breathing and heart
rates also increase. This 'startle reaction' is obvious when the
stimulus is strong (fear) but even on weak stimuli (novelty,
unfamiliarity) the same changes take place, although may be
unnoticed.

Consequently, the more stimuli you are faced with, the
more the body wears out with the orientation response. The

responses described above are mainly neural responses but in addition, when repeated stimuli are offered, the pituitary gland comes into play and releases chemicals called corticosteroids into the body which speed up body metabolism and increase blood pressure, at the same time turning fat and protein into disposable energy. This particular physical reaction is a much more stressful reaction on the body and stress can be actually measured by the amount of corticosteroids and adrenalin, plus noradrenalin, in the bloodstream. So much so that one can provide an acurate index of wear and tear and stress by purely measuring noradrenalin and adrenalin in the blood or urine system.

More important, the harmful effects of constant stress in terms of triggering both orientation response and the hormonal responses indicated above *last a lifetime*; the wear and tear effects are permanent.

5. *Speed of knowledge growth.* The amount of information currently becoming available is literally incredible. Dr. Robert Hilliard has said :

> At the rate at which knowledge is growing, by the time the child born today graduates from college the amount of knowledge in the world will be 4 times as great. By the time that child is 50 years old it will be 32 times as great and 97 per cent of everything known in the world will have been learned since the time he was born.

The stream of data and knowledge has meant that the ways of communicating it both physically and indeed in style have had to change. So the method of presentation verbally is itself fast moving. Everything is advertised as high speed adventure, 'fast reading', 'action packed'. We are literally bombarded with information. The average person is exposed to as many as 50,000 words a day, of which he probably assimilates between 10,000 and 20,000. A study in America indicated that the average adult has 560 advertising messages reaching him per day – of which he notices only 76. The so-called 'block-out' procedure is the only thing that saves his sanity !

What all this is leading to is that speed of change around us compels us to re-learn our environment continuously. This places extra demands upon our nervous system.

That the world is moving at an ever increasing pace is of course unarguable and that human beings are ill equipped to adjust to the pace is almost as unarguable. We took hundreds of thousands of years to evolve into even an agrarian-based society. We have taken approximately 150 years to move to an industrial society and we are trying to move to a super industrial society in little more than a generation. To the majority of people, the developments appear out of their personal control and as their environment becomes more unfamiliar so the sense of dislocation, of disorientation and of alienation increases.

Some, of course, are not only adjusting satisfactorily but welcoming the sheer speed and they are generally the sort of people who are migrating to California, New York, the big cities. Others, probably the majority, have a vague feeling that they should 'stop the world, because they want to get off'.

6. *Impermanence – producing alienation and loneliness.* The trilogy, novelty, speed and impermanence are the fundamental sociologically disruptive elements.

Our relationships are increasingly impermanent. Alvin Toffler must be quoted again – 'We live in the throw away society' – with automobiles of built-in deliberate obsolescence, changed regularly, with disposable lighters, pens, bottles, even buildings. We are living in a rental society in which many of our possessions are rented. We even rent people – witness the booming 'temp secretary' business.

We have an increasingly impermanent relationship with our homes and even our countries. The average American moves once every five years, the average Englishman once every nine years. Certain nations are supplying whole workforces that live and work in other countries, and this is illustrated by the northward migration of the Jugoslavs, Algerians, Turks, and Spanish. The inevitable effect of this transience is a sense of detachment, of not belonging and even not wanting to belong. Indeed because our lifestyle dictates that we meet an increasingly large number of people for a smaller and smaller duration of time it is just not possible to get involved. We are forced to have relationships with them that are largely 'de-personalized'.

This ever faster turnover in jobs, friends, even marriage, has created stress, and nowhere is this stress higher and the pressure of transience higher, than in the scientific and technological communities – the very jobs of the future.

7. *The concentration of economic power in larger and larger corporations and authorities* who cannot know and inevitably must care less about the impact of their activities on the local environment in which their subsidiaries operate. This also involves the profit motive which has become too crude a measuring stick of social contribution.

The bigger the economic entity the more the man tends to be used to fuel the system, not the system used for man's benefit. We talk, for example, of consumers, as if the primary function of individual humans is to provide an outlet for production. Production itself has for the moment and for the majority, become more routine and dehumanized. This in turn makes it difficult for a man to feel a sense of being of some worth and having some identity of his own.

It may well be true, as the article on computers and electronics, in particular, suggest that the liberation of the man from mindless routine is a real prospect, but it certainly will not happen in the short term, neither does it mean that everyone will suddenly become liberated to a free creative and satisfying job. Few readers of this book are likely to have direct personal experience of the dulling and frustrating impact of the assembly-line without which through the technical genius described in *Science Fact* would not have come about.

Science and economics is currently too much concerned with the large scale – with statistical abstractions – and the macro effect of planned programs, forgetting that the individuals who sum to make up those macro-statistics are increasingly subject to alienation, loneliness, fear, stress and frustration by the system. The danger is that we simply treat people as numbers and not as human beings.

8. *We have no price system that reflects the 'irreplace-ability' of finite resources.* Thus the simple price of aluminium is broadly devised on the basis of the cost of acquiring land, mining, processing and profit on capital employed. Nowhere do we cost in a future scarcity factor.

9. *We have been using science and the acquisition of knowledge as if the latest developments* – however sensational and creative was an end in itself. Unless we think out sane and reasonably clear objectives for our progress it is likely to be jumbled, badly co-ordinated and frequently contain undesirable side effects and unforseen consequences. 'The Tree of Knowledge,' said Ortega, 'is not that of Life.' We are, it seems, over-emphasizing the physical and material developments and paying insufficient attention to developing the spiritual, cultural and emotional basis of humanity.

This last point of course links with the other problems. It is useless to exhort individuals to be more moderate in their ambitions of wealth, to voluntarily use less resources, to forego maximum profit in order to organize farms or factories in a less environmentally destructive manner, unless they actually believe such attitudes and actions are right and beneficial to them.

Specifically, it is fair to argue that it must be inefficient for one industrio-scientific community – the USA – to use 40% of the world's primary resources to maintain 6% of the world's population and still fail to achieve a happy, peaceful and well-adjusted society. (1500 estimated suicides in the USA each day; 4% of all adults are alcoholic according to the WHO estimates; one in six of all children appear in court; 20% of all adults have mental hospital treatment.) Yet, if asked, the average American citizen is likely to argue that if anything increasing his happiness is at least partly dependent on further increasing his wealth!

Thus our science-based society has been very successful at satisfying material needs – so much so that man is now looking for psychological satisfaction which is what we are not well geared to do. It could therefore be argued that the price of the technical magic reviewed in this book is the sickness and alienation of the major part of our people.

Unless man's assessment of his real long-term self-interest is changed our muddled and incompatible objectives are likely to lead to :

10. *Totalitarianism.* This could occur as disillusioned people turn to new messiahs who offer imposed solutions and controls to our seemingly insoluble problems and who offer well-defined social objectives. Such a man was Hitler, who offered a clear aim and a defined role and sense of purpose for each individual and the offer was made at a time of high unemployment.

11. *War* – as people realize that 'beggar my neighbor' policies can extend the time period for which they personally are able to maintain their own real material standards, so they are prepared to pursue such policies at the risk of starting a war.

As always, it is easier to state the problems than to find the solutions. Indeed the intention of *Science Fact* is to review latest developments and to pose the questions arising from them. However, the issues raised are so fundamental that out of this book has grown a conviction that only very major changes will allow us to cope with the future and these will be the subject of a follow-up book called *A Life Style for Survival* now in preparation.

Meanwhile, some of the directions that our solutions must take, we believe, are :

1. *A clarification of our space objectives.* Probably the main justification of space research is that it can lead to the colonization of other planets by man.

We are not recommending such exploration as a way of keeping the frenetic race for continuously expanding riches fuelled for ever by raping other planets but as a way of ensuring essential supplies for a 'modest objective' society. Space will also be a much needed challenge to mankind as his physical problems at any rate begin to diminish on earth.

2. *A real attack on the population problem.* We are inevitably abandoning our subsequent generations to war if populations are not reduced. History, many animal experiments and even natural phenomena (e.g. lemmings) have proven that a stable system is impossible for psychological, let alone resource, reasons with overcrowding. It is too easy to postpone decisions affecting another generation – or another continent – but it is irresponsible. If we don't voluntarily find ways to reduce the population, the solution will doubtless come unpleasantly and unexpectedly.

3. *A realization that science is for people not people for science.* An American publication, *Tomorrow through Research*, complains that 'We are in the age of neo-Luddites who seek to smash the technological framework of the future because the machinery of our times is going too fast for their feeble minds.'

Science Fact is in many ways a sort of epic poem dedicated to our achievements and we believe it is impossible – even stupid – to turn the clock back and forgo the very real benefits science has given us, but we must search for a balanced state. If the vast majority of people have, in the elitist phrase, 'feeble minds', who has the right to impose strain on them? Let us not travel at the speed of the slowest but equally why simply allow developments to create the psychological and sociological havoc we can already predict? We must try to control the speed of change and in the absence of a spontaneous popular change in attitude this short term does pre-suppose a supra-national review body. The same body would be very useful in establishing over-all priorities.

4. *We need to redefine 'Standard of Living'.* As science solves more and more of our basic problems of survival and comfort we will need to have a clearer concept of what we mean by 'standard of living'. To express it in possessions or money terms is to over-simplify the problem. A pleasant work environment, unpolluted rivers and air, constitute as much a benefit as cash. To describe a production method which protects the environment as 'less cost efficient' than

a cheaper but potentially more violent system, is obviously short-sighted.

Moreover, we would have sufficient money resources if we solved some of the basic social and psychological problems of society. Huge effort, resources and money are expended by people on distractions from the problems of everyday life – on gambling, trivia, on 'status' enhancing products and on products with contrived obsolescence. A clear thinking, unthreatened society would not have need of these psychological crutches and could 'spend' the saved resources on improved environment, reduced work week, etc. There are precedents which indicate that a less aggressive and less materialistic attitude can be practical, although we need to specify the means by which peoples can be *voluntarily* changed.

One thing above all else is clear, and that is the economic and sociological implications of our present scientific revolution are likely to prove far reaching and we should try to keep our minds open still as to what sort of society will best suit our needs – always assuming one can be found.

5. *We need a revised approach to education.* It is clear that our educational system must be redirected to educate our children to expect and to be able to cope with the speed and depth of change. Not only should adapting to change be taught but we should look at our education not as a once-for-all event at the beginning of the child's life but as a continuous process lasting throughout life. Part-time education over a long period is likely to be far more beneficial than full-time education over a short period; at least this is so after the formative years are passed.

The normal school curriculum is largely geared to past needs and past values. The mere accumulation of fact is likely to be of comparatively small value in the future. Much more important will be the ability to learn how to learn – to be able to learn how to supersede old ideas and to replace them with new ones. Almost all of our children have courses in history, virtually none have courses in 'the future'.

We will have to encourage our children to speculate because only in such speculation will some of the unexpected possibilities become apparent. This means bringing matters of more philosophical importance into the rather fact-orientated educational system. This raises, though, a whole mass of issues that go beyond our present discussion. What we can say is that education is in great need of change, and one cannot but be pessimistic, if only because the bulk of educators seem to be the people most in need of the very education that we are suggesting as necessary for the children.

6. *The need for technically spread democracy.* Alvin Toffler in *Future Shock* quotes an interesting law formulated originally by Ross Ashby based on the mathematically proven principle that

> the whole system based on a number of sub systems tends to be dominated by the sub system that is least stable.

There is every reason to believe that this law is directly applicable to the complex society, made up of sub systems, in which we live. If this is true, society must be made sensitive to its minorities and indeed society is proving increasingly open to blackmail by just such minorities. It is inevitably relatively easy to disrupt a highly centralized and technologically based society.

For this reason it is likely that we will only evolve a stable political system through an expansion of democracy that gives expression to minority needs (so long as the constituents of this democracy are sufficiently well informed to come to a reasoned judgment). There seems to be a real cause for optimism in that the huge advances in telecommunications should enable public opinion to be swiftly and comprehensively sampled. Thus a well-informed society, with good communications, should be able to make its voice felt through plebiscite voting mechanisms, using teletext style equipment.

This may again sound somewhat starry-eyed but something of this sort is likely to be necessary. The alternative is for increasing unrest with minorities who feel alienated and unable to 'affect the system' and who will increasingly turn to disruptive techniques. This in turn may well encourage governments to argue that the freedom of the individual will have to be sacrificed to prevent the increasingly complex society from being brought to a halt for the sake of 'just a few people'. On just such justification, one can envisage the whole armory of electronic surveillance and chemical and electronic brainwashing being used: this is something though that must be fought tooth and nail.

IV SOCIAL AND POLITICAL CHANGE AND PRIORITIES

Let us now change our pace a little, and slow down from the 'speed' of the last section. It is self-evident that we cannot, as a civilization, exploit, even partially exploit, all the ideas that we have and this raises two questions. One is as to the priorities we should have and the other is as to the priorities we will in fact have, since they may well be different. What may make them different is the political and the economic considerations which will accompany scientific and technological advances and social change. This means that in our discussions of social change we must anticipate a new world of greater leisure, better education and greater social complexity. We anticipate a great upsurge in material standards creating a tremendous competition for raw materials and markets. At the same time we might expect increasing nationalism, racial hatred and wars of a regional kind. These political considerations will certainly interact with the social and economic changes envisaged. We shall face a deluge of devices of every kind that increase leisure and reduce work. This also means a society that has far more time to consider life and its implications.

This suggests that people will demand entertainment and education and a maintenance of standards. The latter will be a first priority for governments who will recognize that any failing here – although hardly noticed if it does not occur – would be political dynamite if it does.

Medicine, and the biology which subserves it, are a huge source of optimism. It is clear that we shall continue the story of medical success that has already started in recent times. The story is of the gradual defeat of disease and it is one that is gathering momentum. The bulk of diseases of the nineteenth century have gone and cancer is now almost the only enemy left. Heart disease is very common but ways of defeating it are within our grasp.

The dangers of upsetting the 'natural state' of things are inherent in scientific and technological development and we have to worry about the possibility of new diseases. But in spite of this everything points to a much greater expectation of life. Spare part surgery in all its aspects is a vital part of this process, a process which holds out the possibility of eventual immortality. What though are the shorter term effects of greater longevity? The immediate effect is to increase the population (already increasing fast) and, more important, increase the older age groups in the population.

We can foresee a state where a large part of the population will be inactive or retired. If they commonly live to a hundred years and beyond and still retire at sixty-five the result is clear in terms of activity. Automation and cybernation come in at this point and their role of taking over more and more of the activity involved in the running of society then people at a younger age will further tilt the balance against work and in favour of leisure in an over-populated society. We have already mentioned something of the social significance of these changes : the political ones are mainly in providing the need for new, more flexible, thinking in the fight to maintain an unstable democratic system.

The development of science and technology means that more and more people will become scientists and technologists and this in turn will further increase the steepness of the exponential growth of knowledge. One fear that now arises is that of boredom, and boredom is always a source of political unrest. The problem of boredom is also linked to the developments in drugs and education.

The bettering of material conditions will make social, and therefore political, life very much more complex rather than less. A simple example brings out the difficulty. If we learn to *control* the weather then we have all the terrible problems of agreeing when and where it should rain and where we should have sun and how much. Without weather control the problem does not arise and we make the best of what we have. It is because science increases the control of our environment that we have the growing threat we have.

Financial considerations

The world of western competition should keep up the financial investment in science where the battle is likely to be between those who want it all to go to the 'welfare state' and those who want it to produce profit. The latter will increase scientific development more quickly, and the decision between the two will influence the kind of science carried out.

The capitalist approach has the best hope of motivating people (and thus helping to remove boredom) but for perhaps the wrong reasons, although right and wrong are no longer very clear-cut concepts. The lesson of science seems to suggest that the socialist approach will help destroy the world, although the motives behind socialism may be good. The point here is that it is not sufficiently recognized that people thrive best in adversity. If there are no evils to fight and goals to be achieved then motivation subsides and boredom can set in.

It is possible that education, especially for leisure, is, as we have already suggested, important but no amount of education in the ordinary sense will help if we have no motive to learn or belief in living. This will bring us to a discussion of religion and philosophy, but before we get there, let us consider drugs.

Our great advances in drugs could allow us to replace alcohol and existing hard drugs by harmless ones that reduce our anxiety and offset our boredom. We are now in a slightly difficult position; we face the same difficulty as the raconteur faced in *Brave New World*. The world of conditioning – directly analogous to our drug-induced world – is wholly distasteful for an outside observer. Should it on that account be abandoned? The answer probably lies in the fact that such a state will come about as a combination of voluntary self-oblivion encouraged – possibly demanded – by political authority.

It is not easy to be sure, in forecasting, of the interaction and causal relationships of the future. To what extent will mining the seas, controlling our resources, avoiding pollution and doing all the other 'right' things avoid the threat of boredom and political control? It seems likely that they can make things far better in the short term but not necessarily in the long term. A failure to carry out the necessary controls would ensure disaster but to be successful will present other problems.

It is here that we see one principle emerging clearly. The nature of man is to accept a higher standard easily and in doing so turn his thoughts quickly to what is unsatisfactory. The development of science is a means of greatly amplifying our ability to express ourselves and since our expression is primarily one of dissatisfaction, the dangers emerge as predominant over the advantages.

If man is designed by evolution to solve problems and overcome difficulties, he is not, by the same token, well suited to the business of living in a state of equilibrium. He cannot (or finds it very difficult to) live without challenge, since the leisure available is conducive to the boredom we fear. The great hope is that space research will provide the means of colonizing space, since then we shall be effectively in the same state as we were before we colonized the earth. This would provide all the threats necessary to evolution and *could* be part of the necessary expansion of a basically capitalist economy.

The role of religion

Let us turn next to religion and its associated aspects. The 'paranormal' hope of discovering the soul may or may not be directly concerned with religious beliefs, but much development of that kind could add evidence to (or subtract from) our confidence in the ontological assertions of most religions. On balance, I believe that science has been the main cause of undermining religious belief. This is so because the stories and allegories surrounding most established religions are placed in doubt as we discover more about reality. This is not a logical reason in itself for disbelief in the possibility of God or Gods, nor for that matter of belief. It is, however, a psychological reason for disbelief.

The problem of religion is twofold. One is that religion was used as a 'social cement' which exercised a measure of control on, and provided a sense of purpose for, society in the past. This basis for ethical and purposeful behavior stood or fell on the credibility of the religionist 'story'. As that has come to lose credibility, so the useful aspects of religion started to fail and paved the way for anarchy and confusion. What are the chances of a recrudescence of religious faith and on what basis could it arise? The answer is, as usual, speculative. But I believe that a great deal of thought should be given to the role of religion and although it seems impossible for existing religious organizations to adapt sufficiently quickly, there are good grounds for an all-out research effort on 'religious thought'. The chances of success though seem slim and in this case we are pushed back on to our own resources. Life has no purposes other than those we impose on it.

Priorities

One of our outstanding problems is that of priorities and how they are decided upon is closely interrelated with political decisions and social change. The situation can be summarized briefly in the following terms.

If we held the view that what was good for most people,

in an immediate 'welfare' manner, was paramount, all efforts would be geared, in all probability, to medicine and education and to any other scientific development which alleviated human hardship and pain in the short term.

If we held the view that what was good for most people was to further economic competition, then we might find the emphasis being placed more on various resources such as foodstuffs, oil and therefore on mining of seas, land and space. The idea then being to promote a greater expansion of wealth for reinvestment in medicine and education. This view would be akin to present 'mixed economy' thinking.

If nationalism became uppermost, whether or not a mixed economy was the foundation, the emphasis would switch to space travel and defence (and offence) weapons of every kind.

These priorities are by no means absolute, rather they represent differences of emphasis. If, as we expect, we are going to move through an expanding, then anarchic, then totalitarian phase, we might expect the resources (the second view) to be followed by more nearly the first view and followed in turn by the last view. The Soviet Union is a living example of how a totalitarian nationalistic system can make huge 'military' developments even in a poor financial climate.

Indeed what we said earlier about financial dependence only really applies in the current 'democratic phase' of development.

All this must then be viewed in a background of dramatically decreasing raw materials and fast increasing populations. This particular problem is perhaps the most pressing of all in the short term and it is this that needs to be solved one way or another regardless of all else. Longevity is simply an attempt to swim against the tide, if we are running out of food.

Control of raw materials and population is one answer, and synthesis of raw materials and mining them wherever they are available – under the sea or in space – is the other. The danger is the exponential danger and this can, as Robert

Waller says, escalate so quickly that a safe situation can be made dangerous extremely quickly.

V THE CONTROL OF SCIENCE

If it is asserted that we have already lost control of science, our question is : 'When then can we regain that control?' If we have not gone that far along the line the question becomes : 'Can we avoid losing control?' Which question is appropriate is a matter for debate and one that we cannot easily put to the test. But let us consider what is involved.

We find it difficult to control science already, and have done for a long time in the limited sense that we have come to depend over much on it.. It is a frequent occurrence to see an office or factory reduced to impotence because a computer is out of action. Even more obviously could a civilization receive an unacceptable setback if suddenly all medical and dental science should cease. In this sense we certainly vitally need science.

When it comes to controlling it we think at first in terms of physical and biological pollution of one kind or another. This is quite right and proper and it is true that research in germ warfare, microbiological – especially genetic – experimentation and the like could get out of hand and it is this we had in mind earlier when we mentioned Dr. Paul Berg's appeal for a moratorium. We could extend that notion and have a moratorium over any branch of science – say certain experimental drugs – which threatened dangerous spin-offs. Such a moratorium would be designed to allow us to proceed more carefully and in a well-controlled manner prior to allowing full-scale implementation.

The biggest threat to our immediate future is not physical pollution, but information pollution. The creation of data banks and their integration is bound to cause widespread confusion and injustice. The same processes of law and order which worked well with limited information ironically fail as they become more efficient. The policeman is encouraged to detect crime and hand the criminals over to our justice. The same policeman, aided by a data bank system, would turn his thoughts to preventing crime. This is at first sight a good idea, but at second sight is utterly intolerable because of the increased surveillance it entails. What has happened here is that quantitative change has become qualitative change, and this change is symptomatic of the general impact of science on society.

The danger is that the power interest in data banks is left wholly in the hands of a government or some such body who can then quickly pressurize all opposition and carry out a 'takeover' of society. Society is by this time highly automated and highly organized and has already begun to face the implications of increased 'leisure'. It is, in effect, ready for change and like so many others before will tend to respond to the most improbable overtures. It will in fact 'grasp at straws'.

The totalitarianism threatening society in this way is greatly aided by science and technology for reasons I have already made clear – amplification of effective power will nearly always be used for what most people regard as the worst ends – they at least make the problem of maintaining a balance much more difficult.

We have next to ask ourselves whether there is something 'natural' in the evolution of a totalitarian system – whether left-wing or right-wing is of no concern. The answer must be 'no' and there is a very clear goal before us to try to see that all the enormous and vastly exciting developments of science be carried through and to do so in a very much more stable social and political state than we have yet attained. But to realize what it would have been like fighting a Gestapo, armed with a computerized data base chain, is worth thinking carefully about.

A machine species

Now we return to the other considerations of artificial intelligence. From these developments we have, regardless of the intervention of totalitarianism, a different sort of development – the emergence of a machine species. We can make a machine to construct itself and that machine we currently call a computer. It is of no consequence whether the machine in question will be a computer in the sense of a current fourth (or fifth) generation computer or a machine more specifically made for the 'humanlike' purpose. Nor is it of any consequence whether or not the machine is manufactured of biological or metallic fabric, it can still reproduce itself. If necessary it can do so by using the same manufacturing processes as we use in our factories.

Since the self-reproduction of machines (and machine species) is not explicitly dealt with in the text, it behoves me to mention this important work, due mainly to the mathematician, Prof. John von Neumann.

He was able to show, mathematically, that an artificial system could reproduce itself. The theory is highly mathematical and somewhat similar to a genetic theory, and we can only here depict the basic idea. If a machine has within it a description or model of itself and instructions as to how that model can be constructed externally, it can carry out that construction process.

It is rather like a drawing of a building or a component, or a drawing in a meccano set. The instructions for the construction must now be made clear and as a result the new machine can be built. The components needed for the construction can be either the same raw materials which go into our production factory or the foodstuffs needed to provide a biological production system analogous to an embryo – basically a seed that grows into a fully grown adult entirely on its own.

The next step is to recognize that such an artificially intelligent species can be given complete autonomy, and given that autonomy can no longer 'have its plug pulled out'. In other words there is no point at which we can say that we have allowed as much freedom as we should to our artificial species. It is like having children. We manufacture them and set them free and then can no longer control them. Precisely at what point you say 'I have given the child enough rope and must now clip his wings' is something which is quite impossible to determine. Similarly if a system is given a level of intelligence it is not possible to then say that this or that is the point at which we must stop if we are to avoid being taken over by our own creation. We simply cannot gauge the implications of our decision with that degree of precision.

The whole process will be speeded up in the event of totalitarianism because the power vested in 'the few' will be used more effectively to maintain control over the majority and that power will be immediately available to the machine species at the time of their 'takeover'. Their own development having been hastened by the politburo' in their fear of reprisals from human beings and their determination to use all available technology to suppress people.

I have asserted that machines *can* reproduce themselves, and this, as I have said, was *proved* by the eminent mathematician, Prof. John von Neumann. The methods, rather like genetics, involve maps of people – like blueprints – from which those people are constructed. If you label all the parts a,b,c, ... and then provide the map – like a jigsaw of how they are put together, then the basic idea is clear enough.

Let me summarize this very heart of the matter of control over science. We need to avoid where possible starting processes we cannot stop. Unfortunately, we cannot always tell which these processes are and how we identify them, as in the case of cybernetics and artificial intelligence, it may still be very difficult to exercise the necessary control.

The analogy here is with social evoltion. If you introduce slaves into society, at what point does your grip on them fail? In the case of our machine slaves – versions of I. J. Good's UIMs (Ultra Intelligent Machines) – it is when they are too intelligent and too powerful for us.

There can be only two safeguards and both pre-suppose a thorough awareness of the dangers. One is to 'build in' controls into our machine species, rather as a human being's brain may be controlled by brain washing, psycho-surgery or miniaturized transmitter-receivers in the skull. The problem remains that such a species may, through its own medicine, learn to cut out such man-made controls in much the same way as our surgeons remove a cancer.

The only other safeguard is to try to limit the extent to which artificially intelligent machines are allowed to be autonomous. This means the need for international control of science, since otherwise parochial national interests would lead to the breaking of such legislation.

Perhaps the most important things to be said about the control of science is the need for a thorough awareness of the problem and this is where this book comes in and a determination for the need of international control.

VI A LIFE STYLE FOR TOMORROW

This brings us to the whole question of 'life style'. In the face of all these predictions, what style of life should we adopt, or start planning for?

The first answer is that we should plan first of all to survive. To do that implies the regaining of control of science and at the same time regaining control of civilization. Control is the main ingredient of science and therefore we are talking of control of control.

But a life style is more than just surviving and continuing to try to control our environment, which also means knowing when to leave well alone, it involves philosophy and religion.

So we come to philosophy. We believe in effect that religion will not help us : Politico-economic considerations will support scientific developments but ultimately at the expense of destroying the stability of our existing social system. What is technologically possible is exciting beyond belief, but we have to remember that the changes engendered by the scientific revolution change the nature of the society that engenders it.

So next we have to ask again – what do we want? What makes – or could make – our life worth living? The answer is that we may not be able to tell, since the *Brave New World* syndrome intervenes. If we do not evolve within the new system we shall not know ourselves the psychological changes that will occur and therefore cannot know what our needs will be. But the fact remains that we can see that by conditioning, or being drugged or being taken over, we can avoid the problems that are provided by social motivation: this is one clearcut possibility.

The question can be put in a slightly different form. If we asked ourselves what we would choose the world to be like, what would we say? Science is increasingly making this offer to us and we are unprepared for the answer.

The first part of our answer must be that we accept that no static Utopia is of any use. As soon as we have achieved a goal we are no longer interested in the problem, but turn to something else. 'As soon as we have it, we no longer want it!' So man is designed to be adaptive and evolutionary and any satisfactory Utopian recipe must also be dynamic.

We shall press on with the control of disease, the benefits of medicine and the search for space and resources and hope at the same time to place all these advances in the proper context of change and *sufficient* control. The balance struck here is all important and must be done under the banner of Internationalism.

Sociological and political research to achieve these extremely sophisticated objectives could provide much of the thought necessary to avoid boredom in the coming decade. Now we must also hope that education, entertainment, drugs (of a suitably harmless kind) and other research activities produce personal panaceas. We need an organization of the 'Marriage Guidance' type on a much wider and larger scale which deals with every kind of social problem.

In this background, we can set about ensuring the communication links of the world act effectively to retain the precarious balance of power between peoples and ideals. This can be done by making sure all information is public or publicly accessible and that no 'secret treaties' can be entered into. It is because exponential changes are involved more and more in the future that instability becomes an ever-increasing threat. A total awareness and a total dedication on the part of the whole of civilization will be required.

We have not discussed education in detail in this book, although it is implicit in much that has been said. We should now make one or two points of major importance to the future. The subject was touched on in psychology, and the point there was that there are various ways of making learning more effective. These methods vary from the use of pelmanism, acronymic devices, systematic rehearsal and telepathy. Some of the language teaching that seems to be achieved in so-called 'Suggestics' laboratories could, if in widespread use, rapidly transform the whole educational world. An increase of five to fifty times in the speed of learning has apparently been achieved (a language learnt in six weeks!), and these sorts of speeds provide the basis for exponential developments in the growth of learning – a learning already likely to be speeded up by the computer controlled teaching systems mentioned earlier.

This in turn suggests that a super-species of human beings might yet arise and rival the machine species man is already creating. Whether a super-human species would lead to a super-machine species makes interesting speculation. But, in education, in spite of what was said of its ineffectiveness in a purposeless context, must be a major source of future hope.

Here at last we see the main key to our future. The vast bulk of the time will be spent philosophizing and legalizing and performing acts of cultural integration. What we must never do is sit back and contemplate eternity and ask about 'the meaning of it all'. Its only meaning is what we can impose upon it, and this means that learning and education are paramount.

Thomas Erskine's words, quoted by Arthur Garratt are also most appropriate to our purpose. I will not repeat the whole quotation but merely the last sentence :

> Let reason be opposed to reason, and argument to argument, and every good government will be safe.

The trouble is that this pre-supposes the ability to have a

'good' government, which means trying to retain the inherently unstable type of western democracy.

Two further very considerable lines of hope though can be added to education. One is that in the short term physical conditions will be far better than ever before and that, in vacuo, cannot be anything but good, and secondly that we are forced to reason and think today in a way never either necessary or possible before.

If there are huge threats that inevitably arise from the way in which society is being made more complex, then we and all our helpmates – machines of every possible kind – are becoming more and more adept at dealing with such complexity.

We must be always on our guard, but wholly optimistic about the future, but there are in the end a series of questions that we have to put to ourselves yet again. In the first place we have to ask seriously how long it will be before the present developing avalanche of science and technology will descend upon us? How much time is there? If we answer none, then we must take action : what action?

Immediate action must be in terms of national and international control. Should we not do everything we can to alert everyone about what is happening? Ask for sensible natural control of science and technology, and greater social freedom? How do we formulate our detailed plans? Much here will hinge on whether we see ourselves heading for socialism, totalitarianism or retaining a mixed economy on the present so-called democratic lines : which should we aim for? In answering, can one see all the consequences of our decisions?

Even allowing for these future events and the fact that our own influence will, in any case, be small, what should our social aim be? Should it be to seek a smaller group – perhaps a sort of tribal state – perhaps in Britain, the small town life comes nearest, while retaining the family unit. At the same time should you not resist control and interference here? Make clear that control – even international control – is excellent for science and world economics, but not for individuals, families or tribes : here control should be minimal?

Should we not take Prof. Ubbelohde's advice and return to a materially simpler life with more thought and more contact with nature? No one wishes to dismantle cities or stop entertainment – even automated entertainment – but life could be simpler and more ruminative. What decision do we finally make about religion? Should we have more time to think about it? If we have learnt one thing it is that science offers us almost unlimited opportunities – but it is amoral. It is we who have to take the moral and philosophical choices.

The plain fact is that different people will answer these questions differently and it is quite right and proper, but the more important thing is that these, and many other questions too, should be asked and go on being asked as long as civilization exists. Remember that the problem belongs to human beings. As Milton puts it quite simply :

> Accuse not Nature, she hath done her part;
> Do thee but thine.

INDEX

544

550